Additional praise for Aaron T. Beck's
Prisoners of Hate

"A provocative and most timely report." —*Kirkus Reviews*

"Aaron Beck draws on a lifetime of scientific research and therapeutic experience. . . . He has devoted his career to showing how the rational side of human nature can be trained to overcome the merciless legacy of irrelevant evolutionary imperatives and the tragic result of individual traumas. By reducing conflict arising not from clashes of interest but mistaken judgments and unprocessed impulses, this book will help both laymen and professionals put human rationality to its most important use."

—Ian S. Lustick, Department Chair and
Professor of Political Science, University of Pennsylvania

"This magnificent synthesis crowns a lifetime of achievement in cognitive behavior therapy. The same formulations that account for violence in the individual are found to apply also to collective violence. During the Cold War the West had a convenient distant enemy; now that NATO dominates the world, close neighbors are turning on one another. This is a timely book, closely argued and vividly illustrated with telling examples."

—Sir David Goldberg, Professor of
Psychiatry, Institute of Psychiatry (London)

"The leading authority on depression here turns his clinical radar on the cognitive sources of anger and hostility. As Dr. Beck demonstrates, hate and violence bring pain not only to the victims but also to the perpetrators. Hostility and rage can become habitual and, like other bad habits, these can be broken. This fine book can help."

—David T. Lykken, Ph.D., Professor
Emeritus of Psychology, University of Minnesota

"A brilliant book, deeply needed in today's world. Dr. Beck brings to bear both immense knowledge and creative intelligence to synthesize this amazingly practical yet never too simple book of advice and insight." —Edward M. Hallowell, M.D.,
author of *Worry* and *Connect*

ALSO BY AARON T. BECK, M.D.

Depression: Causes and Treatment

The Diagnosis and Management of Depression

Prediction of Suicide (with Harvey L. Resnik and Dan J. Lettieri)

Cognitive Therapy and the Emotional Disorders

Cognitive Therapy of Depression
(with A. John Rush, Brian Shaw, and Gary Emery)

Anxiety Disorders and Phobias (with Gary Emery)

Love Is Never Enough

Cognitive Therapy of Substance Abuse
(with Fred Wright, Cory F. Newman, and Bruce S. Liese)

Cognitive Therapy of Personality Disorders
(with Arthur Freeman & Associates)

Cognitive Therapy with Inpatients: Developing a Cognitive Milieu
(with Jesse H. Wright and Michael E. Thase)

The Integrative Power of Cognitive Therapy
(with Brad A. Alford)

Cognitive Therapy in Clinical Practice
(with Jan Scott and J. Mark Williams)

Scientific Foundations of Cognitive Theory and Therapy of Depression
(with D. A. Clark and B. A. Alford)

PRISONERS OF HATE

THE COGNITIVE BASIS OF
ANGER, HOSTILITY, AND VIOLENCE

Aaron T. Beck, M.D.

HARPER

NEW YORK • LONDON • TORONTO • SYDNEY

HARPER

A hardcover edition of this book was published in 1999 by HarperCollins Publishers.

PRISONERS OF HATE. Copyright © 1999 by Aaron T. Beck, M.D. All rights reserved. Printed in the United States of America. No part of this book may be used or reproduced in any manner whatsoever without written permission except in the case of brief quotations embodied in critical articles and reviews. For information, address HarperCollins Publishers, 195 Broadway, New York, NY 10007.

HarperCollins books may be purchased for educational, business, or sales promotional use. For information, please e-mail the Special Markets Department at SPsales @harpercollins.com.

First Perennial edition published 2000.

Designed by Kyoko Watanabe

The Library of Congress has catalogued the hardcover edition as follows:
Beck, Aaron T.
    Prisoners of hate: the cognitive basis of anger, hostility, and violence / Aaron T. Beck.—1st ed.
        p.   cm.
    ISBN 0-06-019377-8
    Includes bibliographical referneces and index.
        1. Violence—Psychological aspects 2. Hate. 3. Anger. 4. Hostility (Psychology).
        5. Cognitive therapy. I. Title.
RC569.5.V55B43   1999
616.85'82—dc21
99-22651

ISBN 0-06-093200-7 (pbk.)

17 18 19 20 ❖ /RRD 30 29 28 27 26 25 24 23 22 21 20

CONTENTS

ACKNOWLEDGMENTS

I am grateful to the many people who contributed to the preparation of this volume. First of all, I am deeply grateful to Barbara Marinelli, who coordinated the numerous operations from beginning to end and assisted in typing the manuscript. I also enormously appreciate Jessica Grisham's help in the preparation of the literature review and her editing; she also typed a large portion of the manuscript. I was particularly fortunate to have the professional help of my colleague Kevin Kuehlwein, who contributed his exceptional clinical insight and encyclopedic knowledge in his review and critique of the manuscript. His feedback greatly enhanced the manuscript. I also want to thank Frances Apt and Sally Arteseros for their editorial corrections and suggestions. My professional colleagues have been most helpful in their critiques and suggestions: Jack Bush, Gillian Butler, David M. Clark, Barry Garneau, Paul Gilbert, Joan Goodman, Ruth Greenberg, Robert Hinde, Samuel Klausner, Bob Leahy, Ian Lustick, David Lykken, Joseph P. Newman, Christine Padesky, Jim Pretzer, Paul Salkovskis, Irving Sigel, and Ervin Staub. I also want to thank my research assistants for their help in obtaining reference books and articles: Amy Tapia, Samantha Levy, Rachel Teacher, and Sunny Yuen. Finally, my thanks to my agent, Richard Pine, and my editors, Hugh Van Dusen and Sally Kim.

A word about sexist language: for the most part, I have used male or female pronouns to be compatible either with the subject matter or with the relevant case examples. In the sections on violence, most of the literature and my own personal experience was based on work with male offenders. In the earlier chapters on anger, I drew more on my experience with female patients, since they represented the majority of our clinical population. Also, chapter 8 on intimate relationships is based loosely on the characters and concepts of chapters 2 and 3 of *Love Is Never Enough*.

INTRODUCTION

My approach to interpersonal and social problems can be traced to my work in psychotherapy. In the earlier years—almost four decades ago—I made a series of observations that turned around my understanding and treatment of patients' psychiatric problems. While conducting classical psychoanalysis with patients, I discovered—almost by chance—that they had not been reporting certain thoughts they were experiencing during free association. Although they believed—and I assumed—that they were following the cardinal rule of disclosing everything that went through their minds during therapy, I discovered that they had certain highly significant thoughts at the fringe of their consciousness. The patients were barely aware of, and were certainly not concentrating on, these preconscious thoughts. Based on repeated observations, I suspected that the experience of an emotion or an impulse to do something was generally preceded by such thoughts.

When I coached a patient to focus on these thoughts, I realized that they helped to explain the emotional experience in a more understandable way than the more abstract psychoanalytic interpretations I had been offering. A young woman, for example, was able to access the thought, "Am I boring him?" just prior to spurts of anxiety during therapy. Another patient would have thoughts such as, "Therapy can't help. I'm only going to get worse and worse," prior to a sad feeling. In each instance, there was a logical and plausible connection between thought and feeling. I used a simple technique to capture these fleeting automatic thoughts. When a patient would appear sad or anxious, or report having those feelings, I would ask, "What is going through your mind *right now?*" The patients quickly learned to focus their attention on these thoughts, and it was clear that the thoughts were responsible for the feelings.

Focusing on the thoughts provided a wealth of information that served as a database for explaining not only the patients' emotions but also other psychological phenomena. I discovered, for example, that the patients consistently monitored their own behavior, as well as that of other people. They issued orders to themselves to direct or inhibit activities. They experienced self-critical thoughts when they fell below expectations and self-congratulations when they succeeded.

The themes of their thoughts helped to clarify the specific psychological patterns that produced particular emotions. For example, thoughts (or cognitions) that diminished the patient made him feel sad. These included thoughts of having failed, having been rejected, or having lost something of value. Thoughts of gain and self-enhancement led to feelings of pleasure. Thoughts of danger or threat led to anxiety. Relevant to the topic of this book was the observation that ideas of being wronged by somebody else produced anger and a desire to retaliate. A rapid sequence of thoughts such as, "I should get even," and, "It's okay to hit her," could even culminate in physical violence.

An interesting feature of those thoughts was their fleeting nature. I was surprised to note that even a relatively brief peripheral thought could produce a profound emotion. Moreover, the cognitions were involuntary—the patient could neither initiate nor suppress them. Although they were often adaptive and would reflect an actual loss, gain, danger, or transgression, they were frequently disproportionate or inappropriate to the particular circumstances that triggered them. An anger-prone person, for example, would blow a minor slight or inconvenience out of proportion and want to punish the offender severely.

I also noted, to my surprise, that these patients showed a regular pattern of thinking errors (cognitive distortions). They would greatly magnify the significance of a noxious incident. They exaggerated the frequency of such events: "My assistant *always* messes up," or, "I never get things right." They would attribute what was clearly an accidental or situational difficulty to the other person's bad intentions or character flaw.[1]

The patients characteristically accepted their exaggerated interpretation or misinterpretation at face value—it seemed completely credible. However, when patients learned to focus their attention on these interpretations and to evaluate and question them, they generally realized that they were inappropriate or erroneous. The patients were able to gain perspective on these reactions and, in most instances, to correct them. An easily provoked mother, for example, first observed that she became angry with her children for very minor infractions. When she was able to recognize and respond to her critical thoughts ("they're bad kids") with the idea that they were "just behaving like normal kids," she found that her anger did not last as long. With repeated corrections of her critical, punitive thoughts, they became less frequent.

I was puzzled, however, by this question: why didn't the patients in analytic therapy report these thoughts spontaneously—especially since they were conscientious in expressing whatever came to mind, no matter how embarrassing? Hadn't they been aware of these thoughts in their everyday lives? I came to the conclusion that these thoughts were different from the kind that people generally report to other people. They were part of an internal communication system oriented to the self, a kind of network that was geared to providing ongoing observations about themselves, interpretations of their behavior and that of others, and expectations of what would happen. For example, a middle-aged patient who was engaged in an angry conversation with his older brother had the following sequence of automatic thoughts, which he was able to access despite being involved in the heated interchange. "I'm talking too loudly. . . . He's not listening to me. I'm making a fool of myself. . . . He's got a lot of nerve ignoring what I'm saying. Should I tell him off? He would probably make me look foolish. He never listens to me." My patient was becoming increasingly angry, but on reflecting about the conversation later, he recognized that his anger was not due to the argument but to his overriding interpretation: "My brother does not respect me."

A wife would have the fleeting thought, "My husband is late because he prefers to go out with the other guys," and would feel bad. That is what she communicated to herself. To her husband she would blurt out, "You never come home on time. How can I prepare dinner for the family if you are so irresponsible?" In actuality, her husband would have a beer with the other men in order to unwind after a hard day at work. Her scolding obscured from her husband and herself her feelings of rejection.

The intercommunication system also includes the expectations and demands that people place on themselves and others—something that has been labeled "the tyranny of the shoulds."[2] It is important to recognize these injunctions and prohibitions because rigid expectations or compulsive attempts to regulate the behavior of others are bound to lead to disappointment and frustration.

I was also intrigued by the observation that each patient had his or her own unique set of responses to specific situations and consistently overreacted in an excessive way to certain stimuli but not to others. I was able to predict which interpretations or misinterpretations a particular patient would make in response to a given situation. These overreactions would be apparent in his or her automatic response to specific situations. The patient would characteristically distort, overgeneralize, or exaggerate certain situations but not others that other patients might overreact to.

Certain patterns of beliefs would be activated by a specific set of circumstances and thus generated the thought. These formulas or beliefs constituted a specific vulnerability: when activated by relevant situations, they would shape the patient's automatic interpretation of the situation. The beliefs were highly specific: for example, "If people interrupt me, it means they don't respect me," or, "If my spouse doesn't do what I want, it means she doesn't care." The beliefs provided the meaning of the situation, which was then expressed in the automatic thoughts.

I previously described the angry mother who held the belief, "If kids do not behave themselves, it means they are bad kids." The hurt came from a deeper meaning yielded by the belief, "If my kids misbehave, it shows I'm a bad mother." The overgeneralized belief led to an overgeneralized interpretation. The mother diverted her attention away from the pain of the negative images of herself by blaming her children. Each patient had his or her own specific set of sensitivities.

A similar kind of automatic thought and action occurs when a person is engaged in a routine activity like driving a car. When I'm driving along a city street, for example, I slow down for a pedestrian to cross, steer around a pothole, and pass a slow car ahead of me—all while carrying on a serious conversation with a friend. If I shift my attention to my automatic thoughts about my driving, I become aware of a very rapid sequence—"Watch out for the pothole . . . swerve around it. . . . That guy's going awfully slow . . . is there enough space to pass him?" These thoughts are completely divorced from my conversation with my friend but are controlling my behavior at the wheel.

A NEW THERAPY

As my observations centered on the relations between patients' problematic thoughts—or cognitions—and their feelings and behavior, I developed a cognitive therapy of psychiatric disorders. Applying the theory, I found that helping patients to modify their cognitions resulted in improvement. I consequently applied the term "cognitive therapy" to my therapeutic approach. Cognitive therapy addresses the patients' problems in a number of ways. First, I attempted to give the patients more objectivity toward their thoughts and beliefs. I accomplished this by encouraging them to question their interpretations. Does your conclusion follow from the facts? Are there alternative explanations? What is the evidence for your conclusion? Similarly, we would evaluate the underlying beliefs and formulas. Were they so rigid or extreme that they were used inappropriately and excessively?

These therapeutic strategies helped the patients avoid overreacting to situations. Around the same time that I was formulating my theory and therapy, I was pleased to discover the writings of Albert Ellis. His work, which antedated my own publication by several years, was based on observations similar to my own. I derived a number of new ideas regarding the therapy from his writings. Several of the strategies just described were adapted from Ellis's work.[3]

I observed that these findings were not restricted to people with common, "garden-variety" psychiatric disorders, such as depression and anxiety. The same kinds of erroneous beliefs influenced the feelings and behavior of people experiencing marital problems, addictions, and antisocial behavior.[4] Other therapists who were specialists in these areas developed and applied cognitive theory and therapy to their specific area of specialization. A large body of literature has evolved on the cognitive therapy of various forms of antisocial behavior: spouse battering and child abuse, criminal assaults, and sexual offenses. We observed a common denominator across these various forms of harmful behaviors: namely, that the victim is perceived as the Enemy, and the aggressor sees himself as an innocent victim.

Because I believe that people have the same mental processes when they are engaged in either individual or group violence, I explored the literature on such social ills as prejudice, persecution, genocide, and war. Although there are large differences in the sociological, economic, and historical causes, the final common denominator is the same: the aggressors have a positive bias regarding themselves and a negative bias toward their adversary, often conceived as the Enemy. I was struck by the similarities between a spouse's image of her estranged husband, a militant's image of a racial or religious minority, and a soldier's image of a sniper shooting at him from a tower. Words such as *monster, evil,* or *bastard* are frequently used by these individuals to designate the dangerous Other. When they are in the grip of these extreme patterns of thought, their evaluations of their supposed foes are warped by hatred.

I have prepared this volume with the goal of clarifying the typical psychological problems that lead to anger, hatred, and violence. I have also tried to clarify how these problems manifest themselves in conflicts between friends, family members, groups, and nations. Sharpening our insights into the cognitive factors (interpretations, beliefs, images) can provide some leads in remediating the personal, interpersonal, and social problems of modern society.

In preparing a volume such as this, certain questions naturally arise. What is new and useful about this approach? What is the evidence that the approach is valid and not simply a statement of opinion? I had to ponder similar questions when I first proposed my cognitive theory and therapy of depression, first in 1964 and then in extended form in 1976. Since then my colleagues and I have reviewed almost one thousand articles evaluating specific aspects of the theory. These articles have been largely supportive of the empirical basis and validity of the theory and therapy.[5] A substantial portion of the assumptions validated in these studies also form the basis for the concepts offered in this volume.

In addition to the clinical material, a substantial component of the volume rests on a body of knowledge regarding the cognitive aspects of anger, hostility, and violence in the literature of clinical, social, development, and cognitive psychology. Many formulations regarding broader issues such as group violence, genocide, and war were developed in part from the literature of political science, history, sociology, and criminology.

I have planned the book to introduce the interlocking concepts in a sequential fashion, although they are all part of the same matrix. I begin with the clarification of hostility and anger in everyday life, a subject that readers may be able to relate to their own experiences. I then move on to topics of crucial societal importance: family abuse, crime, prejudice, mass murder, and war. Even though these phenomena are far removed from the personal experiences of most readers, the underlying psychology is similar. Finally, I offer suggestions regarding the application of these insights to personal and societal problems.

PART 1

The Roots of Hate

1

THE PRISON OF HATE

How Egoism and Ideology Hijack the Mind

> It is a magnificent feeling to recognize the unity of a complex of phenomena that to direct observation appear to be separate things.
>
> *Albert Einstein (April 1901)*

The violence of humans against humans appalls us but continues to take its toll today. The dazzling technological advances of our era are paralleled by a reversion to the savagery of the Dark Ages: unimaginable horrors of war and wanton annihilation of ethnic, religious, and political groups. We have succeeded in conquering many deadly diseases, yet we have witnessed the horrors of thousands of murdered people floating down the rivers of Rwanda, innocent civilians driven from their homes and massacred in Kosovo, and blood flowing in the killing fields of Cambodia. Wherever we look, east or west, north or south, we see persecution, violence, and genocide.[1]

In less dramatic ways, crime and violence reign in our countries and cities. There seem to be no limits to the personal misery people inflict on one another. Close, even intimate, relationships crumble under the impact of uncontrolled anger. Child abuse and spouse abuse pose a challenge to legal as well as mental health authorities. Prejudice, discrimination, and racism continue to divide our pluralistic society.

The scientific advances of the age are mocked by the stasis in our ability to understand and solve these interpersonal and societal problems. What can be done to prevent the misery inflicted on the abused child, the battered wife? How can we reduce the medical complications of hostility, including soaring blood pressure, heart attacks, and strokes? What guidelines can be developed to address the broader manifestations of hostility that tear apart the fabric of civilization? What can the policy makers and social engineers—and the average citizens—do? Sociologists, psychologists, and political scientists have made concerted efforts to analyze the social and economic factors leading to crime, violence, and war. Yet the problems remain.

A Personal Experience

Sometimes a relatively isolated experience can expose the inner structure of a phenomenon. I received a clear insight into the nature of hostility many years ago when I was its target. I had received the usual laudatory introduction at a book-signing event in a large bookstore and had just completed a few introductory remarks to an audience of colleagues and other scholars. Suddenly, a middle-aged man, whom I shall call Rob, approached me in a confrontational manner. I recalled later that he seemed "different"—stiff, tense, his eyes glaring. We had the following interchange:

Rob (*sarcastic tone*): Congratulations! You certainly drew a large crowd.
ATB: Thanks. I enjoy getting together with my friends.
Rob: I suppose you enjoy being the center of attention.
ATB: Well, it helps to sell books.
Rob (*angry tone*): I guess you think you're better than me.
ATB: No. I'm just another person.
Rob: You know what I think of you? You're just a phony.
ATB: I hope not.

At this point it was clear that Rob's hostility was rising, that he was getting out of control. Several of my friends moved in and, after a brief scuffle, led him out of the store.

Although this scene might be dismissed as simply the irrational behavior of a disturbed person, I believe it shows, in bold relief, several facets of hostility. The exaggerations in the thinking and behavior of clinical patients often delineate the nature of adaptive as well as excessive human reactions. As I reflect on the incident now, I can note a number of features that illustrate some universal mechanisms involved in the triggering and expression of hostility.

First, why did Rob take my performance as a personal affront, as though I were in some way injuring him? What struck me—and was obvious to the other observers with a background in psychology—was the egocentricity of his reaction: he interpreted the recognition I received as having *diminished* him in some way. Such a reaction, although extreme, probably is not as puzzling as it may seem. Others in the audience may have been thinking about their own professional status—whether they deserved recognition—and may have experienced regret or envy. Rob, however, was totally absorbed in how my position reflected on him; he personalized the experience as though he and I were adversaries, competing for the same prize.

Rob's exaggerated self-focus set the stage for his anger and his desire to attack me. He was impelled to make an invidious comparison between the two of us, and in accord with his egocentric perspective, he presumed that others would consider him less important than I, perhaps less worthy. Also, he felt left out, because he was receiving none of the attention and friendship that were being given to me.

The sense of social isolation, of being disregarded by the rest of the group, undoubtedly hurt him, a reaction commonly reported by patients in like situations. But why didn't he simply experience disappointment or regret? Why the anger and hatred? After all, I was not doing anything to him. Yet he perceived an injustice in the proceedings: I was no more deserving of recognition than he was. Therefore, since he was wronged, he was entitled to feel angry. But he carried this further. His statement, "You think you're better than me," shows the degree to which he personalized our interaction. He imagined what my view of him would be, and then projected it into my mind, as if he knew what I was thinking (something I call the projected image). In essence Rob was using (actually overusing) a frequent and mostly adaptive device: mind-reading.

Reading other people's minds, to some extent, is a crucial adaptive mechanism. Unless we can judge other people's attitudes and intentions toward us with some degree of accuracy, we are continuously vulnerable, stumbling blindly through life. Some authors have noted a deficiency in this capability in autistic children, who are oblivious to other people's thinking and feeling.[2] In contrast, Rob's interpersonal sensitivity and mind-reading were exaggerated and distorted. His projected social image became a reality for him, and with no evidence at all, he believed that he knew what I thought of him. He attributed derogatory thoughts to me, which inflamed him even more. He felt a pressure to retaliate against me because, according to his logic, I had wronged him. *I was the Enemy.*[3]

The egocentric monitoring of events to ascertain their significance, as demonstrated by Rob, is discernible throughout the animal kingdom and is apparently embedded in our genes. Self-protection, as well as self-promotion, is crucial to our survival; both acts help us to detect transgressions and take appropriate defensive actions. Also, without this kind of investment in ourselves we would not seek the pleasures we gain from intimate relations, friendships, and affiliation with groups. Egocentricity is a problem, however, when it becomes exaggerated and is not balanced by such social traits as love, empathy, and altruism, the capacity for which is probably also represented in our genome. Interestingly, very few of us think to look for egocentricity in ourselves, although we are dazzled by it in others.

Once an individual becomes aroused to fight in an ordinary dispute, all of his senses are focused on the Enemy. In some instances this intense narrow focus and mobilization for aggressive action may be life-saving; for example, when one is subjected to physical attack. In most cases, however, the reflexive image of the Enemy creates destructive hatred between individuals and between groups. Although these individuals or groups may feel liberated from restraints against attacking the supposed adversary, such people have actually surrendered their freedom of choice, abdicated their rationality, and are now the prisoners of a primal thinking mechanism.

How can we enable people to recognize and control this automatic mechanism so that they can behave in a more thoughtful, moral way toward each other?

THE HOSTILE FRAME

These egocentric components of anger and aggression have been confirmed by my professional work with patients, but the experience with Rob was my most dramatic public experience. I have wondered for many years whether the insights into human problems gleaned from the psychotherapy of troubled individuals could be generalized to apply to societal problems of violence within families, communities, national groups, and states. Although these domains appear to be remote from each other, the themes underlying anger and hatred in close relationships appear to be similar to those manifested by antagonistic groups and nations. The overreactions of friends, associates, and marital partners to presumed wrongs and offenses are paralleled by the hostile responses of people in confrontation with members of different religious, ethnic, or racial groups. The fury of a betrayed husband or lover resembles that of a member of a militant group who believes his cherished principles and values have been betrayed by his own government. Finally, the biased, distorted thinking of a paranoid patient is akin to the thinking of perpetrators embarking on a program of genocide.

When I was first concerned with the psychotherapy of distressed couples, it became clear that, at least in severe cases, simply coaching people on how to change their distressing behavior—in essence how to "do the right thing"—would not provide a durable solution. No matter how committed they were to following a proposed constructive plan of action, reasonable communication and civil behavior vanished when they became angry with each other.[4]

A clue to their inability to adhere to prescribed guidelines when they felt hurt or threatened lay in their misinterpretation of each other's behavior. "Catastrophic" distortions of each other's motives and attitudes led each partner to feel trapped, injured, and depreciated. These perceptions (or rather, misperceptions) filled them with anger—even hatred—and impelled them to retaliate or to withdraw into hostile isolation.

It was clear that the chronically feuding couples had developed a negative "frame" of each other. In a typical case, each partner saw himself or herself as the victim and the other partner as the villain. Each partner blotted out the favorable attributes of the other as well as the pleasant memories of more tranquil days, or reinterpreted them as false. The process of framing led them to suspect each other's motives and to make biased generalizations about the deficiencies or "badness" of the mate.[5] This rigid negative thinking was in marked contrast to the many ways in which they could flexibly think through solutions to problems encountered in relationships outside their marriage. Their minds, in a sense, were usurped by a kind of primal thinking that forced them to feel mistreated and to behave in an antagonistic manner toward the presumed foe.

There was a bright side to this clinical picture, however. When I helped the partners to focus on their biased thinking about each other and to reframe their negative images, they were able to judge each other in a less pejorative, more realistic way. In many instances they were able to recapture their previous affectionate feelings and form a more stable, satisfying marital relationship. Sometimes the vestiges of their biased perspectives were so strong that the partners decided to separate—but in an amicable way. We could then attain a kind of balanced partition of the family. Relieved of their hatred for each other, the former partners could work out a reasonable settlement of custody and financial issues. Since this approach to couples' problems focused on biased thinking and cognitive distortions, I labeled the treatment "cognitive marital therapy."[6]

I noted the same type of hostile framing and biased thinking in encounters between siblings, parents and children, employers and employees. Each adversary inevitably believed he or she had been wronged and the other persons were contemptible, controlling, and manipulative. They would make arbitrary—often distorted—interpretations of the motives of those with whom they were in conflict. They would take an impersonal statement as a personal affront, attribute malice to an innocent mistake, and overgeneralize the other's unpleasant actions ("You *always* put me down. . . . You *never* treat me as a person").

I observed that even people who were not psychiatric patients were susceptible to this kind of dysfunctional thinking. They routinely framed outgroup members negatively, just as they framed their everyday friends or relatives with whom they were in conflict. This kind of negative framing also appeared to be at the core of negative social stereotypes, religious prejudice, and intolerance. A similar sort of biased thinking seemed to be a driving force in ideological aggression and warfare.

People in conflict perceive and react to the threat emanating from the image rather than to a realistic appraisal of the adversary. They mistake the image for the person.[7] The most negative frame contains an image of the adversary as dangerous, malicious, and evil. Whether applied to a hostile spouse or to members of an unfriendly foreign power, the fixed negative representation is backed up by selective memories of past wrongs, real or imaginary, and malevolent attributions. Their minds are encased in "the prison of hate." In ethnic, national, or international conflict, myths about the Enemy are propagated, giving the image a further dimension.

Insights about harmful behavior can be gleaned from a variety of clinical sources. Patients being treated for substance abuse, as well as other patients who receive the diagnosis of "antisocial personality," provide rich material for an understanding of the mechanisms of anger and destructive behavior.

Bill, a thirty-five-year-old salesman, was addicted to a variety of street drugs and was particularly prone to rage reactions and to physical abuse of his wife and children, as well as to frequent fights with outsiders. As we collaboratively explored the sequence of psychological experiences, we found that when another person (his wife or an outsider) did not show him "respect," as he defined it, he would become so enraged that he wanted to punch or even demolish that person.

Through a "microanalysis" of his rapid-fire reactions, we found that between the other person's statement or action and his own flare-up, Bill experienced a *self-demeaning thought* and a *hurt feeling*. His typical self-deflating interpretation leading to this unpleasant feeling occurred almost instantaneously: "He thinks I'm a wimp," or, "She doesn't respect me."

When Bill learned to detect and evaluate this intervening painful thought, he could recognize that his interpretation of being put down did not necessarily follow from the actual comment or behavior of the other person. I was then able to clarify the beliefs that shaped his hostile reaction. A primary belief, for example, was, "If people disagree with me, it means they don't respect me." What then provoked Bill to attack the offender were his afterthoughts, which were conscious and compelling: "I need to show them they can't get away with this, so they'll know I'm not a wimp, that they can't push me around." It was important for Bill to recognize that these punitive afterthoughts were the result of his feeling hurt, which was covered over by his anger. Our therapeutic work consisted of examining Bill's beliefs and helping him to understand that he could obtain more respect from his family and acquaintances by being "cool" and controlled than belligerent and irascible.

The clues obtained from analyzing the reactions of Bill and other anger-prone persons indicate that these individuals attach a high value to their social image and status. Their personal system of beliefs defines their conclusions regarding the supposed offender. The psychologist Kenneth Dodge found that these kinds of beliefs and the consequent interpretations of events are common among a broad range of individuals who are prone to engage in harmful behavior. For instance, the same type of aggressive beliefs expressed by Bill are held by young children who later become delinquent and include the following:[8]

- The offender wronged them in some way and was thereby responsible for their feelings of hurt and distress.

- The injury was deliberate and unjustified.

- The offender should be punished or eliminated.

These conclusions are derived, in part, from the rules of conduct they impose on other people. Those demands and expectations are similar to the phenomenon labeled the "tyranny of the shoulds" by the psychiatrist Karen Horney.[9] People like Bill believe:

- People must show respect for me at all times.

- My spouse should be sensitive to my needs.

- People should do what I ask of them.

The kind of framing that occurs in angry conflicts may be observed in exaggerated form in the angry paranoid patients. These patients consistently attribute malevolent intent to others and experience urges to punish them for their supposedly antagonistic behavior. Some paranoid patients suffer from persecutory delusions precipitated by traumatic events that lowered their self-esteem—for example, failure to receive an expected promotion.[10] Their persecutory delusions seem, in part, to be an explanation that protects their self-image, as though they are thinking, "You caused my problems because you are prejudiced," or, ". . . because you are conspiring against me." Most of these patients become fearful; others become angry and want to attack the supposed victimizers.

OF HATE AND THE ENEMY

We have often heard people use the expression "I hate you" when they are simply expressing anger. At times, however, intense anger may indeed swell into a state that we can reasonably label "hatred," even though it is transient.

Examine the following interchange between a father and his fourteen-year-old daughter:

Father: What are you up to?
Daughter: I'm leaving now. I'm going to a rock concert.
Father: No, you don't! You know you're grounded.
Daughter: This isn't fair . . . this is a prison.
Father: You should have thought of that before.
Daughter: I can't stand you . . . I hate you!

At this point, the daughter would like to eliminate her father, whom she envisions as a beast hovering over her and blocking her from doing what she "needs" to do. At the peak of a hostile confrontation, individuals see each other as combatants, ready to attack. The father is threatened by his daughter's apparent willfulness, and the daughter by her father's apparently unjust domination and interference. Of course, they are actually disturbed by over-simplified projected *images* of each other. In most such parent-child conflicts, however, the child's hatred eventually subsides along with the anger. When there is a background of continual abuse or frustration by the parent, the child's episodes of intense anger can become transformed into chronic hatred. The child has a fixed image of the parent as a monster and herself or himself as vulnerable to permanent torture.

Similarly, a parent who perceives her child as unreliable, devious, or rebellious may feel acute or chronic anger, without hatred. But once the parent feels vulnerable and views the child as an implacable foe, then she may feel hatred. The hatred experienced between parents and children, between divorced partners, or between siblings can persist for decades, even permanently. The experience of hatred is profound and intense and is probably qualitatively different from the everyday experience of anger. Once the hatred is crystallized, it is like a cold knife poised to plunge into the back of an adversary.

In severe conflicts the adversary may be perceived as ruthless, malicious, and even murderous. Consider the statement of a wife involved in a child custody battle with her husband. "He's irresponsible. He has a terrible temper. He always takes it out on the kids and me. I know he will abuse them. I can't trust him . . . I hate him. I'd like to kill him."[11] Although such negative perceptions of one's former mate may sometimes be accurate, in most cases they are exaggerated.

Because the imagined Enemy may *appear* dangerous, vicious, or evil, the supposed victim feels compelled either to escape or to eliminate the threat by incapacitating or killing the Enemy. In civilian conflicts the actual danger is usually—but not always—greatly exaggerated. The threat is frequently directed not to people's physical being but to their psyches—their pride, their self-image—particularly if they believe their adversary has gotten the upper hand. The sense of vulnerability is generally out of proportion to the adversary's actual transgression.

In some cases the interplay of malevolent images on both sides can lead to homicidal impulses. A jealous husband has fantasies of taking revenge against his ex-wife, who has obtained sole custody of the children and is now living with another man. He feels powerless, trapped, hopeless. He thinks, obsessively, "She has taken everything away from me—my children, my honor. I'm nothing." He believes that he cannot tolerate his anguish or continue to live with this horror, so he formulates a plan to shoot his wife, her lover, and then himself. By this deed, he will presumably settle the score, alleviate his suffering, and regain a sense of power before he shoots himself.

If the husband is receiving treatment at the time, the therapist can demonstrate that the man's major problem is not really his wife but his wounded pride and sense of powerlessness, which can improve as he gains perspective about the situation.[12]

The urge to exact revenge on the supposed tormentor in such a case is so powerful and so primitive that it may be suspected of having evolved from an ancient ancestral setting, where inflicting the supreme punishment for "betrayal" and "treachery" had survival value. Some writers believe that this mechanism is innate in human beings, the result of evolutionary pressures.[13]

The concept of the personal Enemy has its counterpart in warfare between groups. In armed conflict, feeling hatred toward the enemy is adaptive. A soldier who assumes that his image is in the crosshairs of an enemy's telescopic rifle experiences hate as a primitive survival strategy. The powerful framed image of the opponent helps the soldier to fix his attention on his foe's vulnerability and to mobilize his resources to defend himself. The formula "kill or be killed" defines the problem in simplified, unambiguous terms.

The same kind of primal thinking is activated when members of a group are moved to punish supposed offenders. The irrational framing of other people as the Enemy is obvious in incidents of mob violence. Members of a mob engaged in a lynching or soldiers on a rampage of killing innocent villagers are oblivious to the fact that they are destroying human beings like themselves. They do not realize that the impetus for their violent actions comes from their highly charged, primitive thinking. The malevolent images of the victim spread across the group like wildfire. Since they perceive their victims as bad or evil, they are driven by thoughts of vengeance. Inhibitions against killing are automatically lifted by the belief that they are doing the right thing: evildoers *must* be exterminated. Such violent behavior carries immediate rewards, since it relieves their anger, confers a sense of power, and produces satisfaction from the notion of justice having been done.

A member of the marauding mob believes that he is exercising freedom of choice. In actuality, the decision to kill has been made automatically by his mental apparatus, which has been hijacked by the primitive imperative to eliminate a dangerous or loathsome entity. Although the impulse to harm or kill at this point in the hostility cycle is in a sense involuntary, the individual soldier or mob member still has the capacity to control it voluntarily. A more durable remediation of the destructive tendencies needs to be directed toward the primitive belief system that frames the victims as Evil, the system of rules that dictate that they should be punished, and the permissive belief system that waives the rules against harming other human beings.

History is replete with instances in which enmity between families, clans, tribes, ethnic groups, or nations is perpetuated from generation to generation. Some feuds are legendary, like those between the Hatfields and the McCoys, and the Montagues and the Capulets in *Romeo and Juliet*. In act 1, scene 1, of this play, the Prince takes to task his fractious subjects:

> Rebellious subjects, enemies to peace,
> Profaners of this neighbor-stained steel,—
> Will they not hear? What ho! you men, you beasts,
> That quench the fire of your pernicious rage
> With purple fountains issuing from your veins,
> On pain of torture, from those bloody hands
> Throw your mistemper'd weapons to the ground,

And hear the sentence of your moved prince.
Three civil brawls, bred of an airy word,
By thee, Old Capulet, and Montague,
Have thrice disturb'd the quiet of our streets . . .
If ever you disturb our streets again,
Your lives shall pay the forfeit of the peace.

Recent examples of internecine warfare include the Hutus and Tutsis in Rwanda, Jews and Arabs in the Near East, and Hindus and Muslims in South Asia. Nowhere is the creation of the image of the Enemy better illustrated than in the assault by the Serbs on the Muslims in Bosnia. After the breakup of the Communist state in Yugoslavia in 1990, a coalition of nationalists and political and military leaders embarked on a mission to establish a purely Serbian state at the expense of the Muslims. The Serbian leadership stirred up lurid memories of the domination of the Serbs by the Turks and the Bosnian Muslims. Allegedly to preserve their own nation, the power structure promoted a campaign of "ethnic cleansing" and exterminated or displaced thousands of Muslims.[14] More recently, Serbian troops burned villages and massacred civilians in the Yugoslav province of Kosovo in response to a rebellion by ethnic Albanians.

The vivid image of the Muslims as the Enemy, generated by the Serbian leadership, fueled the massacres. Their message was: we suffered through centuries of Turkish domination and will no longer tolerate the yoke of their descendants. Of course, it was the highly charged *image* of "the oppressors" that the Serbian populace could not tolerate. There was little discernible difference between individual Serbs and Muslims, who had lived peacefully together for long periods of time.

Dramatizing the image of the Enemy is also a convenient way for national leaders to explain away military or economic reversals. By attaching responsibility for military defeat to a stigmatized minority, a leader can ameliorate the national image of weakness and vulnerability. Hitler, for example, capitalized on the availability of the Jews to explain Germany's World War I defeat, the political humiliation following the Armistice, and the subsequent inflation and depression.[15] By depicting Jews as warmongers, international capitalists, and Bolsheviks, he projected an evil image onto this vulnerable group.

The scapegoating was empowering for the Nazis. The act of degrading and persecuting the Jews helped to intensify the demonical image. The logical consequence was to eliminate the Enemy so they could not again wreak their alleged destruction (causing wars, oppressing nations economically, and polluting the culture). Hitler helped to create empathy and self-pity among his followers by portraying *them* as the victims of the Jews' control, subversion, and corruption. The "victims" thus became the victimizers—with all the power of an efficient wartime bureaucracy to carry out the Final Solution.[16]

The leaders of a nation taking their country into war may have a sober view of the Enemy. A war of conquest does not require that the leaders actually hate their opponents. But military action is more likely to succeed when the soldiers and the civilian populations view the opponents as an Evil to be destroyed. A military adventure may be a political gamble for the leaders of the government, but it is a life-or-death encounter for the soldiers, who see their personal sacrifices as acts of heroism.

It is possible to detect a unifying theme running through the spectrum of anger, hostility, and antagonistic behavior, from personal verbal abuse to prejudice and bigotry to war and genocide. Anger is anger, whether provoked by a rebellious child or a rebellious colony; hate is hate, whether evoked by an abusive spouse or a ruthless dictator. No matter what the external causes of antagonistic behavior, the same internal or psychological mechanisms are generally involved in its arousal and expression. And as with destructive interpersonal action, cognitive distortions incite anger and prompt the hostile behavior. Thus, unwarranted personal attacks that arise from prejudice, bigotry, ethnocentrism, or military invasion involve the primal thinking apparatus: absolute categorical cognition, on the one hand, and obliviousness to the human identity of the victims, on the other.

If certain cognitive commonalities exist, they can simplify the task of developing remedial psychological interventions. They can provide a framework for conflict resolution between individuals and groups and form a basis for workable solutions to the problems of crime and mass murder. Taking a page out of the psychotherapy experience, we can identify the cognitive biases and apply antidotes based on the understanding, clarification, and modification of these processes and the primal belief systems that underlie them.

THE DIFFERENT ROUTES TO VIOLENCE

Various pathways lead to destructive behavior. *Cold, calculated violence,* for example, does not necessarily require any animosity toward the victim. An armed robber holding up a clerk at a convenience store does not necessarily have any hard feelings toward the clerk, or toward the owner. Similarly, an officer pushing buttons on a console to launch a missile does not necessarily feel angry at the civilian target. The Mongol troops that lay siege to and destroyed the cities opposed to their sweep through Europe had no specific hostility toward the inhabitants of the cities. The master plan rationally conceived by Genghis Khan prior to the invasions called for the total destruction of recalcitrant cities in order to intimidate other cities into submitting to them without a fight. The pleasure derived from the pillaging—like violence from any cause—was undoubtedly reinforcing to the troops.

Other tyrants have made cold-blooded decisions to inflame their people to support aggression against a neighbor or a minority group in their country. Hitler used this strategy in 1939 when he spread rumors that the Czechs were persecuting the minority German population in the Sudetenland of Czechoslovakia. He later invaded Poland to implement his plan of clearing out the population to make room for a greater Germany. Stalin, Mao, and Pol Pot rounded up and killed large sectors of their population in order to enforce their corresponding Communist ideologies and consolidate their power. Their violence was *instrumental*—a job to be done for political and ideological purposes. Instrumental violence is particularly dangerous because it is generally based on the doctrine, "The end justifies the means."

Writers have condemned this kind of justification for ages, but it continues to play a major role in the conduct of international relations. Aldous Huxley's volume of essays *Ends and Means* provides the philosophical underpinnings for repudiating this doctrine.[17] Nonetheless, tyrants (like Saddam Hussein and Stalin) have invaded weak neighbors (Kuwait and Finland), and national groups (like the Serbs) have massacred vulnerable minorities (the Bosnians) to achieve supposedly worthwhile ends. Guards in the Nazi death camps believed they were being model citizens when they sent countless Jews to their death. While world opinion decries such acts, the problem exists: Evil deeds are in the eye of the outside observer—not in the eye of the perpetrator.

Hot, reactive violence is characterized by hatred of the Enemy. The thinking apparatus of individuals who participate in massacres and lynchings becomes focused on the Enemy and produces progressively more extreme images. First, the members of the opposition are *homogenized;* they lose their identities as unique individuals. Each victim is interchangeable, and all are disposable. In the next stage the victims are *dehumanized.* They are no longer perceived as human beings for whom one can feel empathy. They could just as easily be inanimate objects, like mechanical ducks in a shooting gallery or targets in a computer game. Finally, they are *demonized:* the embodiment of Evil. Killing them is no longer optional; they *must* be exterminated. Their continued existence becomes a threat. The abstract notions of Evil[18] and the Enemy are transformed into a concrete image of an entity or force that appears to threaten the existence or vital interests of the aggressors. These reified concepts are projected onto the victims. We attack the projected image, but harm or kill real people.

Hot violence is reactive in nature: an external situation, such as a perceived threat, shifts the leader and his followers into the fighting mode. The external circumstances operate at multiple levels. For example, the arms race produced instability in Europe prior to World War I. As coalitions of European states lined up, each side saw the other as the dangerous Enemy. This produced increasing fear and loathing on the part of the leaders and followers that led to a preemptive strike by Germany.[19]

In a completely different domain, an unstable marital relationship sets the stage for an exchange of insults because each partner sees the other as a mortal Enemy, culminating in blows and the husband battering the wife into submission. Wives are also capable of violent behavior during these altercations.[20] Alcohol often accentuates the loss of objectivity and releases the inhibitions of the violent partner.

In domestic violence the aggressor is locked into the primal thinking mode, which by its exclusive focus on the Enemy precludes empathy for the victim and concern about the long-range consequences of violence. In many instances the batterer is genuinely sorry later (and presumably more sober)— after having regained perspective. The problem here is not a lack of morality per se, but the vise-like hold of the primal thinking, oriented to fighting. The ultimate remedial approach, to be described later in this volume, is the clarification and modification of the belief system that predisposes the individual to overreact to supposed threats, the development of strategies to catch the hostile sequence in its earliest stage, and the abandonment of violence as an acceptable weapon.

In addition to the deliberate, planning type of thinking associated with instrumental (cold) violence and the reflexive thinking of reactive (hot) violence, we can identify a kind of *procedural thinking* that is involved in carrying out destructive assignments. This kind of "low-level" thinking is characteristic of people whose attention is fixed totally on the details of a destructive project in which they are engaged. Procedural thinking is typical of functionaries who fastidiously carry out the destructive assignments, apparently oblivious to their meaning or significance. These individuals can be so focused on what they are doing—a kind of tunnel vision—that they are able to blot out the fact that they are participating in an inhuman action. It seems likely that if they do think about it, they regard the victims as disposable. This kind of thinking was evidently typical of the bureaucrats in the Nazi and Soviet apparatus.[21]

Where can a reformer begin in order to attach responsibility for these two forms of hostile aggression? Obviously, accountability for the cold violence lies in the author of the grand design, ideology, or political dictum that claims the desired goals justify whatever means are necessary to achieve them. Nonetheless, no acts of group violence are possible without the cooperation of the followers, the bureaucrats, and in many cases, the average citizens. Exercising complete freedom of the will, tyrants like Genghis Khan or Saddam Hussein deliberately follow a clearly formulated plan to acquire material wealth through plundering weaker nations. By the same token, the Crusaders during the Middle Ages implemented a well-articulated ideology when they massacred "infidels" on their way to do "God's work" in the Holy Land. And Stalin and Mao fortified their political and economic revolutions through the death of millions of their citizens.

In these instances the authors of the master plan were psychologically free to reflect on the human consequences of their goals. They were capable of considering the victims of their actions in deliberating the costs and benefits of their plan. They could have superimposed a higher level of morality that proscribes killing, but they chose not to.

The international community needs to make it clear that those who carry out destructive instructions are as accountable as those who issue them.

Recent international trials of perpetrators in Bosnian and Rwandan massacres represent an important step in fortifying this principle.

GUILT, ANXIETY, SHAME, AND INHIBITION

Although writers like Roy Baumeister propose that guilty feelings are the main deterrents to harmful behavior, these feelings are seldom experienced during the hostility sequence.[22] People may feel guilty *after* they have done something that they consider wrong. When they evaluate their actions and conclude that they have hurt another person and that it was unwarranted, then they may feel guilty. The memory of this incident may influence their behavior the next time a similar situation arises. The memory acts as a deterrent since it prods the individual to avoid doing something he realizes will make him feel guilty later.

Say I am excessively critical of an assistant. I realize after the fact that I have hurt him, and I feel guilty. This event leads to the formation of a rule: "In the future, show more restraint in your criticisms." The next time my assistant makes a mistake and I feel inclined to blame him, the past memory and the newly minted rule regarding restraint give me a twinge of guilt; I restrain the critical impulse and go no further with it. I also generalize the rule to my critical reactions to other people.

Empathy for the object of the hostility is often sufficient to inhibit the aggressor from inflicting an injury in the first place. In cognitive therapy, we have successfully utilized techniques of empathy training to facilitate an angry person's identification with the target victim (see chapter 8).

Certain precepts that are drummed into us at an early age provide the framework for constructing rules that may influence our later behavior. Even young children recognize that it is wrong to hurt a playmate or get him into trouble. When the impulse to hurt another child is strong, however, they permit themselves to break the rule: the power of the impulse can find a justification for its expression (for example, "she hit me first"). Similarly, adults generally label deliberate physical attacks on other people as immoral. The capacity to feel empathy for others helps to stamp into our minds: "This act is wrong."

There have been many reports of soldiers or police who found it impossible to execute prisoners at point-blank range.[23] Christopher Browning describes how some members of a German police battalion charged with killing Jews in Poland became nauseated and had to withdraw from the scene.[24] Unfortunately, soldiers and secret police who at first sight are automatically revolted by torturing or killing become desensitized to gruesome acts after several exposures. Indeed, some begin to enjoy the feeling of power and righteousness. This reaction suggests that the initial revulsion is associated with empathetic identification with the victim rather than with feelings of guilt. When the identification with the victim wears off, so does the revulsion.

The experience of anxiety in anticipation of the consequences of harmful behavior triggers another important automatic inhibitory mechanism. As a person becomes mobilized to abuse another person, the fear of retaliation by the other person or of punishment by an authority can defuse the hostile impulse. For example, an older child about to hit a younger sibling might experience a flash image of an angry parent and stop himself. The fear of being shamed by other people also inhibits us from crossing the boundary of reasonable behavior in our dealings with rivals or opponents. Our public image exerts powerful control over our actions because of its capacity for evoking pain and shame.

In addition to the negative deterrents to antisocial acts, there are positive factors that promote benign behaviors. We generally like to think of ourselves as mature and kind persons. Impulsive acts connote immaturity, whereas self-control makes us feel proud of ourselves. Self-restraint also helps to bolster our own self-image of being worthwhile, desirable human beings. We all have ideals, values, and standards and expectations of ourselves, often encapsulated in our system of injunctions and prohibitions as "shoulds" and "should-nots." We are generally pleased with ourselves when we live up to our ideal self-image, and displeased when we deviate from it. We may reflect that a harmful act was not worthy of us, feel guilty, and repent. Finally, we make a considered decision to control a hostile impulse, not because of shame, guilt, anxiety, or self-criticalness, but because we know it is personally unacceptable.

Although the mechanisms of anxiety, guilt, and shame may retard the expression of hostility, they do not get at the factors that instigate it in the first place. Further, commandments not to kill or harm others may put the brakes on hostile impulses, but they do not extinguish them. It is of crucial importance to examine the permissive beliefs and justifications that enable us to disregard these commandments.

THE MORAL PARADOX:
A COGNITIVE PROBLEM

On February 25, 1994, Dr. Baruch Goldstein opened fire with an automatic weapon and killed or wounded at least 130 worshiping Moslems in a mosque atop the tomb of the Patriarch in the Palestinian city of Hebron. He believed that he was following the will of God and was hailed as a hero by other hardline Israeli settlers. The Islamic fundamentalists convicted of the bombing of the World Trade Center Towers in 1993 shouted in unison, "God is great," at the time of their sentencing. Michael Griffin, a "pro-life" activist, believed he was fulfilling a Christian mission when he murdered the attending physician at an abortion clinic in Pensacola, Florida.[25]

These destructive acts present a paradox. Judaism, Islam, and Christianity, the religions invoked by these extremists to justify the destructive behavior, are committed to love and peace. Yet militants from each of these religions have considered acts of violence like these the *consummation,* not the contradiction, of their faiths. Interestingly, their destructive actions seldom achieve the good results they envision. Rather, their acts of violence often turn the opinion of the broader community against them, even though their own subgroups hail them as heroes.

The perpetrators demonstrate typical dichotomous thinking—perversely branding the victims as the criminals and glorifying the offenders as the saviors. Such dualistic thinking is characteristic of the belief systems of world cultures and permeates these mainstream religions.[26] The Bible and the Koran divide the moral universe into the absolute categories of good and evil, God and Satan. Through a reverse twist of logic, the faithful who engage in murder (and thus violate the basic tenets of their religions) view their victims as the Evil Ones. In Islamic holy wars (jihads) and the Christian crusades, countless individuals of different faiths were massacred in the name of Allah or Jesus. Even Hitler justified the mass murder of Jews in the name of the Lord.[27]

It is evident that the institutions of religion have, at best, succeeded only partially in solving the problem of either individual or mass violence.[28] What does the understanding of the psychology of the individual have to offer?

Formulating the psychological factors that lead to violence can provide the framework for understanding anger, hostility, and violence. This framework, in turn, can suggest strategies to people for dealing with their own hostile reactions as well as provide a basis for conflict resolution between groups and states.

The failure of moral codes to reduce antagonistic behavior can be analyzed in terms of the cognitive structures that drive and justify harmful behavior. Understanding primal thinking and beliefs can be a first step to solving the moral paradox. When a person perceives that either he himself or a sacred value is threatened or abused, he reverts to categorical, dualistic thinking. When this primal mode of thinking is triggered, he automatically prepares to attack—to defend his highly invested value. This hostile mode takes over the thinking apparatus and crowds out other human qualities such as empathy and morality. The same kind of thinking is activated whether the perpetrator is reacting as a member of a group or as an individual. Unless interrupted, the hostility sequence proceeds from *perception* (of transgression) to *preparation* to *mobilization* to *actual attack.*

SOLUTIONS TO THE COGNITIVE PROBLEM

The basic psychological problems that contribute to anger, hostility, and violence will be expanded in later chapters. Briefly, there are two phases in solving the problem of hostility and hate in interpersonal conflicts. The first centers on deactivating the hostility mode when it is triggered. A variety of methods can be used to institute a cooling-off period when conflicts heat up. Distraction also helps to deactivate the primal mode. When sufficient time has elapsed and the parties are able to review their reactions in perspective, they can modify their misinterpretations of each other's behavior.

A more enduring approach is aimed at the basic way that people perceive themselves, their group, or their fundamental values as vulnerable. People in general, as well as their leaders, need to become more aware of the rigid thinking that gains control of their minds when they are threatened. They need to know that they are not exercising rational judgment when absolute notions of good versus evil and holy versus unholy dominate their interpretations of others' behavior. They need to be able to evaluate the behavior of other groups according to more objective criteria and resist the tendency to assign them to absolute categories, such as aliens or the Enemy. And above all, they have to become aware that they can be abysmally wrong in their characterizations of other people and their motives, often with tragic results if they pursue actions based on these characterizations.

Recent work in cognitive and social psychology has added substantially to our understanding of the nonconscious processing of biased information, especially in relation to phenomena like unacknowledged prejudice and in the priming of hostile attitudes.[29] In addition, contemporary work in anthropology, sociology, and political science provides a broader perspective for the analysis.[30]

New work in evolutionary psychology offers an expanded time scale for speculation about human behavior. A number of writers, starting with Charles Darwin, have suggested that much social and antisocial behavior is based on a biological infrastructure. Certain types of "antisocial" behavior, such as deception, cheating, robbing, and even murder, may be derived from primitive patterns that facilitated survival and reproductive success in prehistoric times. These writers also provide an evolutionary explanation for such prosocial behavior as cooperation, altruism, and parenting.[31] But they have not presented a formulation of a topic to be discussed in this volume— the evolution of cognitive patterns, especially primal thinking.

Other writers, such as Paul Gilbert, have emphasized the importance of social connectedness in Paleolithic times.[32] Presumably the social danger of rejection or loss of status had a bearing in those times on survival and reproduction. Selective pressures would have fostered the evolution of social anxiety to discourage behavior that might have interfered with participation in group acceptance and mating. An individual excluded from the band because of undesirable behavior would have been deprived of protection against predators or ready access to the food garnered by the group. Vulnerable to fatal attacks by human or nonhuman enemies or to starvation, that person would have been less likely to mate and have progeny.

Having a built-in mechanism to cause fear of abandonment or depreciation may have been an important factor in promoting group solidarity. The evolved emotional reactions of shame, anxiety, and guilt contribute a solid basis for moral behavior within the group. Yet this mechanism, possibly adaptive in prehistoric times, is to a large degree inappropriately overactive in contemporary situations.

Some writers also believe that evolutionary pressures have helped to develop socially desirable characteristics.[33] There seems to be an innate program that reinforces sociable behavior. Because people feel pleasure when they are cooperative and altruistic, educators, religious leaders, and social engineers can utilize these forces to counter hostile behavior and promote moral behavior.

2

THE EYE ("I") OF THE STORM

The Egocentric Bias

What triggers hostility? In general, whether we feel anger, anxiety, sadness, or joy in a particular encounter depends on our interpretation, the *meaning* we assign to it. If we did not interpret the meaning of events before reacting, our emotional responses and behavior would occur willy-nilly, without relation to specific circumstances. When we select and process information properly, we are likely to extract what is relevant to the conditions. Consequently, our feelings and behavior are appropriate. If the meaning is, "I am in danger," then I feel anxious; if it is, "I am wronged," I feel angry; if it is, "I am alone," I feel sad; if it is, "I am loved," I feel joy.

If, however, I attach an incorrect or exaggerated meaning, I may feel anxious when I should feel calm, or joyful when I should feel sad. When information processing is affected by a bias (or when our information itself is incorrect), then we are prone to react inappropriately.

The biases may impinge at a very early—unconscious—stage of information processing.[1] A hypersensitive woman interprets a hearty compliment by a male acquaintance as a slur; a second later she snaps at him angrily. Her interpretation of his remarks is, "He is putting me down." Because she is tuned to expect men to reject her, she misinterprets innocent remarks as demeaning.

ON BEING THE "VICTIM"

Consider the following scenarios. The driver of a truck curses a slow driver for holding up traffic, a manager berates an employee for not turning in a report, a large nation attacks a smaller, resistant neighbor for its abundant supply of oil. Interestingly, although there is clearly a difference between victimizer and victim in these examples, the aggressor in each case is likely to lay claim to being the victim: the truck driver for being impeded, the manager for being disobeyed, the invading nation for being opposed. The aggressors are firmly entrenched in their belief that *their* cause is just, *their* rights have been violated. The object of their wrath, the true victim (to disinterested observers), is seen as the offender by the victimizers.

Aggressive, manipulative people generally believe that their entitlements and rights override those of others. An aggressive nation operating under slogans like "The Need for Lebensraum" (living space) (Germany) or eminent domain (the United States) views opposition by the weaker country in much the same way the aggressive driver views the slow driver: as interfering with its legitimate goals.

As members of a group, people can show the same kind of biased thinking they bring to personal conflicts. Hostility—whether experienced by a group or an individual—stems from the same principles: seeing the adversary as wrong or bad, and the self as right and good. In either case, the aggressor shows the same "thinking disorder": construing the facts in his favor, exaggerating the supposed transgression, and attributing malice to the opposition.

As part of our survival heritage, we are very much aware of events that could have a detrimental effect on our well-being and personal interests. We are sensitive to actions that suggest a put-down, imposition, or interference. We monitor other people's behavior so that we can mobilize our defenses against any apparently noxious actions or statements. We are inclined to attach adverse personal meanings to innocuous actions and exaggerate their actual significance to us. As a result, we are particularly prone to feel hurt and angry with other people.

The tendency to overinterpret situations in terms of our own frame of reference is an expression of the "egocentric perspective." When we are under stress or feel threatened, our self-centered thinking becomes accentuated, and at the same time the area of our concern expands to irrelevant or remotely relevant events. Out of the tapestry of the multiple patterns contributing to another person's behavior, we select a single strand that may affect us personally.

We are particularly drawn to a self-centered explanation for another person's apparently adverse behavior. A wife is preoccupied when her husband returns home from work, and he decides, "She doesn't care for me," even though this explanation discounts the fact that she is tired and absorbed in her duties as a homemaker after a day at work outside the home. He assumes that her lack of affection is caused by her withdrawal of her love for him.

We all have the tendency to perceive ourselves as the lead actor of a play and to judge other people's behavior exclusively in reference to ourselves. We take the role of the protagonist and the other players are our supporters or antagonists. The motivations and actions of the other players revolve around us in some way. As in an old-fashioned morality play, we are innocent and good; our adversaries are villainous and bad. Our egocentrism also leads us to believe that other people interpret the situation as we do; they seem even more culpable because they "know" that they are hurting us but persist in their noxious behavior anyhow. In "hot" conflicts, the offender also has an egocentric perspective, and it sets the stage for a vicious cycle of hurt, anger, and retaliation.

This self-centered orientation forces us to focus our attention on controlling the behavior and supposed intentions of the other players. We have implicit rules, such as, "You should not do anything that distresses me." Since we may apply the rules too extensively and rigidly, we are constantly vulnerable to the behavior of others. We are incensed by our perception that our rules have been violated, and since we identify ourselves with our rules, we feel violated. The more we relate irrelevant events to ourselves and exaggerate the significance of relevant events, the more easily we are hurt. Our own self-protective rules are inevitably broken since other people operate according to their own egocentric rules; even if they know our rules, they would not want to be controlled by them.

The impact of the self-centered perspective is clearly seen in close relations, particularly in distressed marriages. Nancy, for example, was enraged because Roger made himself a sandwich and did not ask whether she wanted one. Roger had violated one of Nancy's silent rules: "If Roger cared for me, he would always offer to share things with me." Nancy did not actually want a sandwich, but his not offering her one meant, according to her rule, that he was inconsiderate, that he did not care about her wishes. Even though Roger responded to her subsequent complaint by offering to make her a sandwich, it did not help; he had already "shown" that he did not care. As a result, she clammed up.

By expanding her expectations, Nancy made herself more vulnerable and thus more subject to anger and hurt feeling. Roger, on the other hand, did not care whether Nancy anticipated *his* desires but was very sensitive to any indication that another person was trying to control him. He became angry when she withdrew—thinking that she was attempting to punish him. In Nancy's "morality play," she was the victim and Roger the villain. In Roger's, he was the victim and Nancy the villainess.

The paradox is that the rules the person forms to protect himself from being hurt actually make him *more* vulnerable. A more adaptive rule for Nancy would have been, "If Roger doesn't sense my wishes, I'll inform him." Such a procedural rule, if successful, could accomplish her goal of making Roger more considerate. Roger, on the other hand, had to learn that Nancy's withdrawal was a result of her disappointment in him, not a subtle form of retaliation.

The tendency to see ourselves on center stage and to refer the actions of others to ourselves is pronounced in some psychiatric disorders. As a patient's egocentric perspective is intensified, it may cloud the true characteristics of other people and interactions. He is likely to attach erroneous, even fantastic, meanings to their behavior. This predisposition appears in exaggerated form in the paranoid patient, who refers other people's irrelevant behavior to himself (self-reference) and believes without question the validity of his representation of their attitudes toward him.

Tom, a twenty-nine-year-old computer salesman, was referred for evaluation because he had been in a constant state of agitation for several months. He complained that people he passed on the street stared at him and made derogatory comments. When he approached an animated group of strangers at a corner, he interpreted their laughing as a sign that they were plotting to embarrass him. While Tom's self-reference may seem remote from our own experiences, it dramatizes the human inclination to relate people's behavior to ourselves.

The egocentric perspective can be observed in other clinical conditions such as depression. Depressed patients relate irrelevant events to themselves, but they interpret these events as signs of their own unworthiness or defectiveness. The typical hostile person, in contrast, does not always believe that people are deliberately trying to hurt him but does infer that they interfere with his goals through their stupidity, irresponsibility, or stubbornness. In his drama, fools thwart him—the hero—in his mission to follow his rainbow. The more hostile he is, however, the more likely he is to interpret others' interfering behavior as deliberate attempts to hurt him.

The suspicious person interprets other people's behavior as indicative of their desire to thwart, deceive, or manipulate him. The phrase "paranoid perspective" has been applied to members of political or religious organizations who perceive that their values and interests are being suppressed by an aggressive government or corrupted by other groups. In *The Paranoid Style in American Politics and Other Essays,* Richard Hofstadter delineates the psychology of hate groups in terms of their single-minded belief that a corrupt government is deliberately trying to violate their constitutional rights.[2]

A major problem in human relations is that our words and actions convey unintended meanings to other people, just as their words and actions have unintended meanings to us. Tact and diplomacy involve sensitivity to the possible explanations that people will attach to what we do or to what we do not do. People who want to maintain equilibrium in close relations find that they must steer carefully through the shoals of others' expectations and interpretations. This principle applies to intergroup as well as to interpersonal relations.

INDIVIDUALISM AND EGOISM

The popular notion of egocentricity as simply "being wrapped up in ourselves" underestimates its important role in the interpretation of our experiences and the protection and promotion of our vital interests. Since each person is the vehicle for transferring his or her genes to the next generation, evolution has placed a priority on self-serving bias, acquisitiveness, and self-defense. This centrality and definition of the self is reinforced not only by physical pleasures and pains but by psychological ones as well. Our delight after a triumph, for example, reflects a rise in our appraisal of our own value, a phenomenon that we label "self-esteem." The pain after a failure, on the other hand, results from reduced self-esteem.

The experience of such pleasure and pain reinforces our sense of personal identity, which is further consolidated by other people's defining and rewarding or punishing us. As other people mark *their* boundaries, they also serve to define *our* sense of being separate individuals. The anger that we evoke when we encroach on the domain of another establishes the perimeter of our personal domain. We have a specific mental representation of ourselves, including our sense of personal identity and our concept of our physical and psychological characteristics. We recognize our strong investment in ourselves as well as in the "external" components of our domain, other individuals and institutions that we value, and tangible possessions. Our extended domain may in fact stretch to include all our affiliations—race, religion, political party, government—and we may construe any attack on our domain as an attack on our personal identity. Maintaining such an extended domain, unfortunately, can cause us to be hypersensitive to a variety of possible insults.

As a rudimentary sense of self is established (probably by the second year of life), individuals think and plan in terms of their self-interest. This programmed orientation can be—and is— overridden by social pressures to conform to the rules and regulations of the community. Our self-esteem acts as an internal barometer that pressures us to expand our resources and our domain and registers the fluctuating evaluations of our domain. We feel pleasure when our valued domain is expanded, and pain when it is constricted or devalued. When we have been hurt, we draw on a variety of strategies to bolster our self-esteem. When we are blocked in our attempts to attain a goal and expand our domain, we may take offense and attack or punish the offending persons.

THE PRIMAL BELIEFS

Our beliefs and information-processing systems play a decisive role in determining our feelings and behavior. We interpret and misinterpret signals from others according to our values, rules, and beliefs. When we overemphasize the significance of personal success, or national superiority, we slip into the trap of regarding individual competitors, members of outgroups, or citizens of other nations as less worthy than ourselves. Primitive mechanisms for processing information, retained from our evolutionary experience, prejudice our judgments about people who differ from us. Cognitive biases may also lead us to indiscriminately attribute malice to anyone whose actions or beliefs conflict with ours. As the vise of our cognitive apparatus tightens, we tend to squeeze these people into the Enemy category: the angry spouse, the member of a religious or racial minority, the outspoken political revolutionary. It becomes increasingly difficult to observe others reflectively, objectively, and with perspective.

The predilection to become excessively or inappropriately angry and violent may be understood in terms of this primal thinking. The patterns are primal not only in the sense of being basic but also because they probably originate in primordial times, when they would have been useful to our animal and human ancestors in dealing with dangerous problems with other individuals or groups.

People generally believe that their anger is their first response to an offense; their initial interpretation preceding their angry response, however, is so rapid and often so subtle that they may not be aware of it. Upon reflection and introspection, however, they can almost recognize that their initial emotional response is a distressed feeling rather than anger. With training, they can generally "catch" the meaning of an event leading to their distress.

The hostility sequence thus proceeds from the interpretation of a transgression to anger and then to hostile verbal or physical action. For many years I believed that the anger followed immediately after the interpretation of being wronged. Several years ago, however, I observed that patients who focused on their feelings following a noxious experience noted a fleeting hurt or anxious feeling *before* they experienced anger.[3] Close examination revealed the common theme leading to the distress that preceded the experience of anger: the perception of being diminished in some way. If the individual evaluates this distress as wrongfully caused by another person, his behavioral system is mobilized in preparation for counterattack. A simplified version of the stages in the development of hostility can be represented in this diagram:

EVENT → DISTRESSED → "WRONGED" → ANGER → MOBILIZE TO ATTACK

If we perceive that a threat or loss is simply due to an impersonal situation—for example, sickness or an economic crisis—we feel upset or unhappy but not angry. If we conclude that some person or group is at fault, however, then we feel angry and are impelled to retaliate to undo the wrong. We may even feel angry at an inanimate object when we feel that it is unfairly impinging on us (such as a chair that shouldn't be there, or a glass that shouldn't have fallen from our grip). Our subjective feeling ranges in quality and intensity from being miffed to being enraged. Although the term "anger" is often used in common parlance to express not only a person's feeling but also his or her destructive behavior, I reserve the term simply for the *feeling*. I will use the term "hostile aggression" to refer to the *behavior*.

When we are mobilized to fight or to counterattack, we may inhibit our actions out of concern for the consequences. Nonetheless, as long as our image of the transgressor persists, our biological attack systems remain activated, expressed physiologically by an increase in the heart rate, a rise in blood pressure, and growing tension of the muscles. Our mobilization to fight also includes our display of intimidating facial expressions, such as scowling and staring.

Problems arise in interpersonal relations when we misinterpret or exaggerate what seems to be a transgression. Say we believe that somebody has disparaged us, deceived us, or challenged a cherished value. This violation rouses us to counterattack in order to terminate the damage and punish the offender. We all have specific vulnerabilities that predispose us to *overreact* to situations that impinge on them. These vulnerable areas actually consist of problematic beliefs, such as, "If somebody doesn't show respect, it means I appear weak," or, "If my wife doesn't express appreciation to me, it means she doesn't care," or, "If my spouse rejects me, I am helpless."

To protect ourselves from discrimination, coercion, injustice, and abandonment, we construct rules regarding equality, freedom, fairness, and rejection. If we perceive that we are receiving unfair treatment or that our freedom is curtailed, not only are we diminished by it, but we become vulnerable to further disparagement. We may seek to retaliate and punish the violator, even if we have not sustained any damage, in order to reinstate the balance of power. Whether we have been damaged or not, we determine the nature of the offense, weigh the pros and cons of desired retaliation, and decide on what form of remedial action to take.

We use these formulas to monitor and evaluate our interpersonal transactions, but because they are exaggerated and rigid they lead to unnecessary distress. Faulty beliefs are embedded in a network of self-protective compensatory demands: "People must show me respect," or, "My wife should consistently demonstrate she cares." If these injunctions are violated, then another set of coercive retaliatory beliefs is activated. "I should punish anybody who doesn't show me respect," or, "I should withdraw from my wife if she is not responsive." Beliefs that protect what we consider vital to our existence or our identity assume a primal form, such as, "Somebody who slurs my honor is my enemy."

The primal beliefs are often extreme and can lead to violence. Hank, a construction worker who believed, "If somebody doesn't show respect, I should beat him up," got into a number of fights on the job, at bars, and at other social gatherings. Sometimes he extended the same rule to his wife and struck her when she scolded him. One such event led to his entry into couples' therapy, where he recognized that his sense of vulnerability underlay his drive to preserve his macho image at all costs. When he realized that giving in to his impulse to hurt people was generally regarded as a sign of weakness, not strength, he was more motivated to control his behavior. We have found clinically that a number of abusive people have a defective image of themselves, for which they compensate by attempting to intimidate others.

A similar set of beliefs may promote anger and hostility between groups and nations. This is not surprising, since group behavior represents the cumulative effect of the individual members' thinking. Even stronger is the core belief that individuals in the outgroup are dangerous, which leads to a sense of vulnerability and defensiveness. When a conflict between the two groups arises, these beliefs are activated and shape the perceptions of both groups. As these beliefs become more intense, the adversarial group is seen as the Enemy, and one of the groups may be driven to make a preemptive strike. Beliefs like these also appear to underlie large-scale violence, such as race riots, wars, and genocide.

Although the enemies of our prehistoric past, such as animal predators or bands of human marauders, are no longer a threat to our everyday existence, we are encumbered by the legacy from our ancestors, who were exposed to and feared these dangers. We unwittingly construct a phantom world composed of individuals who are poised to dominate, deceive, and exploit us. We are overly suspicious of actions that hint of manipulation or deception, and we may transform trivial or innocuous events or mild challenges into serious offenses. These automatic, exaggerated self-protective processes lead to unnecessary friction and pain in our contemporary lives. It probably was useful in our evolutionary past to react in an either-or fashion in discriminating friend from foe, prey from predator. It may have been adaptive to be on guard against the intrusive behavior of other members of the clan when our own survival was at stake, but we generally no longer need the margin of safety provided by these archaic mechanisms in our ordinary interactions.

What we learn by studying people in individual sessions can apply to the collective thinking of individuals in a group. The tendency of an individual to show bias against a competitor may be reflected in the combined biases of all group members against members of an outgroup. We know that we can modify the thinking of individuals in therapy to alleviate their self-defeating anger and hostile behavior.[4] Can we apply the same principles to the problems leading to group conflict and ethnic strife?

The counterparts to the rules we have erected to protect ourselves from the incursions of other people are the laws erected by a community to protect its constituency. We acknowledge these laws because we recognize their protective value, even though we may not personally obey them. However, when we observe somebody else breaking a law, we are likely to become irate and punitive toward the violator. If you run your car through a stop signal, you make me vulnerable; I could be the victim the next time you do so. Therefore, we collectively agree that rules must be enforced, but primarily to control *other people*. (We like to think we have a special exemption from these communal laws.)

Although we are inclined to be tolerant of our own self-serving behavior, we tend to be judgmental toward others who show similar behavior. The designation of avarice, vanity, and sloth among the Seven Deadly Sins results from the attempts of social and religious institutions to curb people's inherent tendencies toward self-inflation and self-indulgence at the expense of others. What is advantageous to a single person may be disadvantageous to the rest of the group. I may get gratification from expanding my domain (avarice) or conserving my energy (sloth), but these "natural" tendencies would interfere with the interests of the community, so they are discouraged. Sanctions are imposed to produce shame (if not guilt) and resultant behavioral change in the offending person.

POSITIVE BELIEFS AND FEELINGS

The self-serving, egoistic attitudes and behaviors represent only one side of human nature, however. They are softened and balanced by the powerful evolutionary forces of affection, kindness, and empathy. Thus, we manifest our basic ambivalence—self-indulgence, self-adulation, and selfishness in one situation, and self-sacrifice, humility, and generosity in another.

Relations between individuals can be characterized by the contrasting metaphors of fission and fusion. While individualism and egoism can separate people and lead to hostility within a family or group, the craving for affection, nurturance, and solidarity can bring them together. This latter bonding process is also a product of the evolutionary master plan. The affiliative, or sociophilic, tendencies are obvious in various close relationships, such as those between parent and child, lovers, spouses, relatives, and friends. Intimacy, rapport, and camaraderie are reinforced and maintained by feelings of pleasure. Strong, sometimes transient, bonds also form among members of clubs, political organizations, schools, and ethnic, racial, and national groups when they have a common purpose. The enthusiastic Earth Day celebrations are an example of this. Group solidarity can be enhanced by esprit de corps, as well as by the experience of common loss, which promotes communal grieving and mutual support.

Bonds of loyalty to other individuals in the group, as well as to the group as an entity, give the group cohesiveness, definition, and boundaries. However, as Arthur Koestler pointed out, this very cohesiveness has drawbacks for our species, because we define other individuals and groups as outsiders, potential adversaries, or even enemies.[5] The combination of sociality among group members and the individualism of the group members lays the foundation for aggressive competition, intolerance, and hostility toward outsiders. When people identify their own individualistic and sociophilic strivings with the goals of the group, they are subject not only to the benefits of group identification but also to its dark side: xenophobia, chauvinism, prejudice, and intolerance. They also exhibit the same thinking toward other groups that they show toward individuals within their group who have offended them. This involves such errors as overgeneralization and dichotomous thinking, as well as the fixation on "single-cause" explanations—seeing the outgroup as the sole cause of their distress, otherwise known as scapegoating.

Ingroups often fall into the trap of equating feeling good with being superior (that is, more worthy). It's easier to hurt others and demean them if you believe that they are not worth much. Biased thinking becomes embedded in the form and content of memories of past conflict with the "alien" group. These memories may be passed on (through cultural media) from generation to generation. Inevitably, the twin-headed creature of individualism and sociality (labeled "Janus" by Koestler) twists thinking into rabid nationalism, religious crusades, and political battles.

People in a confrontational or evaluative mode are particularly prone to view their adversaries in a negative way. One reason that religion and less formalized codes of morality have not been more successful in neutralizing our inherent egoism and acquisitiveness is that they do not eliminate the flaws in people's information processing or substantially modify their beliefs about outsiders. Indeed, with their emphasis on absolute (dualistic, overgeneralized) evaluative judgments, many religions often reinforce people's tendencies to judge themselves and other people in biased ways: good versus evil, benevolent versus malicious. Such thinking obviously creates problems in interpersonal as well as intergroup relations.

Rather than relying exclusively on codes of morality and religious canons to eliminate excessive anger and violence, we need to clarify the cognitive aberrations and faulty beliefs that drive interpersonal and intergroup conflict. That knowledge can then serve as the basis for appropriate intervention at an individual or group level, and the remedy can focus on the problems that lie at the core of anger, hostility, and hatred.

Sociophilic tendencies are not necessarily confined to the self-aggrandizing goals of the group; they can also form the substrate for cooperation, understanding, and empathy among group members. Moreover, these prosocial tendencies can provide a bridge between groups (for example, interfaith worship services). People have used codes of morality, ethics, and religious principle to attempt to transcend individual and group boundaries and to neutralize hostility. But ironically, tenets such as universal brotherhood have sometimes forced "unbelievers" to either accept a group's faith or be expelled or even annihilated if they resist. The Puritans, for example, left England because of hostility from other religions, but then showed the same kind of intolerance that they had experienced toward dissenters in the New World.

THE GENESIS OF HOSTILITY

How does the notion of the egocentric bias fit in with the theories of hostility? The explanations range in focus from innate factors to environmental factors to the interaction between the two. The most prominent theory of intrinsic causes was elaborated by Sigmund Freud. Disillusioned with the nature of mankind after the apparent irrationality of World War I, he elaborated the theory of Eros (love) and Thanatos (death). The death instinct was sufficiently powerful to overwhelm the defenses erected against it and consume the supposed adversaries. This theory also fit with Freud's well-known hydraulic theory, which posited that, like water in a reservoir, hostility could build up over time and overflow. Another psychoanalytic theory holds that people project their hostile fantasies onto other people and then react with rage against these projections.[6]

An evolutionary thesis advanced by Konrad Lorenz views hostile aggression as an instinct released by certain external stimuli. Nonhuman animals, according to Lorenz, presumably have a built-in inhibition against killing their own species, which has not evolved as yet in humans.[7] Biological theorists have considered a variety of neurochemical factors to account for violence: excessive hormones, like testosterone, or a deficiency of neurotransmitters, such as serotonin or dopamine.

Another school of thought places the blame for hostile aggression on circumstances or situations. Among the most popular has been the concept that people can be induced to acquiesce to the dictate of an authority to harm designated individuals. The results of a series of experiments by Stanley Milgram were proposed as an explanation for mass murder of large groups of people.[8] The implication of these "situationist" theories is that practically anyone could, under the right circumstances, be induced to engage in antisocial destructive behavior.

The third school emphasizes the interaction between external circumstances and the intrinsic capability for violence. These theories regard hostility as an adaptive response to specific noxious circumstances. W. B. Cannon elaborated the concepts of the fight-flight reaction, which serves as an appropriate attack or escape strategy to a threat.[9] Leonard Berkowitz has emphasized the importance of frustration as a cause of hostility.[10] Albert Bandura provides a detailed scheme according to which people engage in aggression in order to attain certain goals.[11] My own formulation combines elements of the theories of Cannon, Berkowitz, and Bandura but emphasizes the crucial importance of the meaning attached to our interactions as a key factor in arousing anger and hostility. I view the hostile responses as a strategy that was adaptive in early stages of our prehistory but is mainly maladaptive today.

The sensitive operation of the fight-flight mechanism undoubtedly prolonged the lives of our prehistoric ancestors. The capacity to fight off or escape from an enemy was finely honed by natural selection. But it is the hyperreactivity of these defensive strategies that poses problems in contemporary society, where the perceived threats are for the most part psychological rather than physical. Disparagement, domination, and deception, which represent threats to our status in a group and diminish our self-esteem, do not in themselves constitute dangers to physical well-being or survival. Yet we often react as strongly to a verbal attack as we would to a physical one, and become just as intent on retaliating.

The forerunner of our profound sensitivity to such psychological encounters may have been primordial encounters with group rejection, which led to deprivation of food resources and loss of protection by other group members. Those group members who developed a fear of rejection and abandonment presumably were more likely to cultivate cooperative relations with the rest of the group, thereby improving their chances of survival in an unpredictable environment. Socially desirable traits, selected by evolutionary pressures and the relevant genes, would have been passed on to future generations. A further reason, observed at the present time in "the code of the streets" in the inner cities and the "culture of honor" in the South, is the need to be able to respond promptly to any insult, real or imagined, to forestall being regarded as too soft to resist a more aggressive attack (see chapter 9).[12]

In addition to survival and social strategies, perpetuating our genetic lineage depended on crucial cognitive skills: discriminating prey from predator, friend from foe. Just as our individual cells detect and destroy intrusive foreign bodies and our immune systems detect and destroy toxins and microbes, so our cognitive and behavioral systems identify and ward off intruders. Primitive patterns of thinking were probably adaptive in those prehistoric conditions under which survival depended on an individual's reacting, often instantaneously, without time for reflection, to apparent threats from strangers—or even from certain members of the same band. A premium would have been placed on identifying enemies even at the cost of misidentifying a fellow group member as a foe (false positive). A single false negative (misidentifying an enemy as a friend) would have been fatal. Encounters with others had to be rapidly categorized as either threat or non-threat, with a distinct boundary between them. There was no latitude for ambiguity. This crude either-or categorization is the prototype of the dichotomous thinking that we see in chronically angry, hypercritical, or hyperirritable individuals. It also shapes the reactions of members of feuding groups and warring communities and nations.

Our understanding of this aspect of behavior has also been facilitated by recent developments in psychology, labeled the "cognitive revolution." This relatively recent focus has provided rich information and elegant theories about how people think, form concepts, and develop beliefs. The studies extend to processes like reading others' intentions and forming representations of ourselves and others.[13] Most relevant to the problems of society are the clinical observations: biased perceptions and thinking become set in a mental vise in response to threat, real or imaginary. This rigid frame, the prison of the mind, is responsible for much of the hate and violence that plague us.

Recent developments in psychology, biology, and anthropology have expanded our understanding of people's innate tendencies to be kind and cooperative. The "New Look" in Darwin's biology and psychology has directed attention not only to the genetic programs promoting personal survival and reproductive success but also to the evolved strategies facilitating adaptation to the social group. We understand that the ability to fight was necessary to achieve "reproductive success" in an ancient environment of often vicious competition for resources; our ancestors sought to ensure their opportunities for mating, often at the expense of other members of their species. "Nature, red in tooth and claw," operating through natural selection, provided important strategies for survival.

But these ancestors were also directed by selective pressures to develop social traits for accommodating to group living. Among these were skills useful in forming close attachments that enabled them to share food, information, protection, and parenting duties. It is apparent from the empathy we feel when we observe a child in pain, the pleasure we enjoy when we help somebody, and the happiness we experience when we have an intimate relationship that prosocial reactions are wired into our makeup as much as antisocial reactions. Studies of hunter-gatherer societies and chimpanzee colonies have provided valuable insights into the evolution of our cooperative social behavior.[14]

3

FROM HURT TO HATE

The Vulnerable Self-image

> We are all serving a life sentence in the dungeon
> of the self.
>
> Cyril Connolly, The Unquiet Grave (1995)

Recall different situations in which you have felt pain. Somebody you trusted cheated you; a person you relied on let you down; a friend spread gossip about you. What purpose was served by the suffering you experienced?

Humans seem fated—but are also constructed—to endure "the slings and arrows of outrageous fortune." Suffering appears to be a ubiquitous feature of the human condition. When we are hurt, however, it rarely occurs to us that psychological pain may serve a useful function. In contrast, it is easy to accept the notion that acute *physical* pain alerts us to the presence of bodily damage and speeds us to terminate or repair the damage. We know that people who lose their pain sensitivity as a result of neurological disorders are vulnerable to serious, possibly fatal, damage, so we recognize that sensitivity to physical pain provides a crucial protection. But what conceivable purpose can be served by the psychological hurt that often casts a pall over our lives? What do we gain by feeling sad, humiliated, lonely?

We typically experience psychological pain, such as hurt feelings, sadness, distress, even anxiety, in the context of our interactions with other people. These unpleasant reactions have a special function: they prod us to take corrective action with ourselves or with them, or to review the circumstances leading to our distress. Although we may recognize intellectually that a particular interpersonal experience is damaging to us, we may not be sufficiently motivated to do anything about it unless we experience some distress. Without the sting of hurt feelings, we would be putty in the hands of other people. Anyone would be able to take advantage of us, control, manipulate, or betray us, absent our robust efforts to stop such actions.

Psychological pain is often necessary to rouse us to overcome our natural inertia and focus our attention on the wrong and the wrongdoers. We are impelled to do something to remove the source of the pain—either by righting the wrong or by getting out of the situation. Pain mobilizes our entire system to get away (flight) or to remove the source (fight). The experience of anger is the catalyst for attacking an external agent; anxiety causes us to escape from it or to avoid it.

Pain—whether physical or psychological—also has long-range functions. We are born with the ability to associate certain situations with pain or hurt, and after the linkage has been established we are prepared to deal promptly with similar situations in the future. Having learned to distinguish an intentional insult from a friendly tease, we can react almost reflexively to malevolent behavior. We also learn adaptive patterns to prevent avoidable hurt. "Once burned, twice shy" typifies the almost automatic avoidance of repeating a traumatic situation. Of course, we learn ways of dealing with adversity other than backing off. We can use our more mature social skills to defuse a potentially harmful situation and apply our problem-solving skills to difficulties before hurt or psychological damage occurs.

We often learn from experience that our own behavior may unwittingly provoke other people. As our awareness of other people's feelings develops, we discover that we are not the only ones who have needs and sensitivities: we can unintentionally hurt other people by our actions. Criticism and punishment help us incorporate a social code of conduct and formulate our own code of behavior. With these social skills established, we can enlist the cooperation of other people.

The study of psychological distress teaches us about important aspects of human nature. We identify emotional hurt partly by the associated distressing feelings in the body, such as a sinking feeling in the abdomen, a lump in the throat, a clutch in the chest. Using these and other markers, therapists can help patients to pinpoint their thoughts before and after the hurtful experience and thus clarify their excessive and self-defeating reactions.

Theorist and therapist alike can profit from identifying people's thoughts and images as well as their pain. In this way, the totality of behavior—thinking, feeling, and acting—can be understood.

THE MEANING OF MEANING

Although we seem to respond almost instantaneously to assaults, whether physical or psychological, we do not always experience anger. Whether we do so depends on the context of the injury and the explanation for it. A young child subjected to an injection by the family doctor will fight and scream to protect herself from an inexplicable infliction of pain. An adult receiving such an injection, and experiencing the same kind of pain, may have some anxiety but will not typically respond with anger.

The obvious difference between the child's and the adult's reactions lies in the *meaning* of the event. For the child, there is no comprehensible explanation for having to undergo the frightening and painful procedure except that the doctor is overpowering and cruel. Moreover, her typically benevolent parents have betrayed her by facilitating the assault. For the adult, the procedure, although painful and possibly anxiety-producing, is warranted and acceptable. Responding with anger would be illogical, because he is voluntarily submitting to a beneficial procedure. Unlike the child, he has learned to discriminate between malevolent and benevolent injuries, between acceptable and unacceptable infliction of pain. He has expanded his *construct* of pain to include experiences that, while painful, are ultimately positive.

This example shows the importance of meanings, attributions, and explanations in determining how we respond to our experiences. When somebody hurts us, our natural reaction is to feel anxious and try to escape, or to feel angry and try to fight back. If the threat is overwhelming, we are disposed to get out of the situation. Whether or not we become angry depends on whether we judge that we have been wronged or victimized: we are likely to become angry if we believe the other person was unjustified. If we attribute a benevolent motivation to the act, we do not generally become angry. Unless we are specifically "primed" to explain assaults as benign, however, our immediate reaction is to regard unpleasant actions as intentional and malevolent and to prepare to punish the offender or to escape.

Picture the following scene: I am waiting at a bus stop. A bus comes by and doesn't stop. First I feel distress at being inconvenienced, then a sense of helplessness as the bus speeds by without even slowing down. I think, "He (the driver) deliberately ignored me," and feel angry. But then I notice that the bus is full, and my anger subsides. The key to my angry reaction was my interpretation that the driver arbitrarily chose to ignore me. The actual inconvenience is minor compared with the presumed offensive behavior. Once I reframe the situation, the "offense" fades away and I regard the incident as simply an inconvenience. I can then turn my attention to ascertaining when the next bus is due or considering other ways of getting to my destination.

Delays and frustrations do not in themselves necessarily produce anger. The crucial element is the *explanation* of the other person's action, and whether that explanation makes the other person's behavior acceptable to us. If it does not, we become angry and want to punish the offender. For the most part we regard behavior that offends us as intentional rather than accidental, as malicious rather than benign. Inconveniences and frustrations come and go, but the sense of being wronged persists.

An illustrative clinical example of how anger is aroused comes from the files in our clinic. Analyses of clinical cases are particularly illuminating: since the reactions tend to be magnified, they are more clearly delineated and understood.

Louise, a personnel supervisor in a large employment agency, found that she was almost continually angry at her subordinates or superiors, as well as at family and friends. A few of her angry reactions demonstrate the mechanisms involved in the triggering and the expression of her hostile response. On one occasion, her boss corrected a memorandum that she had prepared. Louise had these automatic thoughts following her boss's "criticisms": "Uh-oh, I've made a mistake." Then: "He really thinks I did a bad job. . . . I messed it up this time." Her self-esteem was damaged, and she felt bad. Louise's reactions demonstrate the typical dichotomous thinking triggered by threats to the self-esteem. If feedback is not all positive, it becomes totally negative: a mistake becomes a really bad job, a criticism becomes total rejection.

Later, as she mulled over the event, she became increasingly angry and had a different set of automatic thoughts: "He had no right to treat me that way after all I've done for him. . . . He's unfair. He never shows appreciation for my work. All he does is criticize. . . . I hate him." By shifting the explanation for her hurt to her boss's "unfairness," she was able to salve the hurt to her self-esteem. In essence, her focus shifted from, "He disapproves of me; he considers me inadequate," to, "He was *wrong* to have criticized me." Assigning responsibility to another person for unjustly "causing" an unpleasant feeling is a prelude to feeling angry. The persistence of a sense of threat and the fixed image of a malicious person leads to at least a temporary feeling of hate. It is much easier to sustain anger or aggression when we drift from *specific actions* (he criticized this memo in two places) to *overgeneralizations* (he always criticizes me) or *labels* (he's unfair). The drift is often outside awareness; people may hold grudges about matters they no longer recall.

THE MALEVOLENT TRANSFORMATION

The more one regards a distressing act as intentional or due to the negligence, indifference, or deficiency of the offender, the stronger the reaction. Louise's experience illustrates the chain reaction from vulnerability to pain to anger. However, people who are particularly prone to react to situations with anger have little awareness of either the transient hurt feeling prior to the anger or the rapid "automatic thoughts" preceding both the pain and the anger. The thoughts that precede the distress may be self-deflating ("I made a bad mistake"), self-doubting ("Can't I do anything right?"), fearful ("I may lose my job"), or disappointing ("He doesn't respect or appreciate me"). I have referred to these kinds of thoughts as "hidden fears" and "secret doubts".[1]

We are generally influenced by how others perceive us—or how we *think* they perceive us. Conclusions such as, "She disapproves of me," affect our image not only of our critic but of ourselves as well. We form an idea about the impression we make on other people, our self-presentation. When we are involved in interactions with others, we tend to project this image onto them and assume that this is the way that they see us. If they mistreat us, our projected interpersonal image may be that of a "pushover" or "slob" or "misfit." Our self-image may shift from, "I appear to be a misfit," to, "I *am* a misfit." Since the way people perceive us is associated with how much they value us, the debasing of our social image produces psychic pain. The effect of criticism or insult is analogous to that of a physical attack: we become aroused to fend off the attack or to retaliate. In this way we minimize the psychological impact of the blow. If we can discredit the "attacker," the effect on our self-esteem is lessened.

Louise's initial construction of her boss's "criticism" led to her feeling hurt. Her later *reconstruction* of her boss's behavior led to anger, and even hatred. Between her hurt and anger came a shift of the focus to the person who "caused" her pain and his presumed fault in not appreciating her more. Thoughts like, "He has no business treating me this way," and, "He has a lot of nerve after all I've done for him," made him the wrongdoer and prompted the transition from downgrading herself to downgrading him. The change in her construction allowed her to move from feeling hurt to feeling angry. Her angry feeling, while still distressing, was far more acceptable than her hurt and replaced her sense of vulnerability with a sense of power. In a way it served the same function as a verbal counterattack, even though the retaliation was confined simply to thinking and imagining revenge.

A different kind of "offense" can lead to anger and a desire to punish the offender. Louise became angry not only when she considered herself unfairly treated by a superior but also when her subordinates did not live up to her expectations. One day she flared up at her assistant, Phil, for not having attended promptly to an assignment. Even in this situation, Louise was initially distressed. When she noted her assistant's omission, she had a medley of rapid thoughts and feelings: "He let me down; I'd counted on him." These were followed by a pang of disappointment. The next rapid sequence, "What will he do next? I can't trust him," led to a spurt of anxiety. Louise's anxious and hurt feelings were overshadowed by the next set of thoughts—"He *should not* have made the mistake. . . . He should have been more careful. . . . He's irresponsible"— which led to anger.

Here again we observe the sequence:

LOSS AND FEAR → DISTRESS → SHIFT OF FOCUS TO THE "OFFENDER" → FEELING OF ANGER

A clue to the angry response is the intrusion of the imperatives "should" and "should not," which impose responsibility for the difficulty on the other person. Louise believed strongly that Phil "should have been more responsible" and needed to be scolded for his transgression. Her thwarted expectations and her sense of being wronged prompted her angry reactions.

We all set up expectations of other people: they should be helpful, cooperative, reasonable, fair. These expectations are often elevated to the level of rules and demands. When a person whom we count on breaks a rule, we become angry and are disposed to punish him. In the background is the sense that the breach of the code makes us more vulnerable, less effective. Punishing the wrongdoer, however, helps to restore our sense of power and influence.

Having scolded her assistant, Louise felt relieved and moved on to other things. There is generally no "staying power" from punishing other people, however. The increase in self-esteem and gratification does not provide any protection against being upset by the next unpleasant encounter. Phil was hurt and angry and complained to other people in the office about being mistreated. Louise was surprised to hear of this, since she believed she had treated him fairly. Bothered by the image he presented of her, she then became annoyed all over again.

The incident illustrates other principles about the expression of anger. In this encounter both people felt wronged and vulnerable—Louise by Phil's error of omission, and Phil by Louise's scolding. Each regarded himself or herself as the victim and the other as the wrongdoer. The disciplinarian or the punisher is often oblivious to the long-term impact of the pain inflicted on the "offender." We feel that once we have gotten our complaint "off our chest," we have restored harmony to a relationship. The target of our reproach, however, is hurt and builds up a grievance against us. Restoring balance to the relationship for me upsets the balance for you. In this case the balance was upset again by Phil's complaint, and a typical vicious cycle was created.

THE SHOULDS AND SHOULD-NOTS

Louise had several other typical (for her) angry encounters in the course of the next day. In one, Louise had asked Clare, a close friend for whom she had done a number of favors, to buy her a special brand of perfume during Clare's trip to a shopping center. Louise fully expected Clare to comply with this request, and when she learned that Clare had forgotten, she felt let down and then angry. Clare had violated the "reciprocity rule": "Since I will do anything my friends request, I expect them to follow my rare request." As suggested earlier, her boss had broken the "fairness rule" ("If people criticize me, they are unfair"). Her subordinate had broken the "reliability rule."

The rules provoking Louise's anger are contained in her thoughts of "shoulds" and "should-nots": "My boss *should not* have criticized me"; "Phil *should* have completed his work in time"; "Clare *should* have cared enough to remember." We all live by rules that we use to judge whether another person's behavior is favorable or unfavorable to our interests. If favorable, we feel good; if unfavorable, we feel hurt and, often, angry. Involved in these rules, of course, is a crucial question: Do people respect me? Do they care about me? If the rule is broken, it means that the person doesn't respect me or doesn't care about me.

People are not usually aware of either "shoulds" or "should-nots," however, until one of their rules has been broken. These imperatives are the derivatives of the broken rule. "Shoulds" or "should-nots" are evoked automatically as a response, and to reinforce the inviolability of the rule. Sometimes we couch our reprimand in the form of the rhetorical "why": *Why* was he critical? *Why* didn't she do what I asked? The "why" is generally an accusation rather than a true inquiry, as though to say, "*Why* are you so uncaring, so disrespectful, that you treat me this way?"[2] The insinuation is that by disobeying the rule, the other person has behaved improperly and must be coerced into behaving properly in the future.

The rules and standards of behavior provide us with a framework within which we interact with one another in a relatively smooth, balanced fashion. Social pressures impel us to appear to be fair, reasonable, and just in our dealings with others. One of us may offend another, wittingly or unwittingly. If our self-esteem is hurt, we draw on a relevant rule by which to judge the event. If we decide that the culprit has been arbitrary, unreasonable, or unjust, then we regard her as wrong or bad, and we feel angry. Those people who have fragile self-esteem will try to protect themselves with a thicket of rules that are destined to be broken and to lead to further upset. Since people vary in terms of sensitivity and hypersensitivity, what is a breach of acceptable conduct to one may be perfectly permissible to another.

The interactions of Louise with her employer, subordinate, and friend underscore important problems in contemporary life. Why do we become so upset over criticism, even when it is intended to be helpful? How can we distinguish between useful corrective feedback, intended to improve our performance, and disparaging criticism?

Many people react to constructive criticism as though it were a personal attack. It is readily observable that even constructive criticism can contain an element of disparagement, sometimes reflecting the frustration of the critic (parent, teacher, superior). Moreover, even when a correction or criticism is objective (it could, for example, be the marking of a wrong answer on an exam), it affects our self-esteem ("I'm not as good as I thought I was," or, "She doesn't think much of me"). When our self-esteem is injured, we are prone to conclude that we are being judged unfairly and may then become angry to protect our self-esteem.

The problem is heightened by the fact that we are programmed by a combination of innate factors and life experiences to overinterpret others' comments as put-downs. We wonder, "Is she trying to help me or show me that she's smarter?" or, "Is he implying that I'm stupid?" Depressed people frequently accept unfair criticism as reasonable and correct because the judgment is consistent with their own negative view of themselves and is therefore justified.

Caryn, a twenty-nine-year-old airline supervisor, provides another example of how an individual's sense of vulnerability can lead to self-defeating anger. Although Caryn's sensitivity may be more pronounced than that of the average person, her vulnerability captures the essence of much of the anger that we can observe. She was very successful in her career, but not in her relations with other people. She was generally regarded as standoffish by her associates, and as "stuck up" by the men with whom she had romantic relationships. Since she kept most people at arm's length, they regarded her as completely wrapped up in herself, but her appearance of aloofness belied the insecurity she felt. Her signal to her male suitors, "Don't get too close," was derived from her fear of being hurt. In her dealings with associates and subordinates, she felt it necessary to maintain a "cool" front to conceal her feelings of nervousness.

Caryn had had a series of love affairs, all of which ended in a flurry of angry recriminations on her part. In the affairs she showed an either-or (dichotomous) reaction by following the formula, "Either my lover shows complete, unambiguous signs of affection and acceptance at all times, or he is really indifferent and deceiving." Her reaction to what she perceived as a lack of affection from the man was to feel hurt and then angry. The formula was derived from her basic belief about herself: since I am not lovable, I can't trust any other person's protestations of affection. Her compensation for this belief was the rule: my lover must show affection all the time.

When a conflict arose, she would rarely discuss the problem for fear of confirming the "awful truth": namely, that her partner did not really love her (and therefore she was unlovable). She thought, "He's just using me. He's just led me on." By feeling justified in her anger, she was able to diminish her sense of rejection. She would then withdraw.

A review of Caryn's history casts some light on her reactions. During her early childhood, both parents were loving and accepting. When she was six years old, her father died suddenly, and her mother became withdrawn and irritable. Practically the only remarks that Caryn recalled were her mother's criticisms of her. In retrospect, Caryn realized that her mother was probably depressed over the death of her husband, but at the time her critical behavior was inexplicable and produced a great sense of vulnerability in Caryn.

As time went on, Caryn built up a wall—actually a cluster of self-protecting attitudes—around herself. Operating on the principle of not exposing herself to rejection, she appeared cold and detached in most of her encounters. Nonetheless, she continued to experience an inordinate need for demonstrations of affection. This, then, posed a real conflict: her great desire for love versus her intense fear of being rejected. At the deepest level, Caryn had an image of herself as basically unlovable, a self-representation resulting from her mother's critical attitudes and, probably, the loss of her approving father.

Caryn did attract a number of suitors. Each time she became romantically involved, she first had to overcome a reluctance to commit herself. She was always haunted in her relationships by her fear of rejection. When a suitor's interest seemed to fluctuate, she blamed him for "being manipulative" and thus was able to evade the possibility that she was being rejected because she was unlovable.

Caryn's rules were ultimately self-defeating rather than self-protective because they prevented her from attaining what she really wanted: a love relationship. An additional consequence of overprotecting herself was that she could not achieve the security that would have allowed her to discard her protective armor. Caryn illustrates a universal dilemma. We want love, affection, friendship, but we are afraid that if we expose ourselves we may be rejected and consequently hurt.

When we reflect on our vulnerability, sensitive self-esteem, and readiness to become annoyed at other people, we may wonder what function is served by these traits. Since they cast a shadow over our lives and often damage other people, especially those with whom we are the most intimate, how can we understand the powerful grip of our self-image and self-esteem?

SELF-ESTEEM

An individual's self-esteem represents the value he attaches to himself at a given time—"how much I like myself." Our self-esteem serves as a kind of barometer of our success in satisfying our personal goals and in coping with other people's demands and restrictions. It automatically quantifies how worthwhile we regard ourselves to be at any one time. A global self-evaluation— or more importantly, a change in self-evaluation or self-esteem—generally triggers an emotional response: pleasure or pain, anger or anxiety. People judge their worth according to a scale that indexes the difference between what they "should be" and how they currently regard themselves. Depressed patients generally describe a big discrepancy between "What I should be" and "What I am." Consequently they frequently characterize themselves as "worthless." Anger-prone people see a discrepancy in the opposite direction. They feel that other people should attribute more value to them.

The impact of an event on our self-esteem varies according to the degree of importance of the personality characteristic involved. Having one of our "important" traits devalued will obviously affect our self-esteem more and produce more hurt and anger than would devaluation of an unimportant trait. If the impact of a negative event (for example, rejection or failure) is strong, then our view of ourselves may shift to a more absolute, categorical self-appraisal (for example, we are weak, unlovable, useless), and we will experience a pronounced drop in self-esteem.

Of course, we learn in time to buffer the effect of many adverse events by applying a number of techniques designed to deemphasize their importance: placing them in perspective, finding "face-saving" explanations, disqualifying the validity of a criticism, or depreciating the person who supposedly devalued us. In a similar way, self-enhancing events activate our positive self-image, leading to an elevation of our self-esteem and to positive expectations, which in turn stimulate us to engage in further expansive actions.

Our self-esteem is affected not only by our personal experiences but also by boosts or blows to our inner social circle (family and friends). It is easy to recognize a phenomenon akin to a "collective self-esteem," which is raised or lowered according to whether a favorite sports team or political party wins or loses. The same phenomenon may be observed in people's reactions to national victories or defeats. It is important to note that, although the individuals in the same group (winners or losers) experience a similar fluctuation in self-esteem, the degree will vary from person to person, depending on how much he or she identifies with the group and its aspirations.

Our feelings are especially influenced by the change in our self-esteem from a previous rating or comparison with other people. For example, Ted was feeling good about getting a raise in salary until he heard that his friend Evan had also received an increase in salary. Ted felt deflated even though his job was totally different from Evan's. To Ted, Evan's success meant: "I am not as well regarded as I thought I was. If the boss also gave Evan a raise, he doesn't regard me as very special." In his mind, Ted lost all of his previous success gains; this was reflected in a drop in his overall self-esteem. The relation of self-esteem to our ratings of others is like a seesaw. When somebody else's value goes up, ours goes down. Of course, when we can identify with another person's good fortune, our self-esteem increases and we feel good.

Liz, for example, annoyed other people because, no matter what the topic, she turned conversations into monologues about her own experiences and opinions. One time her best friend told her that many people disliked her because she talked only about herself. The alleged unfavorable social image made Liz feel bad; she thought, "I'm socially out of it," and her self-esteem went down. Her secondary reaction was one of anger at "not being appreciated." As her pain subsided, however, she was able to view her difficulties as the outcome of a correctable form of behavior rather than as a reflection of an offensive, irremediable personality defect. She resolved to be less egocentric in her conversations with other people. Thus, the event and the subsequent drop in self-esteem led to a useful learning experience. If Liz had simply become angry, visualizing her friend as an enemy, and plotted revenge, her self-esteem might have been salvaged temporarily, but the opportunity to convert the unpleasant experience into a constructive learning experience would have been lost.

The psychological pain or distress that one suffers in response to, say, being excluded or, like Liz, being criticized, serves a function similar to the physical pain from an attack. The physical pain serves to mobilize a person to cope with a problem that demands correction and can also serve as a learning experience to be applied to similar situations in the future.

Similarly, psychological pain from a presumed put-down can prompt the person to deal with the problem. If an individual experiences a flash of anger, he may "solve" the problem by attacking the cause of his pain—namely, the other person—rather than by trying to clarify the other's intent. In this case the psychological pain is generally due to a lowering of the individual's projected social image, the picture he assumes that others have of him. While a counterattack might ameliorate this image and restore his self-esteem, it would not necessarily solve the interpersonal problem.

Say a wife believes that she was tricked by her husband into doing something she did not want to do. The initial effect is a drop in the level of her self-esteem, and consequent pain. She perceives herself as powerless and vulnerable. As she focuses on her husband's wrongdoing, she becomes angry and wants to strike back. The sequence may run something like this: She thinks, "He took advantage of me. He was wrong to do this. I must look like an idiot." She feels pain, and then anger. She decides, "I must punish him." Her projected image of herself—"looking like an idiot"—lowers her personal esteem and produces pain. Note that she will feel angry only if she considers his behavior unjustified and experiences a negative image of him.

Fighting back neutralizes the damage to her projected social image and to her self-esteem and can temporarily relieve the pain. The retaliation may serve to equalize the balance of power between them (for example, serve as a "learning experience" for the husband). But it may become yet another round of hostile interactions, depending on a number of factors, such as the quality of the couple's relationship and the husband's receptivity to criticism.

Many people who "fly off the handle" easily or have a "short fuse" actually have a shaky self-esteem. Their hypersensitivity is often based on a core image of themselves as weak, vulnerable, and malleable. They have developed ways of protecting themselves from the incursions of other people, however, by being alert to any possible encroachment on their vital interests and by a readiness to perceive their opponents as wrong or bad. Their psychological "defenses" may be strong enough to prevent any damage to their self-concept.

Occasionally, an insult or reprimand may penetrate their defenses, and their self-esteem will drop. By mobilizing a defensive strategy, construing an opponent as "the enemy," and counterattacking, a person can rapidly switch his self-image from that of a helpless victim to a strong, successful retaliator. The change in self-image temporarily repairs the damage to self-esteem, but the memory of being vulnerable and helpless is stored away and reinforces the basic image of weakness and vulnerability. Partly to compensate for his negative self-image, the victim may elaborate a negative image of the "victimizer" as persecutor and conspirator.[3] Such an image of the oppressor or enemy is dramatized in the fantastic delusions of paranoid schizophrenic patients.

A person's portfolio of self-images tends to be consistent over time, each image specific to (or activated by) a particular class of situations. Their consistent selectivity for specific events supports the notion that these self-images represent durable aspects of a more global mental organization. The self-concept, which integrates the various self-images, is multifaceted and multidimensional and not completely accessible at any one time. Like a filing cabinet, the self-concept incorporates the various self-images. It includes representations of the individual's major and minor characteristics, internal resources and assets, and liabilities.

PROJECTED SOCIAL IMAGE

Much more of our life is regulated by our self-images than we realize. When we perceive ourselves as powerful, efficient, and competent, we are motivated to tackle difficult tasks. When we have a self-image of helplessness and powerlessness, as in depression, we feel sad. Our feelings and motivations are also influenced by the way we believe others perceive us: our projected social (or interpersonal) image. Our prevailing social image influences how we react to other people. If we perceive other people as unfriendly and critical, we adopt strategies to protect ourselves. The person with an avoidant personality simply minimizes her social interactions in order to protect self-esteem. The hostile personality is hypervigilant to disparagements, fancied or real, and stays poised to attack. In both instances, the person projects an unfriendly image onto other people and tries to protect his vulnerable self-image. Of course, the projected or fantasized images can become self-fulfilling. People may become critical of and ignore the avoidant personality and become angry at the hostile personality.

Consider this example. Al refuses Bob's invitation to go to a show with him. Bob's reaction is: "He thinks I'm not good enough," and he feels bad, because his negative social image has been triggered. Underlying this is the schema: "People do not regard me as worth being with." The hurt comes from Bob's debased social image rather than from the deprivation of Al's company. For example, if Bob had ascertained that Al was too sick to go to the theater, he could have gone alone, with only a minimal sense of loss, and he would not have felt devalued.

Because our images of other people generally have a definite frame and we see only those characteristics that are consistent with the image in the frame, we screen out other characteristics. It is parsimonious to simplify and homogenize a phenomenon that is at once complex, variable, and unstable, but our interpretations of another person may be distorted by the frame we have constructed. We similarly frame ourselves and obtain at best an incomplete, and at worst a distorted, perception.

How we generally feel about ourselves is, to a great degree, a function of our prevailing self-image. If Bob thinks that Al considers him undesirable, he feels bad because Al's presumed disapproval becomes factored into his self-esteem. But if he thinks, "Nobody likes me," and accepts the explanation, "I must be unlikable," then the negative personal self-image is activated and there is a significant drop in self-esteem. If, on the other hand, Bob has a stable positive image of himself, then Al's refusal will not affect his self-esteem. Bob may, indeed, explain Al's refusal as a sign of Al's selfishness and become angry with him. By viewing Al in a negative light, Bob can protect his self-esteem. Alternatively, if Bob's self-esteem is not threatened by Al's refusal, he can disregard it, without the need to defend himself. In short, the personalized meaning and its relevance to self-esteem issues determine the reaction.

It is clear that in any interpersonal encounter there are at least six images involved: my image of me, my image of you, and my projected image (what I visualize as your picture of me), your image of me, and your projected social image (what you imagine is my picture of you), and your image of yourself. The interaction of these images is reflected in each individual's behavior. If I perceive myself as weak and you as powerful, and you perceive me as weak and yourself as powerful, one likely outcome is that you will dominate me or at least attempt to do so. Many combinations of these various images are possible and can account, in part, for people's unfriendly as well as friendly behaviors and actions toward each other.

4

LET ME COUNT THE WAYS
YOU'VE WRONGED ME

Have you ever stopped to think of all the threats we face in our lifetime? The social world, in particular, is replete with potential danger. Aside from worrying about the risk of accidents or physical assaults, think of the anxiety that people experience in public speaking, job interviews, and love relationships. To understand the nature of these reactions and their relation to anger and hostility, it is helpful to reflect on their development.

As children we were fearful of thunder and lightning, animals, and high places, but as we grew up we realized that we were more vulnerable to *psychological* injuries: to being controlled, insulted, rejected, and thwarted. We were conditioned to perceive these psychological assaults as no less threatening than physical dangers. For the most part our interpersonal problems involve other individuals who are not necessarily a menace to our survival but can nonetheless cause us considerable psychological pain. Regardless of the form of the danger, or the nature of the pain, we fall back on the strategies that served our ancestors in their quest to survive and to avoid physical injury: fight, flight, or freeze.[1] A small child threatened by a large dog or by a bully in the schoolyard will react with feelings of anxiety and will either freeze or become mobilized to flee and seek help. If he is trapped, however, he may be forced to stand his ground and fight.

ANXIETY OR ANGER?

Whether threatened with physical pain from a cutting instrument or psychological pain from cutting words, the individual automatically prepares to cope with the attack. In the first instance, the pain is localized and circumscribed; in the second, it is unlocalized and amorphous. The common denominator of the assaults is the suffering the individual experiences. Psychological "damage" can produce distress as intense as physical damage. The suffering is illustrated in our language itself in the numerous analogies to physical pain: a bruised ego, hurt pride, injured psyche. Because of the unpleasantness of physical pain, people go to great lengths to prevent physical damage and preserve their physical functions. Similarly, the importance of psychological pain is underscored by the extreme care people take to avoid being humiliated or rejected. The victim may retaliate physically or verbally or may withdraw and nurse the physical or psychological "wound."

What factors lead specifically to anxiety or anger? Our response to threat depends on the way we apply a kind of formula that balances the perceived risk against our confidence in coping with the threat. Through a rapid mental calculation, we evaluate whether the risk of damage in a threatening situation outweighs our resources for blunting the assault. If we estimate that the anticipated damage exceeds our capacity to cope with it, we feel anxious and are impelled to escape or withdraw from the threat. If we surmise that we can ward off the offender without sustaining unacceptable damage, we are more likely to feel angry and be disposed to counterattack. In an emergency our calculations are automatic and can take place in a split second; they are not the product of reflective thinking. Certain standard configurations, like a charging bull, are sufficient to set off an immediate alarm reaction. Other situations, particularly interactions with people, may require somewhat longer processing. Our evaluation of the potential harm or risk and of our coping resources can be carried on in parallel, and then integrated to determine the appropriate coping strategy.

Consider this example. You see a person coming toward you, brandishing a club. If you believe he can harm you (he is larger than you and appears angry), you will experience anxiety. If you are confident that you can handle the situation (he seems smaller and insecure), you focus on his vulnerability and mobilize your resources to disarm or to repel him. Often the situation contains sufficient information to tell you immediately whether the risk of being harmed exceeds your resources for repelling the attack. If you are not concerned about being vulnerable, you can consider the wrongfulness of his behavior. Although you may be experiencing some anxiety, you will feel mostly anger and will probably want to punish as well as disarm him.

Similarly, in an ordinary nonviolent confrontation your judgment of your vulnerability and that of your adversary will influence your response. In addition, you will make a rapid (not necessarily accurate) calculation of the advantages and disadvantages of fighting back and of punishing your opponent. Even if you are mobilized to attack and are confident of winning, you will not necessarily follow through with the fight. You decide whether *it is in your best interest* to fight back (usually verbally) or to inhibit your impulse to retaliate. A wife, for example, decides not to continue in a shouting match with her husband because she knows from past experience that yelling back leads to more arguing and may culminate in a physical assault. So even though her muscles grow tense, her fists are clenched, and she has a strong urge to scream at him, she inhibits these impulses in order to keep the conflict from escalating.

In everyday life the circumstances that are most likely to trouble us cause psychological rather than physical pain. We want to retaliate when we are put down, cheated, or slighted. Circumstances like these can rouse us to do battle. In general, the "wrongs" that most concern us are transgressions against our rights, our status, our personal domain, or our efficacy. We expect our freedom, our reputation, and our needs for intimacy and support to be respected. Interference with or threat to these values constitutes an offense. Many of these alleged wrongs are based not on actual violations or transgressions but on the *meanings* attached to particular events.

Bob, a twenty-five-year-old salesman, periodically displayed an idiosyncratic rage when there was no ostensible offense. Usually gentle and easygoing, he would "go bananas" whenever he felt threatened. Typically he felt charged up at the mere sight of a policeman. He also became tense when dealing with an officious clerk at a store, when his wife questioned him about where he had spent his money, or when surrounded by doctors while he was in a hospital bed.

In none of these circumstances was there much likelihood that the other person would demean him; it was the *possibility* of being talked down to or controlled that troubled him, since he felt vulnerable in the presence of authority. He became aroused to attack others before they could hurt him. Any person assuming an authoritarian role seemed to encroach on his autonomy. First he would have a feeling of being smothered, immobilized, or weakened; then he would become enraged and seek to "defend" himself by attacking the presumed offender.

Bob's trigger reactions illustrate another aspect of our emotional reactions.

To a person like Bob, a policeman represents a threatening authority; to someone else, he is a symbol of protection. Bob's vulnerability was embodied in the core belief: if I allow other people any leeway, they will smother me. Therefore, I must fight them off if they pressure me, or if I think that they will.

Many threats, as well as hurts, may be outcomes of our hypersensitivity. Most of us are especially sensitive to specific behaviors that we consider abrasive but that do not bother our friends or relatives. Like Bob, some people react to an authority figure as if he or she will punish them rather than direct or even offer them protection. Others may believe they are being imposed on or exploited when another individual merely wants to enlist their help or to borrow something from them. Some people are "insult-prone," overinterpreting others' benign teases as verbal attacks. Others are rejection-prone and go through life evaluating every interaction ("She loves me, she loves me not"). Our reactions are often based less on the true intent of the other person than on how that behavior makes us "feel": controlled, used, rejected. The feelings are an expression of the meaning we attach to the event.

It is not difficult to determine the meaning of a particular event. Simply ask yourself: What thought went through my mind immediately after the event occurred and just prior to or at the same time as the hurt feeling? This automatic thought—the interpretation of the event—discloses the meaning of the transgression. Examples of these events and the automatic thoughts are listed below. They were obtained not only from patients in therapy but from other people who were coached in tracking their cognitive responses to situations that bothered them and made them angry.

TYPICAL OFFENSE	AUTOMATIC THOUGHT
Friend doesn't return her call.	She has *no respect* for me.
Husband dismisses her opinion.	He thinks I'm *unimportant*.
Guards wave him on without notice.	They think I'm *disposable*.
Waiter delays taking his order.	He thinks I'm a *nobody*.
Shopper stalls supermarket line.	She's holding me up.
Friend doesn't return his lawn mower.	He's *using* me.
Wife doesn't respond to his request.	She doesn't *care* for me.
Boss gives him extra assignment.	He's *pushing* me around.
Teacher corrects her.	She is *critical* of me.
Spouse is late for dinner.	She's inconsiderate.

SELF-IMAGE AND SOCIAL IMAGE

Interpersonal conflict is not simply a matter of who comes out on top, who wins or loses. A crucial aspect is the effect on the victim's self-image and his projected social image—what he believes other people think of him. In some encounters the individual is threatened by the prospect of conveying an image of inferiority or even undesirability. We are sensitive to criticism not because of the critical words themselves, but because of what we think they convey of the other person's mental representation of us. Although we dislike being vulnerable, we don't know how to combat helplessness or its consequences.

Take the example of a graduate student undergoing an oral exam by a group of professors. Power resides in the teachers and vulnerability in the student; if the examiners have a biased image of him and are grossly unfair, the student does not have the chance to fight back. Any overt expression of anger may be used against him. His main response to this situation may be anxiety. After the examination, however, when he is no longer exposed to a negative evaluation by the examiners, he can afford to experience anger and perhaps retaliate indirectly by complaining to his classmates or other faculty members. When there is such a difference in power and the person is concerned about the possibility of punishment, he may simply submit. Submission to the dominant figures may remove the threat and the anxiety. This strategy of submission is often observed in the dominance hierarchy of primates.[2]

Although it is important to recognize that both anger and anxiety are potentially adaptive reactions, they can become maladaptive when we exaggerate the degree of danger or the magnitude of an offense. The student who exaggerates his vulnerability during an oral examination may find that his mind goes blank and he performs just as badly as he feared he would.

We are all familiar with people who show exaggerated reactions to disagreements or criticism; we tend to discredit them as "thin-skinned" or "hotheaded." However, the tendency to misread or exaggerate threats or criticisms may represent a cognitive strategy designed to protect oneself. In life-or-death situations, it is better to interpret a neutral act incorrectly as offensive than to miss a real threat by underplaying its importance. In the prehistoric wild, there presumably was survival value in overreacting to any specific noxious stimulus. Since the *appraisal* of a threat is the key to the behavioral mobilization, overreactions are represented cognitively by terms like "magnification" or "catastrophizing."[3]

The most hypersensitive reactors among us are destined to receive a psychiatric diagnosis, which serves as a mandate to receive help in moderating the exaggerated reactions. The distress of these people revolves mostly around intangible psychological issues, such as the way they believe others evaluate them and the way they evaluate themselves. Interestingly, there are other people with "psychological blind spots" who fail to see potential threats or negative reactions of other people and are therefore fleeced, manipulated, and victimized.

TRANSGRESSIONS AND VIOLATIONS

There is a wide range of experiences that we regard as offenses and that, consequently, make us angry. In the most concrete form, a transgression results in actual physical damage, pain, or the threat of physical damage, such as being choked or threatened with a gun. In the ordinary circumstances of everyday life, however, the typical transgressions are damage or threat of damage to our psychological self. The various kinds of offenses that we sustain have a common denominator: we feel *diminished* in some way, and as a result, we feel hurt, sad, or anxious. We construe this experience as an offense if we regard it as unjustified. We then frame the offenders as wrong or even evil for having hurt us.

Being told that we can't do something we want to do may make us feel immobilized; a lack of obvious caring from a partner may make us feel rejected; a criticism may make us feel socially unacceptable. We also feel personally diminished when another person violates our standards or values or fails to live up to our expectations. Sometimes a relatively minor infraction may enrage us—usually because the distasteful event evokes a sense of loss or helplessness.

When somebody does something that reduces our status, self-esteem, or resources, we may feel hurt at first. If we agree with a criticism, justify a rejection, or resign ourselves to being immobilized, we are unlikely to feel angry but may feel sad. Further, if we blame ourselves for the unpleasant event, we may experience a transient depression. In all likelihood, however, unless we are depression-prone, we will experience some degree of anger in response to these experiences.

The broad spectrum of abuses that we may encounter is illustrated by the huge number of negative evaluation words in the thesaurus. There are many more negative than positive verbs, adjectives, and nouns relevant to interpersonal relations. A substantial number of the negative verbs (for example, *debase, humiliate,* and *reject*) reflect different shades of meaning but have a common theme: reducing another person in terms of his self-esteem or social attachment. The vast majority of the adjectives are evaluative, organizing our social world into nuances of the broad categories of good or bad.

Our language clearly pays homage to the vast array of "wrongful" traumas that we may experience and enables us to identify and distinguish them, just as it helps us to identify the vast array of natural dangers that we might encounter. Perhaps the wide variety of negative words indicates the value of being able to pinpoint the precise nature of other people's harmful behaviors. Concepts like love and affection, while obviously important, do not require and are not represented by such a variety of words. A person can function properly if she is limited in the number of words available to describe affectionate or friendly actions, but she would have difficulty if she could not differentiate the wide variety of unpleasant behaviors to which she might be subjected. Obviously, an adaptive response to being trapped will be different from the response to being abandoned, even though the emotion experienced—anger—may be the same. A precise description of an offense helps us to apply a proper strategy: ignore it, retaliate, or restrain the offender.

Although it is not feasible to list all the offenses we confront, I can indicate the various categories, with examples of each. Note that these offenses are directed at aspects of ourselves or our personal domain that we especially value: our functioning, relationships, rights, resources, possessions, and physical integrity. If we mistakenly interpret another person's neutral or benign actions as a transgression, we are just as hurt and angry as we would be following a real offense.

We readily react to gross offenses, such as being dominated, controlled, rejected, criticized, devalued, or abandoned. An offense may be disguised, however; witness how often people complain of being deceived, used, and manipulated. Whether the wrongs are obvious or concealed, we learn to be alert to them so that we can defend our interests and maintain our well-being.

DIMINISHED FREEDOM	DIMINISHED FUNCTIONING	DIMINISHED RESOURCE	DIMINISHED RELATIONSHIP
Controlled	Disabled	Defrauded	Distanced
Dominated	Immobilized	Robbed	Rejected
Intruded on	Tricked	Cheated	Abandoned
Exploited	Weakened	Dispossessed	Isolated
Manipulated	Trapped	Disenfranchised	Displaced

DIMINISHED SELF-ESTEEM	DIMINISHED EFFICACY	DIMINISHED SECURITY	DAMAGED PHYSICALLY
Disrespected	Misled	Intimidated	Assaulted
Downgraded	Thwarted	Imperiled	Attacked
Insulted	Opposed	Exposed	Wounded
Defeated	Undermined	Betrayed	Shot
Slighted	Let down	Threatened	Trapped

It is not difficult to allocate certain offenses to specific categories, although there is obviously some overlap. Being controlled, dominated, and manipulated subordinates us to another person's will and interferes with our freedom of choice and action. Being trapped, immobilized, or disabled reduces our ability to function. We are particularly sensitive to the loss of our resources when someone cheats us, robs us, or otherwise appropriates our domain or money. As members of groups, we can be aroused to battle with authorities who attempt to dip into our economic resources, as exemplified by the tea-tax rebellion of the American colonies against Britain, the Whiskey Rebellion, and present-day resistance to the income tax by militant groups. Much of the theory of political psychology can be addressed in terms of individual psychology. The collective image of being insulted, opposed, and threatened by another nation is similar to the reaction of one individual to another.

The significance of closeness or stability in relationships is reflected in the perception of being rejected, abandoned, or isolated. Being displaced by another person can lead to homicidal rage, as reported in dramatic media accounts of the murder of a wife and her lover by a jealous husband.[4]

A common source of deepening distress and anger is an exchange of insults. You do something that directly or indirectly, intentionally or unintentionally, diminishes my self-esteem, and I am driven to relieve my hurt by punishing you, most commonly by making a comment that will lower your self-esteem. If you are now angered by my insult, you try to restore your social image by retaliating, thus setting up a vicious cycle of mutual recriminations. And so it goes. We are all sensitive to the actions of other people that make us seem less attractive, influential, or adequate, and we rely on our weapon of retaliation to terminate such incursions and prevent their repetition. Unfortunately, the retaliations often lead to a continuing covert, if not overt, hostility between the adversaries.

Any kind of interference with a person's goal-directed activity can lead to anger; low frustration-tolerance is an especially common cause of hostility. A person with a shaky sense of efficacy will feel weakened when he is thwarted. He then is disposed to punish the offender as a way of restoring his sense of power. Another source of distress involves increased exposure to danger—being betrayed, intimidated, or abandoned. And of course, actual physical assault is an obvious catalyst for anger.

Other offenses that are not listed but produce a sense of loss are concerned with personal or general expectations that have not been met or collective standards that have been violated. We tend to lay claim to other people's loyalty and consideration—a kind of social contract—so that when they default on a responsibility or commit an error, we feel let down and become angry, as though they have broken a promise. The violation of community standards, including behaviors that do not directly injure another person but are considered socially undesirable, can make us angry. These include six of the "seven deadly sins": greed, gluttony, lust, sloth, pride, and envy (anger being the seventh). In addition, profanity, vulgarity, boorishness, aggressiveness, and sloppiness evoke a special form of hostility, namely, contempt, which is often associated with a wish to humiliate the violator. These qualities offend us because their egocentric nature smacks of insensitivity to the needs of others and unwillingness to contribute to the welfare of the group. Self-indulgence and self-centeredness are particularly disparaged because they place the individuals' investments ahead of that of the group.

The various forms of supposed abuse are in some instances defined by the culture and in others are idiosyncratic for a given person. In most cases

conditional beliefs or formulas shape the interpretation of a particular interaction as abusive. In the culture of the streets the following formula defines
disrespect: "If somebody does not look me in the eye, he is dissing me." An
idiosyncratic belief in a marital reaction defines disrespect as follows: "If my
spouse disagrees with me, it means she does not respect me." Practically all of
the kinds of psychological injuries listed above are formed by similar formulas and beliefs.

VERTICAL AND HORIZONTAL SCALES

The transgressions and violations that people are exposed to may be considered in terms of their relative positions of power, status, and attachment.
Certain imbalances or asymmetries predispose them to the possibility of feeling abused. These relationships particularly can be represented graphically in
terms of vertical and horizontal scales. The vertical scale is anchored by the
concepts of "superior" and "inferior," the horizontal scale by "close/friendly"
and "distant/unfriendly."[5] Our relations with other people may be viewed in
the simplified form presented in figure 4.1 in terms of the four quadrants created by these axes: superior-unfriendly, superior-friendly, inferior-unfriendly,
and inferior-friendly.

A person automatically assigns herself a position on the scales in reference to another person or group. If she locates herself in an inferior position
and another person in her superior-distant quadrant, she will feel exposed to
the possibility of being demeaned, controlled, manipulated, rejected, or
abandoned.

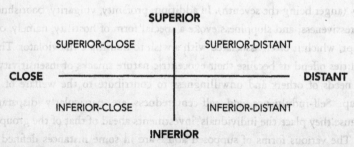

Figure 4.1

A variety of inimical interactions can be delineated according to these two axes. The superior-inferior axis encompasses all the ways in which one person attempts in an unfriendly way to gain an upper hand over another; competition for status, power, influence, and resources produces a winner and a loser. When the struggle produces a top dog and an underdog, a master and a subordinate, the "upper" experiences a sense of triumph, control, and power, and the "lower" a reduction in control, power, and esteem. The underdogs, however, are not necessarily lacking in power to undercut their superiors through subversion, passive resistance, or overt rebellion. The relationship of superior-unfriendly to inferior can also be viewed in terms of victimizer and victim.

In social comparisons, the individual who experiences a loss of position, status, or power relative to another person is likely to feel hurt. If he then attributes this loss to a wrongdoing by the other person, he is likely to feel resentful or angry. Direct, unfriendly actions by another person thus include many of the aforementioned offenses: domination, exploitation, deception, devaluation, intimidation, and immobilization. Punishment (which can include debasement, intimidation, and immobilization) also exacerbates the vulnerable position of the inferior person. When a person views his punishment as unjustified and the punisher as wrong, he or she feels angry and motivated to retaliate.

While at first glance there may seem to be a built-in unfairness in the superior-to-subordinate positioning, this type of hierarchy evidently served our primate relatives well. A hierarchy of this sort ensures a kind of structure among members of a group and generally places limits on individual hostility.[6]

The other axis (close/friendly-distant/unfriendly) is concerned with affiliation. The superior is not necessarily unfriendly. The parent, teacher, leader, or coach can have a friendly-superior relationship with the child, student, or follower, and the subordinate may feel grateful for being nurtured, helped, or educated. If the subordinate is in a learning mode, he can be comfortable in an inferior relation to the instructor. Similarly, when he receives help, he does not necessarily feel inferior. Relationships fluctuate even in normal circumstances. One can be thankful for being helped at one time and resentful at being in the subordinate position at another time. The ascendant person similarly may enjoy the high status, influence, and power at one time but at another time resent the responsibility of taking care of or being imposed on by subordinates. Further, she may be ignored or defied by the subordinate, each reaction a source of annoyance.

Many hurts—and subsequent anger—are related to negative changes in one's relative position along either or both of these axes, rather than the absolute position itself. Changes involving, for example, involuntary movement from a higher to a lower position can produce a sense of loss, weakness, and discouragement. Similarly, movement in a valued relationship from close and friendly to distant and unfriendly can lead to anxiety and sadness. The negatives on the horizontal axis include rejection, abandonment, and withdrawal of affection. If another person or persons can be faulted for the shift, the victim may feel righteously angry and experience a desire to retaliate.

Individuals who have a higher position on the vertical scale (for example, people who are judging) may feel empowered to be overly critical of their vulnerable underlings (for example, those who are being judged). This relationship calls to mind an incident that occurred when I was conducting a workshop for clinical psychology trainees and called for volunteers to perform roles in a play. Only one person, a foreign exchange student, was willing to volunteer. When he came forward, I observed, "You seem concerned. What are you afraid of?" and he replied, "I'm afraid I'll appear nervous and they'll cut me up." To demonstrate how far-fetched his fear was, I asked the members of the audience to raise their hands if they would be critical if he appeared nervous. Practically the entire group raised their hands!

This event taught me several lessons. First, the context of the situation determines what set of attitudes will be active. Psychology students in this particular academic setting were highly competitive. The members of the audience saw themselves as vulnerable to ridicule by the others and so were loath to volunteer. And they knew they would feel superior to anyone else placed in a vulnerable position. Contrary to my expectations, the vulnerability of one of the group members, especially a foreign student, stirred up their critical thoughts rather than their empathy. A show of weakness in themselves or somebody else was something to be scorned. The application of the vertical scale (superior-unfriendly quadrant) explains their reluctance to expose themselves—and their readiness to debase anyone whose weakness ("inferiority") was exposed. Those on the top are in a position to disparage the person on the bottom.

Experiences like this explain why we have reason to feel vulnerable in situations where we can be evaluated. Thus, most people have anxiety over giving a speech in public. It is valid to expect that, under some circumstances, people may be unkind, even cruel or sadistic. And we can anticipate a negative bias. If we have a stable, positive view of ourselves, however, we can shrug off other people's disparaging comments. But if our self-esteem is low or fluctuates according to life events, we can be deeply hurt by the unflattering judgment of others.

Why didn't the members of my workshop feel empathy for one of their colleagues who was distressed? In this setting their competitive-evaluative attitudes preempted caring. Also, since the foreign student was not "really" one of them, it was easier for them to be scornful. He was, after all, part of an outgroup. There is a strange contrast in our reactions to somebody who is "down," depending on whether we identify with him or distance ourselves from him. In the competitive mode (vertical axis), we are likely to distance ourselves from an outsider, particularly if we conceive of him as different from us; at the same time we experience a negative bias. On the other hand, if we project ourselves into his position, we identify with him (horizontal axis). Thus, our mental set influences whether we feel caring or contempt.

The workshop participants could afford to feel superior because they were in the position of passing judgment on their fellow student who was in the inferior-vulnerable quadrant. At other times, however, these same mental health professionals could be caring, nurturing, and sympathetic (the superior-friendly quadrant). Indeed, most of them were therapists who showed those sociophilic characteristics with their patients. This distinction illustrates how different circumstances activate totally different modes and, consequently, different behaviors.

Our relationship to groups is complex, involving not only the vertical axis (superior-inferior), but the horizontal (close-distant). Our relationships may fluctuate from transient to enduring. The transient alliances include a bond formed with others on the same team or group against a common opponent (another team, an ethnic group, or nationality). The natural tendency for each group is to see itself as superior to opponents and to experience a negative bias toward them.

The horizontal axis is most often applicable to relations with other group members, family, and friends. People join in parades to welcome a hero, to celebrate an event. A surprising camaraderie evolves when otherwise distant people are united in response to a disaster, such as fire or flood. More ominously, group bonding can draw people together in gangs to lynch, loot, and rape.

The signals representing solidarity among people bring pleasure and activate earlier patterns of cooperation, mutuality, and reciprocity. The group shows a kind of ripple effect. As the gregarious mode of one person is manifested, it ignites the same mode in others. The cumulative effect of the individual modes operating in synchrony is a kind of "groupthink" that directs the individual into constructive or destructive group action.[7] If camaraderie, solidarity, and group acceptance produce satisfaction, then isolation from the group will cause pain. In some societies, isolation, in the form of shunning, is the official punishment for anyone who dares to violate the expectations of the group.[8] While closeness generally produces pleasure, rejection leads to pain, and often anger. An extreme example of the impact of rejection is the action of a spouse or lover who, aroused by desertion, attacks or even kills a mate.

It seems likely that in ancient times the realistic risks of rejection by the group or a family member posed threats to life because of the interdependence for food and protection. Our progenitors were presumably programmed to respond to social threats with the same type of reaction they had to physical threats. Although the contemporary risk of death from interpersonal rejection is substantially less than it was for our ancestors, we can still react to rejection by our group or family as though our existence is endangered. Being shunned by the group not infrequently leads to depression, a reaction to the loss of a major resource. Anger is usually not pronounced in these circumstances, because the offended person cannot punish the whole group. Rejection by a spouse or lover leads initially to pain but, in many cases, progresses to rage and a craving to punish the offender.

I have applied the term "mode" to describe the composite of beliefs, motivations, and behavior that characterizes enduring reaction patterns. People can function in a variety of modes: in a rejection mode, the person overinterprets experiences as forms of rejection and feels hurt; in a depressive mode, the patient places a negative construction on experiences and feels sad; in a nurturing mode, the individual responds to help-eliciting signals from others. In the hostile mode, she is more likely to read offenses in other people's actions, to magnify the seriousness of slights, to be relatively unmoved by positive events or offers of reconciliation, and to feel angry.

People experience a broad spectrum of modes, some of them self-serving, even antisocial, such as the expansive-exploitative, controlling-dominating mode. While these modes bring gratification to the subject, they often inflict damage on the "target." The modes represent people's psychological states, whereas the axes and quadrants represent the relationships between people in terms of attraction, power, and status. The person's perception of these relationships, rather than the specific circumstances, are responsible for activating the relevant mode.

The discussion of relationships, whether symmetrical or asymmetrical, close or distant, shows the importance of the meaning attached to the relationship. This meaning is responsible for the kind of mode that is evoked.

FITTING THE RESPONSE TO
THE PERCEIVED OFFENSE

We feel less secure, weaker, and more vulnerable when somebody has successfully deceived or manipulated us. Our exaggerated reaction may be, "How stupid I was to be taken in," and to feel enraged at the offender, both for having taken advantage of us and for having lowered our self-esteem ("I'm a dupe"). Or we feel helpless when others control us and unlovable when our beloved rejects us. In each instance our own value is depreciated. If we then focus our attention on how wrong or bad the other person is for treating us this way, we become angry at that person.

Since we are susceptible to such a wide array of assaults on our person and our self-esteem, what methods can we call on to protect ourselves? Of course, the most obvious defense is to become angry at the offender and counterattack. By retaliating we can demonstrate that we are not weaklings who invite abuse; we communicate the message, "Don't mess with me." Fighting can ward off not only present but also future transgressions and can help restore a sense of power and efficacy, crucial components of self-esteem.

When we are hurt, we want to hurt back. The forms of retaliation are usually appropriate to the type of offense that we experience. Unlike their experience of automatic thoughts preceding the feeling of distress, people are very much aware of the possible responses and feel pressure to visualize them.

These are typical verbal counterattacks and the offenses that elicit them. The offenses can be inferred from the response of the "offended" individuals:

- How dare you tell me what to do! (controlled)

- I'll never trust you again. (betrayed)

- You can't talk to me this way. (degraded)

- I thought I could depend on you. (let down)

- You're a rotten cheat. (dispossessed)

- Don't give me the "silent treatment." (ignored)

- You've got a lot of nerve lying to me. (deceived)

- You made me make a fool of myself. (exposed)

- Watch where you're going. (shoved)

These verbal retaliations are intended to hurt the offender and to elicit an apology—certainly to prevent a repetition.

Our social order, apparently cognizant of our supersensitivity to being misused and abused, has provided us with several verbal and nonverbal devices to reassure each other that a behavior should not be taken as an affront. We smile and say "please" when we make a request, so as not to appear that we are imposing or demanding. When people make a comment that could be taken as a criticism, they will often preface it with something like, "No offense intended, but . . . " We often precede a critical comment with a compliment: "You've been doing a splendid job, but . . . " Above all, we learn to apologize or make amends when we realize that, wittingly or unwittingly, we have hurt somebody's feelings. Many successful managers have refined these social skills so that they can express themselves effectively without unduly ruffling other people. They also are adept at inducing others to serve them by skillfully using charm, flattery, and inspiration. Given that there is frequently a disparity of self-interest, it is a marvel that people are able to get along together as well as they do. Of course, the lubricant that reduces the friction is also related to our basic affiliation instincts.

5

PRIMAL THINKING

Cognitive Errors and Distortions

> The human understanding when it has once adopted an
> opinion (either as being the received opinion or as being
> agreeable to itself) draws all things else to support and agree
> with itself.
>
> *Francis Bacon, 1620*

Suppose you see a flying object in the distance. As it comes closer, you decide that it is probably a bird. If you are not particularly interested in birds, your attention wanders to other things, and since this sighting has no relevance to you, you do not waste time and energy in determining the creature's species.

Now imagine that your country is at war and your job is identifying enemy aircraft. Your attention is riveted to the distant flying object. If you think it may be an enemy aircraft, your psychological and physiological systems are totally mobilized. You become hyperalert, you feel tense and anxious, your heart starts to pound, and you breathe faster. Because of the disastrous consequences if any enemy bomber penetrates the defense, you are likely to err in favor of misidentifying a civilian aircraft as a threat—a false positive.[1]

Also, as you move about off-duty, you will be poised to recognize enemy agents who might be mingling in the crowd. Conditioned by government warnings on television and billboards, you will be suspicious of strangers whose appearance or behavior does not match your view of a patriotic countryman. You will focus on small details such as a man's slight foreign accent, his ignorance of certain sports figures in your country, or his secret, suspicious-looking meetings with other strangers. You may build a case against the stranger based on no more than these bits of observation. You will probably be overly inclusive in your identifications. After all, overgeneralization is adaptive when people are at risk, since not recognizing the enemy could be dangerous. In normal times many of our interpretations are based on extracting small pieces of data like these. Because we may string together bits of information often taken out of context, our conclusions are subject to error.

GENERALIZATION AND OVERGENERALIZATION

If you are engaged in a military operation, you will be on "red alert," prepared to process ambiguous stimuli rapidly, to assume that they are directed against you (*self-reference* or *personalization*), and to focus on small details—perhaps out of context—that might be indicative of a threat (*selective abstraction*). You will make *dichotomous* judgments (a stranger is either a friend or an enemy), and you will be overly inclusive in your evaluations (*overgeneralization*). When there is a clear and present danger, these kinds of appraisals are adaptive because they help to prepare us for action that may save our lives.

When we are confronted with a threat, we have to be able to label the circumstances quickly so that an appropriate strategy (fight or flight) can be put into effect. The thought processes activated by threats compress complex information into a simplified, unambiguous category as rapidly as possible. These processes produce dichotomous evaluations, such as harmful/harmless, friendly/unfriendly.

As I explained previously, I have applied the label "primal thinking" to denote these fundamental cognitive processes. This kind of thinking is egocentric in orientation and operates within a frame of reference of "what is good or bad for me (or us)." It is primal in the sense of being absolute—it occurs at the earliest stage of information processing—and also of being apparent in the early developmental phases, when children think largely in global evaluative terms, such as good or bad.[2] Some aspects of primal thinking are similar to the "primary process" form of thinking described by Freud.[3] He described a primitive cognitive process that generally operated unconsciously but could be manifested in dreams, in slips of the tongue, and in the speech of primitive societies.

Such primal information processing operates adaptively in true emergencies, but maladaptively at other times. As we slip into the "danger" or "defensive" mode, this thinking process crowds out our more reflective thinking. When our interpretations are chronically erroneous or exaggerated, we pay the price of persistent discomfort and probable wear and tear on our nervous systems. A persistent dominance of this mode is typical of psychopathology such as paranoia and chronic anxiety, as well as certain cardiovascular disorders.[4]

These primal thinking processes are activated whenever people believe that their vital interests are at stake. The cognitive processes extract the most personally salient features of the situation and are economical in their efficiency. Because of its reflexive nature, primal thinking is adapted for emergencies that do not allow time for reflection and fine discrimination. It is the very simplicity that facilitates the triggering of relevant primal strategies to deal with a threat.

The efficient features of primal thinking are also its disadvantages. The selective reduction of data into a few crude categories wastes much available information. Certain features of the situation are highlighted or exaggerated, and others are minimized or excluded from processing. Personally relevant details are taken out of context, the meanings tending to be excessively egocentric and the conclusions too broad. Consequently, the thinking is unbalanced: it may be satisfactory for true life-or-death emergencies, but it is disruptive to the smooth functioning of everyday life and to the solution of normal interpersonal problems.

Primal thinking is frequently evoked in interpersonal and group conflicts that carry an exaggerated sense of threat. When people become adversaries, their primal thinking may displace their adaptive skills, such as negotiation, problem solving, compromise. The manifestations of primal thinking extend across a variety of situations: adaptive emergency reactions, dysfunctional interpersonal conflict, and intergroup conflict. Emergency mechanisms that are potentially life-saving in dangerous conditions, such as at the battlefront, are often triggered inappropriately in day-to-day interpersonal conflicts. We are therefore not only susceptible to thinking errors but may experience considerable distress, even psychological damage.

We are especially likely to make such errors if we have a negative bias toward a particular person or group, perhaps based on previous unpleasant encounters or on some negative stereotype, ethnic or racial. When we extract or distort data to fit our preconceptions, we perceive offenses where they don't exist and misinterpret innocuous behavior. This kind of selective bias makes us susceptible to drawing the same kinds of arbitrary conclusions as when a real risk was present. However, the seemingly innocuous bias is the basis for much interpersonal conflict, as well as for serious problems between groups, such as prejudice and discrimination (as though it is the replication of previous offenses).

For example, a wife questioned her spouse about his reason for selecting a specific model of a vacuum cleaner. Instead of giving the reason for his choice, the husband exploded and stomped out of the room. His conclusion, "She doesn't have confidence in my judgment," had expanded to, "She never has confidence in me." Her simple request gave rise to the generalization that she not only doubted his judgment but in fact had a low opinion of him. What fed the reaction were associations with previous instances when she had questioned him about his purchases. His anger intensified as he reached back in time and recalled similar events that seemed to support his interpretation of the present event.

Although the wife might actually view her husband's judgment as flawed, it was unlikely that she did so in this particular case. This couple generally had an amicable relationship. It was evident that the husband's interpretation was arbitrary. When people believe they are being challenged, they appear to draw from their memory bank all past difficulties that bear some resemblance to the present one. The memories, unfortunately, are not necessarily accurate but have been organized by a chronic belief, such as, "She has a low opinion of me." This kind of overgeneralization, reflected in absolute words like *never* or *always,* can be observed in the complaints of a youngster to her parent: "You always let Sara have her own way. . . . You never give me what I want. . . . Nobody likes me."

The more a person overgeneralizes, the more upset she will be. It is obviously far more painful for a person to be "always" mistreated than mistreated on a single occasion. The overgeneralized explanation, rather than the event itself, accounts for the degree of anger. A key factor in overgeneralizing is what the "victim" imagines is the offender's view of him: stupid, expendable, undesirable. This factor, the "projected self-image" or the "social image," is often at the heart of problems between people. We are provoked not so much by what people say or do as by what we believe they think and feel about us.

SELF-REFERENCES,
PERSONALIZATION, AND ENGAGEMENT

The application of a personal meaning to events or comments that are essentially impersonal is a common cause of anger and other emotional reactions. Some of the clearest examples of this tendency to refer neutral actions of others to oneself occur when people are driving along a highway. Oscar, for example, would become annoyed when another car passed him on the highway. Once, a large truck passed his car and moved ahead into his lane. The following stream of thoughts went through his mind: "He's trying to show me that he can just cut me off. . . . I can't let him get away with this." Initially Oscar felt belittled, expressed as a transient feeling of weakness. This was followed by anger, accompanied by the decision, "I'll show him!" and a sense of empowerment.

Oscar pressed his thumb on the horn, and while he passed the truck, he observed that the driver was engaged in conversation with a woman companion. It suddenly occurred to him that the truck driver "could not care less" about him but was probably engrossed in talking about when they would stop for lunch or where they would spend the night. Oscar had been engaged in an imaginary confrontation with the truck driver; obviously the latter was oblivious to him. Oscar's subsequent reframing of the incident led to a rapid reduction in anger, the kind of "disengagement" described by the psychologist Irving Sigel.[5]

On another occasion, while Oscar was driving into a parking lot at a hospital, the guard mechanically waved him on in what he interpreted as a perfunctory manner: "He thinks I'm just somebody you can flick off like a fly." When asked to reevaluate this experience, he said, "I guess he wasn't even thinking about me. He was just moving traffic." Again, the recognition that there was "nothing personal" in the guard's behavior nullified his sense of hostility and thus eliminated his anger.

In most situations Oscar was on guard against being belittled. He was chronically concerned that people would regard him as a nobody, as somebody disposable. He would interpret any comments from persons in authority as put-downs. He regarded their comments as directed subtly against him even when they were talking about other people. Oscar also had a pattern of reading personal antagonism into his discussion with others, for example, when he and another person disagreed in a conversation about politics or sports.

In the course of therapy, he came to realize that he was arbitrarily reading contempt, disdain, and disrespect into people's comments. I pointed out to Oscar that he was automatically attaching meaning to the behavior of other people when in fact it was devoid of personal meaning; he had been turning neutral interactions into significant confrontations. He had to learn to detach himself from meanings and accept at face value what other people said.

Oscar's reactions were not unique. It is important to note that the phenomenon of self-reference (or personalization) is observed in normal people, as well as in those who have clinical problems. Many people create egocentric meanings out of their purely impersonal exchanges with strangers, for example, salespeople and other service personnel. They think, "He doesn't like me," or, "She's looking down on me."

When two or more people become engaged in a confrontation, it matters to each what the other is thinking and feeling. However, they can become disengaged when one person decides that the confrontation is not worth the investment of energy and diverts the conversation to a neutral topic or walks away. Similarly, an intervention by somebody else, telling them to "break it up" or "calm down," can lead to the same kind of psychological disengagement and a defusion of hostility. Becoming overly involved in what people think, and personalizing their actions, is in part the expression of an overinvestment in the importance of how one is evaluated by others.

Therapy can help people to recognize the degree of their overinvestment in their social image and their resulting tendency to become engaged with other people in an imaginary or actual confrontation. In most cases analyzing the image that other people, particularly strangers, might have formed has little value, since they have little bearing on one's life. Straightforward focusing on whatever agreements or disagreements might be involved in a business negotiation is more productive than worrying about losing. It is conceivable that in an earlier era, however, being accepted by other members of one's own clan or confronting a stranger might have been life-or-death matters. Residuals of such anachronisms today can interfere with appreciating the subtlety of the give-and-take of ordinary interactions in one's personal or vocational life but can be overcome if recognized.

Intense inappropriate reactions may be expressed when an event touches on a vital issue. Louise, the supervisor described in chapter 3, became furious at an associate who made a minor mistake. It was not the lapse per se that aroused her anger but the personalized meaning she attached to it. The highly charged issue was the question of trust, as indicated by her thought, "I can't trust him to do even a minor thing correctly." Similar "hot" issues revolve around motives of loyalty, fidelity, and honesty. According to the "law of opposites" embedded in the primal mode, if a person is not loyal, faithful, or honest on one occasion, he is the opposite: disloyal, unfaithful, dishonest. This perception destabilizes the relationship and rouses the "victim" to anger and an urge to punish the offender.

Interpreting other people's actions as directed against oneself may lead to perceiving the other person as an enemy, or at least an adversary. For example, Bob, whom I described in chapter 4, presented himself at a medical clinic, only to be told by the receptionist that he could not be seen by one of the doctors because he had not brought his Blue Cross/Blue Shield insurance cards. When informed that he would have to get the cards, he became enraged at the receptionist and started shouting at her. His thoughts were, "She's giving me a hard time; she's deliberately making a lot of rules to put me down." Not until later did he realize that the receptionist was simply following standard procedures and was not being deliberately difficult with him, but by that time it was too late for the appointment for much-needed medical attention. Bob had a long-term problem adjusting to rules and regulations. Whenever a rule was applied to him, he inferred that the person was arbitrarily picking on him. After experiencing a transient feeling of weakness and helplessness, he would become angry and have an impulse to lash out at his assumed antagonist.

DICHOTOMOUS THINKING

Alfred was dubbed "the Last Angry Man" (from a book of that title) by his friends. They remarked that he became angry if he did not get his own way when plans were being made or when he wanted something. It was apparent that his angry reactions were the result of an inner sense that he had no influence over other people. When a friend disagreed with him or seemed to ignore his comments, he thought, "Nobody ever listens to me," and became angry. When his wife did not accept one of his suggestions, his automatic thought was, "She has no regard for my opinion." When he could not persuade the plumber to come out right away to look for a leak in one of the pipes, he thought, "I can't get anywhere with these people." In each instance he felt defeated, weak, and helpless.

Alfred judged each of these situations in terms of being either effective or ineffective; he had either total control or the opposite—no control at all. Such dichotomous thinking is an expression of an underlying belief: "If I can't influence other people, I am ineffective, powerless." He would indiscriminately judge situations according to this formula, and if he decided that he was effective (the minority of instances), he would feel pleased for a short time. When he wasn't able to make an immediate impact on other people, he felt hurt and then angry.

When Alfred compared himself with others, he found them to be more effective than he was in getting people to cooperate and in being in command of a situation. His anger was triggered by his image of other people as oppositional and stubborn. Underlying these overreactions was his basic view of himself: "I am weak."

Why was Alfred angry rather than sad at being thwarted? In effect, he was able to shift the "cause" of his hurt from his own sense of ineffectiveness to the behavior of the others. "They are wrong to oppose me. . . . They never listen to me." Shifting the blame enabled him to soften the pain of his frustration and disappointment and to experience the less unpleasant emotion of anger. Continual anger, of course, takes its toll psychologically and medically, and Alfred had a continuing elevation of his blood pressure. He did not enjoy life very much and was chronically fatigued, probably the result of being frequently "worked up."

To have a less turbulent life, Alfred will have to examine his interpretations of the situations in which he seems to be ineffective. He will need perspective: is he as ineffective in these situations as he believes? Are there situations in which he is clearly effective? How can he reconcile his image of himself as weak and powerless with his obvious achievements in life? He will also have to recognize that situations do not generally fall into either/or categories, that he can have many gradations of influence, efficacy, and power that vary from one situation to another. In addition, Alfred might consider where he could refine his social skills so that he can be more effective. With an improved self-image, he might actually become more effective.

Dichotomous thinking is observable in a variety of interpersonal problems. Sally regarded herself as having a "hang-up on rejection." If she did not receive immediate reassurance, affection, or approval from a close friend or lover, she would feel rejected. When she recovered from this feeling, she would be critical and angry at the other person for letting her down. She would turn her attention from her hurt feelings to the offender, whom she saw as wrongfully rejecting her.

Sally's dichotomous belief was, "If I am not fully accepted or loved, then I am rejected," derived from her core image of herself as being unlovable. She sought reassurance to compensate for this image but needed a constant show of affection and approval to avoid feeling bad. When this was not forthcoming, the all-or-nothing mental set made her conclude that her friend or lover was rejecting her. Since the notion of being unlovable was devastating, she would shift to finding the "rejecter" at fault and would become enraged at him.

Larry was regarded by his friends and family as a "control freak" and a "perfectionist." He insisted on monitoring the performance and conduct of members of his family and his subordinates at work to see whether they met his standards. If they were not up to the mark, he would react with annoyance and anger. An analysis of his reactions revealed a man who had a profound underlying fear of things going wrong. His perfectionism was an outgrowth of his dichotomous belief that if things were not done properly, chaos would ensue. This belief was clear in his acute anxiety when a mistake was made; his initial thought was, "This *could be* a disaster." Then he would shift his focus to the other person and level the blame at him, not for making the mistake but for upsetting him.

Larry's core belief centered on the notion that he was defective, a belief rooted in childhood experiences when, owing to undiagnosed attention deficit disorder, he felt overwhelmed by the academic demands placed on him. His need to control his own behavior and that of others can be seen as a compensation for this belief about himself. When his control over a situation slipped, he would be overwhelmed by a catastrophic fear of being incapable of functioning. His compensation for his fear of being defective was to blame others when things did not work out properly. Since he held others responsible for his distress, he would naturally feel angry at the "wrongdoers" and feel motivated to punish them. In addition, when things were not done well, his fear of chaos was aroused. His largely successful compensation for this fear was to adhere to the highest standards of performance.

The tendency to anticipate the worst possible outcome of imperfect performance or an error is generally highly dysfunctional in normal times. Nonetheless, partly because of inherited predispositions and partly as a result of previous learning, many people are prone to "catastrophize" when problems arise. This mental mechanism is involved in chronic anxiety and hypochondriasis as well as in excessive blaming and anger.[6]

CAUSAL THINKING AND THINKING PROBLEMS

Consider the following scene. While you are walking down the street, somebody puts a cane in your path and you stumble. You instantly decide that this person was deliberately trying to harm you, and you become intent on punishing him. But then you discover that the person who inadvertently placed his cane in your path is blind. You correct your construction of the event, perhaps feeling a little guilty or embarrassed over becoming angry. The offender's *intentions* are more important than the relatively mild jolt you received. Once you understand that the incident was unintentional—owing to neutral circumstances—then nobody is to blame. You no longer feel the need to punish the other person.

Primal thinking plays out in a crucial way in our explanations for unpleasant events, such as being obstructed by another person. When information regarding the *cause* of an aversive situation is incomplete or ambiguous, an individual is disposed to assume that the cause is deliberate, not accidental.[7] In our conflictual relations with other people, however, we may become stuck in the pattern of making erroneous causal explanations because our minds are already closed to contradictory information or alternative explanations. Disagreeable situations often summon up this pattern of thinking automatically, without any conscious reflection on our part.

People's actions and words are important to us, but the reasons and motives—the causes—behind them are even more important. Unpleasant reactions such as hurt, sadness, anxiety, frustration, or suffocation demand an explanation. It obviously makes a big difference whether a particular act was intended to hurt or was accidental. When we are distressed, we are likely to search for causes: "Why hasn't she called?" or, "Why did he talk sharply to me?" Any act by another person that gives us discomfort sparks the question, "Why?"

We fasten our thinking on the cause of an unpleasant occurrence because the explanation is crucial both for our anticipation of what will happen next and for our long-term expectations. However, the way we assign the cause is susceptible to bias. We may incorrectly attribute the act to a malevolent wish or to a character defect in the other person when the most parsimonious explanation is usually that the act was simply an unavoidable mistake, due either to the situation or to chance. A harmful error due to negligence will be judged more harshly than one that was unavoidable.

If another person grimaces and thrusts his finger at you, it is important for you to know whether he is making a mock gesture or a serious threat. If the "cause" is his anger at you, it could be the prelude to a direct attack or a harbinger of unpleasant actions. The anxiety that you experience can be adaptive if it helps you to mobilize for defensive action, since threatening gestures or words could be followed by a physical attack or other hostile actions. You can anticipate the possible consequences of the offensive behavior if you can decipher the offender's state of mind.

Considerable research indicates that people have definite styles in explaining events.[8] Some consider themselves responsible for the good things that happen to them and blame others for the bad things. Most normal people show this kind of self-serving bias. Others have the opposite pattern. People who are depressed or susceptible to depression, for example, attribute their successes to luck and their failure to internal causes, such as their presumed inadequacy.

Although it is relatively easy to label another person's irritating behavior (noisy, late, inattentive), it is not easy to decipher his state of mind—the cause of his behavior—correctly. Indeed, our reckonings are often fraught with potential error. If I have slipped into egocentric primal thinking because of inherent difficulties in understanding another person's mental process or because of my own biased thinking, it is difficult for me to arrive at a reasonable explanation for that person's distressing behavior. Because of the biased thinking that occurs when an event concerns an important problem, people are prone to make erroneous explanations.

Consider some typical problems and the presumed causes.

EVENT	INFERRED MEANING
A husband is late for dinner.	"He would rather be at the office than at home with me."
A student receives a lower grade on an exam than he expected	"The professor doesn't like me."
A friend neglects to keep a promise.	"He's angry at me."

Note that in each case one can insert the word *because* between the event and the interpretation. The *meaning* that one attaches to the troubling behavior of others is generally the same as one's interpretation of the *cause* of that behavior. The automatic explanation drives the alternate, more benign one out of circulation—for example, "He is swamped with work at the office," or, "That's a fair grade," or, "My father forgot."

The tardy husband complains, "Why does my wife get hysterical because I'm a few minutes late?" Again, her overreaction springs from her inferred "cause": "He cares more for his work than for me". Of course, the wife's interpretation may be correct. That is, he may care more about his work. The pain, however, comes from the extreme meanings inferred: "He doesn't love me anymore," or, "He doesn't respect me." In most life situations the problem may be addressed objectively and a more accurate explanation can be offered, or the problem can be compensated for or dismissed. The student could discuss his grade with his professor; the wife could determine whether her husband really had to work late; the disappointed friend could discuss why the promise was broken and find out whether his friend was actually angry.

Although the disappointing situation may be corrected or dismissed, its inferred cause is often unacceptable. Thus, when a person becomes enraged, he wants to inflict pain and stamp out the "cause" more than he wants to correct or compensate for the disturbing event. Inflicting pain is intended not only to change behavior but to change the offender's motivation.

EXCLUSIVE CAUSES

We are driven to account for the cause of others' poor behavior even though that behavior is easily misunderstood. There is a high premium on knowing what other people think of us and feel toward us. Their state of mind will be of utmost importance as we try to anticipate what problems may arise and how to act if they do. We automatically tick off the possible explanations for someone's behavior and form conclusions about that person. Depending on our conclusions, we decide whether to be friendly to him or to avoid him, whether to cherish or punish him. The problem is compounded by the fact that although there are generally multiple causes for a particular event, our primal thinking impels us to focus on "the one single cause" and exclude other possibilities.

In earlier times certain natural phenomena, such as windstorms, rainstorms, and drought, were attributed to a single personal cause—the fickle disposition or wrath of the gods, for example. We realize now that there are "natural" reasons for changes in the weather, one that are so complex, in fact, that we have difficulty in ascertaining the causes. We have even greater difficulty in predicting long-term weather patterns, despite our knowledge of multiple complex factors. Nonetheless, the specific target for a retaliation in human interactions is usually the immediate obvious factor, the proximal "cause." Even when the inference of the proximal cause is correct, however, there may be numerous contributing factors, some subtle and distant, arranged not in a linear fashion but in a weblike fashion.

Consider this case. A woman was furious at her husband because their son had bumped his car into another car while driving to school with his friends. At first she blamed her husband completely, because "he should have been more strict in teaching our son how to drive." On further probing by a therapist, the mother recognized that there were other factors that contributed to the accident: (1) their son was rushing to school because he was late, (2) his companions were egging him on, (3) it was raining and the roads were slippery, (4) the driver of the other car was at fault for not signaling a turn, (5) the son characteristically rebelled against any rules. As she reviewed the other causes of the accident, she divided the blame among the other contributing factors. She realized that their son played only a small part in the cause of the accident.

Fixing on a single cause of an individual's "misdeeds" results in the automatic exclusion of alternative explanations. The offended person jumps to a conclusion regarding the cause of behavior and thus ignores other possible explanations or dismisses them as "excuses."

A teacher became annoyed that her high school class had not performed as well as she had hoped on their examinations. Her first explanation was that they had not been working hard enough and were attempting to show her up, getting even with her for enforcing discipline. Then she became annoyed at the school system and society at large for not providing the proper arrangement for learning. During a therapeutic interview, she was asked for her very first thought when she received notice of her class's performance. She became tearful and said, "I thought it was my fault. I'm responsible. I should have been able to stimulate them. . . . I must be a poor teacher." Her anger at the students and at the school system had covered up her own self-doubts. Either she was totally responsible for the students or the school system was responsible.

Although focusing on a single external cause may seem to protect self-esteem, it only serves to disguise, not remove, underlying self-blame. At the same time, the self-doubts produce discomfort, which fuels the blame assigned to the system and to others.

To give the teacher a sense of all of the relevant factors, I used a cognitive technique called a "pie diagram." The same type of crucial judgment occurs in normal situations, when there is no immediate threat to life. I asked the teacher to indicate in the diagram how much of the responsibility for the disappointing class performance could be attributed to any single factor. She first wrote down 100 percent for "the system," then she changed this and rated herself as the exclusive cause. We then brainstormed to list all other possible causes. After assembling a group of contributing causes, she assigned the following percentages:

The students come from the inner city, where the
role models and general atmosphere do not
stimulate academic achievement.   15%

Most come from single-parent homes and do not
receive encouragement, support, or help at home.   15%

They enter high school with a weak scholastic
foundation and minimal motivation.   15%

The system has let them down—for example, classes are too large.	20%
The teachers are forced to spend too much time being disciplinarians.	15%
The peer pressure, particularly among the boys, is not to do well scholastically but to excel at sports.	10%
I am not as inspiring a teacher as I could have been.	10%

After she made the new analysis, her perspective was dramatically broadened. She realized that there was no one factor that contributed to their poor performance. She also realized that her own teaching deficiency—if present—contributed only a trivial percentage to the outcome. Then she was no longer angry at the students for "letting her down," and she thought of ways in which she could counteract some of the negative factors. In the end she reflected that, in fact, the students had done as well as the previous year's class.

The case illustrates the tendency to assign a single cause to a negative event—in this case, that the students were attempting to thwart the teacher. Her anger was associated with a vague desire to punish them for letting her down. This "cause," as it turned out, was only the superficial reason for the teacher's distress. The students' grades at first evoked the reaction, "I am responsible for their inadequate performance." She was then able to escape from the pain of this attribution by diverting her attention from herself and latching onto another reason for the disappointing performance: her students' recalcitrance. She also added, "I feel that they rejected me." On reflection, she recognized that she had initially blamed herself entirely and, when this was too painful, had redirected the cause to the students' presumed resistance.

The teacher had initially succumbed to the cognitive error of personalization: she took the class's performance as a personal failure. She then shifted the blame to the class—a process technically known as "externalization" or "projection"—and became angry. But her underlying sense of inadequacy persisted, contributing to her unremitting anger.

Additional facets of human nature are illuminated by the example. We frequently find that a parent will hold herself solely responsible for her child's misbehavior at first and then quickly shift the blame to the child and become angry at him. The damage to self-esteem by the self-criticism ("I am a failure") initiates the self-protective reaction of blame ("He's a rotten kid"). But the underlying hurt continues, in a sense fueling external attribution and anger. Much of the anger emanates from damage to self-esteem, but externalizing the blame to relieve self-criticism really is a smoke screen. By shifting the blame, the mother forgoes the opportunity to work through—to reevaluate—her notion of, say, being a failure. Consequently the disguised damage to her self-esteem remains. The self-doubt and self-blame that occurs prior to blaming other people are generally covert and, as described earlier, take the form of automatic thoughts. People can be trained, however, to identify them by observing their state of consciousness prior to becoming angry.

Some people have learned to "catch" themselves when they are slipping into an angry or anxious mode by recognizing some of the features of biased thinking, such as overgeneralization or personalization. They can then apply strategies such as distraction or looking for contradictory evidence to force themselves to correct the primal thinking. Others may become stuck in the primal mode and slip into a clinical disorder.

We are not enslaved by our personal history or by evolution-derived patterns of thinking. We are endowed with the capacity for mature and flexible thinking, which allows for reflection and judgment and can supersede the primitive primal thinking. This kind of reflection is more realistic, logical, and rational and can correct the primal thinking, but it can have the disadvantage of being slower and requiring more effort. In fact, it has been described in the literature as "effortful thinking." When we are not engaged in a hostile encounter, we have the cognitive capacity to view circumstances in perspective. When we do become so engaged, however, it requires a powerful mental effort to transcend our automatic primal thinking. Taking perspective gets easier in time and provides a foundation for constructive problem solving and a more tranquil life.

6

FORMULA FOR ANGER

Rights, Wrongs, and Retaliation

> Our brain has developed a capacity to create for us a world
> of our own making and imagination. Very few of us live in
> the real world. We live in the world of our perceptions, and
> those perceptions differ dramatically according to our per-
> sonal experiences. We may perceive anger where there is
> none. If the distortion is ever enough, we may think we are
> living among enemies even while surrounded by friends.
>
> Willard Gaylin, 1984

Consider this illustration of a rule violation that leads to angry reactions. A
wife discovers that her husband has not attended to certain obligations, such
as fixing the leaky faucet in the kitchen, calling the electrician, and paying
the bills; she becomes enraged, and they have the following exchange.

Wife: You never do what you promise.

Husband: Why are you making a federal case of this?

Wife: Because you never do what I ask you to. You never pull your
own weight.

Husband: There you go again—you never appreciate what I do.

Wife: How can I? You're always watching the games on TV or
hitting golf balls.

Husband: You just can't stand it if you see me enjoying myself. You need
to control everything I do.

Wife: Why don't you shut up and do what you're supposed to do?

Husband: You even want to control my speaking.

It is notable that the partners do not discuss their respective initial hurts, the human feelings that they could have accepted in each other. They certainly could have felt empathy for each other's hurt feelings more readily than with the angry feelings. As the interchange shows, the partners are driven to view each other's behavior in absolute terms (as evidenced by the words *never, always,* and *everything*). The content of their expressed thoughts also shows the presumed violation of the rules.

Let's roll back in time and see this episode from the standpoint of the wife, who reported it to me in treatment. Her chain of thoughts ran like this: "He's let me down again. . . . He never does what he is supposed to do. . . . He's doing this on purpose to undermine me. . . . He's totally irresponsible."

This sequence demonstrates the steps in the generation of anger and hostility: the *thwarted expectation* and disappointment and the consequent *sense of betrayal* ("He has let me down—again"). The initial feeling, often described as a sinking *feeling of weakness,* is the bodily manifestation of the loss of power. The letdown is rapidly superseded by her *framing* of his omissions as a *pattern of undesirable behavior* ("He never does . . . ") and the characterization of his omission as *intentional*. Finally, she *fixes blame* on him and condemns him as "irresponsible." Once his wrong behavior has crystallized in her mind, her anger accelerates. She even has thoughts of attacking him physically.

A further probe of the wife's psyche elicits a fear or exaggerated worry, which takes the typical "what if" form. "What if he doesn't attend to things? Everything could fall apart," or, "We will be hounded by creditors," or, on a less conscious level, "I'll be helpless, unable to cope." Thus, the background for her anger is not only her frustration but also her fears and sense of helplessness. As she shifts her focus to the *cause* of her distress, she inevitably latches onto her husband's supposed obstinacy and irresponsibility.

The following figure illustrates the progression of the chain reaction from the husband's default on obligations to the wife's anger. It is clear that the perceived lack of responsibility would be sufficient in itself to anger her, but in this case, as in many, additional factors feed into her feeling upset and ultimately enraged. She has a tendency to catastrophize and thus to envision

**Figure 6.1   Network of Reactions in Wife Whose Husband
Did Not Attend to Responsibilities**

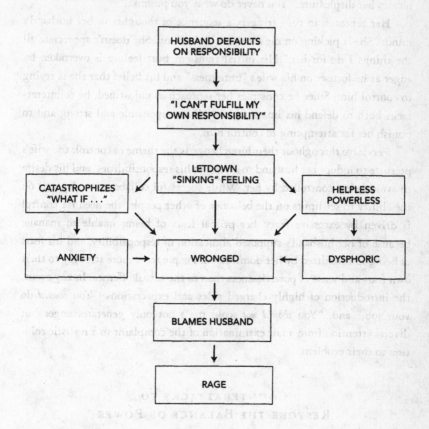

the worst possible outcome, namely, total chaos. Further, his procrastination affects her own sense of competence— or incompetence—and she perceives herself as helpless.

As a result of this constellation of beliefs, interpretations, and feelings, psychiatric symptoms—or even a full-blown disorder—may develop. The wife may become fixed on worry and slip into a generalized anxiety disorder. This is particularly likely if she feels intimidated by her husband and is overly concerned about displeasing him. Another path might lead her from feeling weakened and inadequate to hopeless, resigned, and possibly depressed. If her reaction does not follow the route to anxiety disorder or depression, the wife is likely to become angry or enraged and fully mobilized to punish her husband for his behavior. Energized by her anger, she communicates her displeasure: "You never do what you promise."

Her reproach in turn triggers a sequence of thoughts in her husband's mind: "She is picking on me for every little thing. She doesn't appreciate all the things I do for her." His initial transient hurt feeling is overtaken by anger as he focuses on his wife's "unfairness" and his belief that she is trying to control him. Since he construes her reproach as unjustified, he counterattacks both to defend his image of himself as responsible and strong and to punish her for attempting to control him.

Pervasive throughout their interchange is the theme of control: the wife's pressure to induce her husband to attend to his responsibilities, and his desire to avoid being controlled by her. While successful collaboration depends on the ability to set limits on the behavior of other people, the need for control is driven by excessive fears: her primal fears of being unable to manage because of her husband's supposed abdication of responsibility, and his fears of being immobilized by her domination. People react more strongly to their own fears and sense of powerlessness than to the actual offenses. In any event, the introduction of highly charged rules and expectations ("You *should* do your job," and, "You *should not* scold me") not only generates anger but diverts attention from a fair examination of the complaint to a realistic solution to their problem.

COUNTERATTACKS TO
RESTORE THE BALANCE OF POWER

There seems to be an axiom that when one person perceives a loss of power in a relationship, he or she must strike back, even though retaliation is self-defeating in the long run and may lead to even further attacks and pain as the conflict grows. A number of related factors contribute to this. First, there is a primitive, almost reflexive, reaction to pain, whether physical or psychological: to remove its source. Another important factor is that criticism, even when justified, often produces an upset in the balance of power. In the foregoing example, the husband felt diminished and stripped of power by his wife, who similarly felt powerless as a result of his behavior. His retaliation represents a desire to restore the balance in his power struggle with his wife.

As long as the hostile engagement persists, she is faced with the choice of fighting back, which could exacerbate the fighting, or giving up and feeling worse. Although the conflict centers on the division of labor, it could occur in many other domains, such as child-rearing, social engagements, or contacts with relatives.

Of course, couples acknowledge that the exchange of insults usually does not accomplish the objectives of either person and frequently drives them further apart. The proper route would be disengagement and a focus on solving their problem so that there is neither a "winner" nor a "loser." It is often difficult to follow this route, however, because the ostensible cause of a conflict may mask such personal issues as low tolerance for frustration, a sense of inadequacy, and sensitivity to criticism. If the personal problems are particularly strong, it will be necessary to address them, perhaps in psychotherapy or marriage counseling.

FROM THINKING TO ACTION: THE HOSTILE MODE

It is of interest that a battle often rages in a person's mind before he or she expresses anger vocally or in hostile action. There seems to be a straight-line progression from thinking derogatory thoughts to transforming them into spoken words to taking violent actions. In the previous example, the offended wife gave voice to her thought, "He never does what he is supposed to do." Her automatic thoughts already contained a reproach—as well as a critical explanation for her husband's upsetting behavior.

Let us examine what a person involved in a hostile engagement experiences. When she expresses what she is thinking and feeling, her hostility is apparent not only in her words but also in her cutting tone of voice, her taut facial muscles, staring eyes, clenched fists, and rigid posture. She feels angry and also has a strong wish to punish her husband. All of her relevant attack systems have been mobilized: cognition (derogatory view of husband), affect (anger), motivation (desire to criticize), and behavior (mobilization to attack). The words that were formed initially as negative evaluations progress to criticisms and an urge to reproach him, and they now become instruments both to punish her husband and to pressure him to comply with her wishes. It is interesting that once she is in the attack mode her initial feelings of being hurt, frustrated, and powerless are submerged by a greater sense of personal power and an expectation that she can influence his behavior. Whether she proceeds to attack him physically depends, of course, on her overcoming a number of deterrents and inhibiting factors, such as the fear of worsening conflict and of ultimate physical damage to herself.

In the attack mode, the thinking of each partner reverts to a primal form. The wife perceives her husband as bad and wrong to the exclusion of any positive traits. Her focus is fixed on his culpability, and she interprets his behavior in absolute terms. Her husband frames her as a nag whose behavior is not only ugly but unjustified. As long as they continue in the hostile mode, they recall only past misdeeds of the other and interpret each other's present behavior in a biased fashion. Later, when they have calmed down (technically, disengaged from the hostile encounter), they can see each other more objectively and perhaps proceed to solve their practical household problems, if not their interpersonal problems.

A number of factors, as displayed in the figure on page 95, are involved in the generation of hostility. Although these are presented in a somewhat sequential order, it seems likely that the appraisals occur practically simultaneously, so the individual makes an integrated global judgment.

Several of the factors in this diagram are necessary (for example, a perceived loss or threat) but not sufficient to produce hostility. The presumed *loss* in the diagram usually involves an appraisal of being diminished in some way, such as feeling less desirable or effective or being deprived of a personal relationship or resource. The *threat* may be directed toward the individual's security or values. The weights carried by the factors differ according to the nature of a particular event and the surrounding circumstances. If this loss or threat is viewed as excusable or justified, the progression stops at that point and the person does not become angry. On the other hand, if these factors are present, they add a special force to the eventual experience of hostility.

Although feelings of *distress*—hurt, anxiety, or frustration—usually occur early in the sequence, they are not inevitable in the generation of anger. V*iolation of a rule,* often implicit rather than explicit, is practically always present, however. A history of violations exacerbates the hostility response.

The magnitude of the violation contributes to the degree of hostility, as does the presumed offender's motivation for the transgression—that is, whether the offender acted unintentionally or knowingly. In fact, most people seem to operate on the dubious assumption that, until proven otherwise, any act they perceive as noxious is deliberate. If the offender *could* have behaved otherwise, he *should* have; therefore, he is responsible for the offense. If, on the other hand, the offender is judged to have had no control over the offensive act, or it is seen as involuntary, the attribution of responsibility is reduced.

It is possible to be exposed even to an intentional hurt and not become upset—for example, if the offender is not regarded as "responsible" for the act. Being punched by a small child in a rage or screamed at by a delirious patient does not usually make us angry because we know that neither one is *accountable* for his action. Because the behavior is excusable, we do not become engaged in an adversarial encounter. Further, the action is not a breach of a "should-not" rule (prohibition against an unprovoked attack) because "should" and "should-not" imply the availability of choices or self-determination and control on the part of the offender. If a person's hostile behavior is dictated by an immature or disordered brain, the should-not rule is not generally applied: a person with a brain disorder is *expected* to behave irrationally. Thus, if an unpleasant assault is justified or excusable, the progression toward hostility is arrested. However, a past history of similar transgressions by the offender will cast him more solidly as an offender and will intensify the hostility.

It is important to recall that the experience of anger is generated not by the event but by its final meaning. Constructive, diplomatic criticism from a coach or teacher or a painful injection by a physician is acceptable because the benefit justifies one's pain. Even though another person's actions may seem intentionally provocative, the target person's reaction depends on the meaning he attributes to the act. A depressed person, for example, may respond to a deliberate insult by becoming more depressed rather than angry because his interpretation of the insult may be, "I deserve it. . . . This only proves how undesirable I am."

The verbal rules that govern our interpretations and consequently our feelings and behavior are often complex and appear to form a kind of algorithm. Our information-processing system is sufficiently complex to evaluate each of the features of a situation practically simultaneously, as though through multiple channels. For example, suppose that a friend refuses your invitation to have dinner with you. The rules in the algorithm are framed to answer the following questions, practically simultaneously.

- "Did the refusal diminish me in some way, that is, does it indicate I am not desirable company?"

- "Was this unwarranted or unjustified?"

- "Did she intend to hurt me?"

- "Was this behavior characteristic of her or out of character?"

- "Does she deserve to be punished?"

The rules in the algorithm provide rapid and concurrent answers, which are immediately integrated into a conclusion. As illustrated in figure 6.2, affirmative responses to these questions will arouse anger.

AUTOMATIC FRAMING

Imagine that you enter a store in an unfamiliar part of the city and a sales-person—say, a woman of a different ethnic background—smiles as she approaches you. Your immediate reaction may be, "She seems to be a friendly person," and you automatically return the smile. But suppose that you have had unpleasant encounters with people of her ethnic background or you have heard derogatory remarks about them. Then your positive reactions will be muted. Perhaps the meaning of a derogatory parental warning will echo through your mind: "Don't have anything to do with those people." Or you may have created a picture from past experiences of salespersons as control-ling and self-serving. The memories and beliefs that you bring into the situa-tion will help to shape your interpretation of the salesperson's behavior. You then jump to the conclusion that her smile is insincere—an attempt to manipulate you. Instead of returning her smile, you feel tense and stiffen up.[1]

We frequently have such interactions with other people. How do we interpret someone's communications—their words, tone of voice, facial expressions, and body language (whether stiff or relaxed)? We have a reper-toire of beliefs that we apply to particular situations and consequently make

Figure 6.2   Algorithm of the Factors Leading to Hostility

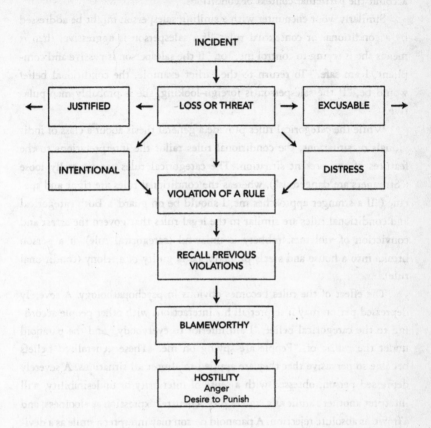

sense of them for the most part. We already have these rules or formulas at our disposal when we enter a situation. Depending on its nature, one or another pattern of beliefs is automatically activated.

The beliefs and formulas tend to be global: "Foreigners are dangerous," or, "Salespeople are manipulative." The global beliefs are applied to fit a particular situation in the form of conditional or "if-then" rules. A global belief, for example, would be, "Tigers are dangerous." However, you will obviously react to a saber-toothed tiger quite differently if you encounter it in the zoo rather than in the wild. The conditional rule activated in this case would be, "If the tiger is caged, then I am safe." The general belief about the danger of fierce animals has been refined to a conditional belief in order to take into account the particular context or conditions.

Similarly, your encounter with a smiling salesperson might be addressed by a conditional or contextual rule: "If a salesperson is aggressive, then it means she is trying to control me," or, "If the salesperson is passive and compliant, I am safe." To return to the earlier example, the conditional belief would be, "If the salesperson is foreign-looking, she is probably manipulative."

While the categorical rules provide a general thesis about a class of individuals or situations, the conditional rules tailor the interpretations to the features of the present situation. The categorical rules are generally loose ("Strangers are dangerous"), whereas the conditional rules are tight and specific ("If a stranger approaches me, I should be on guard"). Both categorical and conditional rules are similar to the legal rules that govern the arrest and conviction of violators: robbery is unlawful (categorical rule); if a person breaks into a house and steals property, he is guilty of a felony (conditional rule).

The effect of the rules becomes obvious in psychopathology. A severely depressed person may interpret all his interactions with other people according to the categorical belief, "I am inferior to everybody," and the paranoid under the rubric of, "People are spying on me." These generalized beliefs become so pervasive that they are applied to almost all situations. A severely depressed person, obsessed with a sense of inferiority or undesirability, will interpret another's smile as a sign of pity, a neutral expression as aloofness, and a frown as absolute rejection. A paranoid person may interpret a smile as a devious attempt to manipulate him and a neutral expression as feigned indifference. Thus, the dominance of the categorical rules can distort the specific features of a situation. The biased beliefs of the depressive or paranoid person produce biased interpretations of reality. Such biased beliefs and thinking occur both in psychopathology and in interpersonal animosity and intergroup conflict.

Overly broad categorical beliefs about strangers or foreigners can lead to erroneous labeling of friendly strangers as dangerous or unfriendly ("false positives"). To the degree that the categorical beliefs regarding strangers are a mixture of our evolutionary and cultural heritage and our idiosyncratic learning history, we are predisposed to react to unfamiliar or different people as being alien to us. During an early developmental stage, for example, children generally respond to the approach of unfamiliar individuals with obvious distress, presumably fear.[2]

Even though most children outgrow the fear of strangers, they may retain this categorical belief in a latent form that is called into action when they are in contact with foreign-looking people or when they hear unfavorable comments about them. At a more conscious level, the reflex aversion to people different from us is apparent in xenophobia—ethnic or racial prejudice. Further, the more general biased beliefs regarding outsiders become activated in the presence of conflict with other groups or nations.

How do we extract the meaning of a particular combination of stimuli, as when we are approached by a smiling salesperson? In interpreting what we see, we draw on an information-processing system based on images and memories as well as on beliefs. The visual configuration of the salesperson is matched against templates in our memory. When a match is made between the external configuration—for example, the saleslady's smiling face—and the relevant template, "recognition" occurs and the associated beliefs and rules yield an interpretation of her motives.

The conditional beliefs flesh out and modify the meaning generated by this matching process. A memory of a particular person who was deceptive or manipulative may override the perception of this person's smiling image and our belief that "smiling people are friendly." The associated rule, "Don't trust her," will modify our reaction to her smile and evoke a question: "Is she trying to manipulate me?" In fact, our visualization may even "change" her smile from an innocent one to a crafty one.

"Could Have" and "Should Have"

One of the vague rules we apply to our own actions and those of others concerns our expectations of favorable outcomes of what we do. It is a truism, however, that we learn even more from our mistakes than from our successes; we are particularly likely to scrutinize our actions when things go wrong. "What happened?" "What could have been done differently?" and, "Why wasn't it?" can lead to a corrective strategy, namely, to identify the sequence of events and reflect on alternative actions. The strategy is an important component of problem solving and "learning from experience." By correlating results with actions, we construct a base for an evaluation of our own actions and those of others. For example, if we select a certain route or plan, we may want to consider whether it led to the intended goal in the past and whether it was the most efficient one possible.

More often than not, however, when we are dissatisfied with a turn of events, we experience an inner pressure to blame ourselves or others for the misadventure instead of compensating for the undesirable outcome or profiting from the experience. We compare the actual disappointing outcome with an imaginary one that is favorable. If a different course of action *could have* produced a favorable result, then the person *should have* followed this course; since that person did not, he should be reproached.

Consider the example of a shopper waiting in line in a market. Because most people like to get their groceries checked out promptly so that they can leave as soon as possible, they select what they judge will be the fastest line. But there is a popular notion among shoppers that they generally end up in the slowest line. They check the other lines and make the subjective determination that those lines are moving faster. When they change lines, then that line appears to slow down.

A typical shopper may then imagine a different scenario. Since the hypothetical scenario would have supposedly produced a better outcome, the shopper blames himself for making the wrong choice: "If only I'd got on the other line. If only I hadn't changed lines." It is interesting that the impatience and frustration arise less from "what was" (the actual time "wasted") than from "what could have been." "I could have chosen a different line" becomes rapidly transformed into "I *should* have chosen a different line"—a self-reproach.

More than we realize, we tend to imagine an alternative and optimal scenario and compare it with the actual happening. The degree of discrepancy will contribute to the degree of dissatisfaction. Much of the stress that people experience results from the cumulative effects of such relatively minor reactions. As a consequence, they may become chronically irritable and may overreact with anger to minor problems with their family, friends, or business associates.

Suppose that our shopper's line becomes stalled because a customer whose purchases are being checked out starts searching for coupons or leaves the line temporarily to get another item. Then the frustration is accentuated by what our shopper sees as the selfishness or thoughtlessness of the other shopper. A mild reaction is the thought, "*If only* she'd had her coupons ready." This merges rapidly with the criticism, "She *should not have* delayed us."

At this point our angry shopper would *like* to shout, "Why didn't you have your coupons ready?" This is an expression of a more optimal scenario, as though it were possible to modify the woman's behavior retroactively. Although social restraints will probably keep our shopper from saying this out loud, he will surely think of a disparaging adjective for the wayward customer: self-indulgent, stupid, self-absorbed, inadequate. But if his self-control is weak and the impulse is strong, he may well mutter these criticisms aloud.

This formulation shows how comparing a chosen strategy with a potentially more effective one can become counterproductive. Demanding that other people *should not have* behaved the way they did can be a source of unproductive anger. This kind of cognitive process has been described as "counterfactual thinking"—imagining a scenario that did not in fact occur.[3] At the most pathological, an individual distressed over an unpleasant event may have obsessive fantasies in which the facts are changed to allow a more favorable outcome.

At times it becomes so obvious that things could not have happened differently that the sense of outrage dissipates. If an offense is reframed as totally unavoidable, then it ceases to be an offense. Consider the following. Parents are concerned that their teenage son has not returned home with the family car long after he was expected. They grow anxious over the possibilities: "What if he was mugged?" "Suppose he got into an automobile accident?" When their son does appear, they are relieved but then angry that he did not leave for home earlier (alternative scenario). When the son offers a reasonable explanation that eliminates the possibility of that scenario, their anger is defused. Perhaps something happened to the car's engine and he was unable to telephone his parents, or one of the passengers was sick and had to be rushed to the hospital. As soon as the parents recognize that their son's actions and their distress were unavoidable, the "should-haves" are no longer operative, and they no longer believe that they were treated badly. It is interesting that when we are distressed by another person's behavior, we tend to assume that the hurt was either intentional or due to his or her negligence. We do not initially consider the reverse: that the distressing incident was accidental or unavoidable.[4]

People are usually aware that they will be absolved of criticism if they can show that a violation was inevitable. Indeed, to forestall being blamed, they may invent explanations to show that they could not possibly have done anything other than what they did. Teenagers are often adept at concocting irrefutable explanations like that. Whether the listener accepts this explanation as reasonable determines whether he or she becomes angry or hostile.[5]

THE TYRANNY OF THE SHOULDS

The compelling rationale for imposing rules and standards of behavior on others is to provide both protection for us and a strategy with which we can meet our "needs." The significant rules that we live by are designed to control the behavior of others as well as our own. Like the laws of government, these rules take the form of injunctions and prohibitions: the do's and don'ts; the "should-haves" and "should-not-haves" that pervade our own speech and thoughts; what we say to others and what we say to ourselves. And the violations of these imperatives are regarded—like legal violations—as punishable offenses.

Psychotherapists and theorists, such as Karen Horney and Albert Ellis, recognized and elaborated on the dominance of these imperatives in patients with a variety of psychiatric disorders. Moreover, they showed that people in general—those without discrete psychiatric disorders—have the same kinds of problems as those with the disorders. Horney particularly focused on the role of the imperatives in people's exaggerated goals that are incorporated into their "idealized self-image." Those individuals who are particularly prone to depression are largely driven by what Horney called the "tyranny of the shoulds." When they are depressed, their thoughts are a cacophony of what they *should not have* done and what they *should* do—but are not doing. Ellis, in a tour de force, showed how the "shoulds" and "should-nots" produce problems for a wide variety of individuals, especially those whose excessive expectations of others lead them to outbursts of anger.[6]

When people start to focus on their thoughts about others (and themselves), they can readily recognize how their emotional reactions and angry feelings are governed by these imperatives. Although the imperatives may occur only as unexpressed thoughts, they are readily verbalized:

- "He should have known better."

- "She should have listened to me."

- "You should have been more careful."

- "They should have worked harder."

The common denominator of these statements is a subtle demand to reshape people's actions, wrapped in an accusation that they have violated an imperative or rule. The language of the imperative may assume a somewhat different form with the use of certain "loaded" words that still contain the same critical message—namely, that the speaker's rights have in some way been violated.

- "She has no right to treat me this way."

- "I have the right to an honest answer."

- "You've got a lot of nerve talking to me that way."

Implicit are the "shoulds" and "should-nots," which are designed to protect one's rights, to fix the responsibility for their violation, and to punish the offender.

Since these imperatives are such frequent intruders into our consciousness and play so important a role in shaping our emotions and behavior—often to our detriment and to that of people whom we care about—it is important to understand their functions and their applications to interpersonal conflicts. It is also of interest to speculate about their origins as adaptational strategies.

IMPERATIVES:
PROTECTING RIGHTS, FULFILLING NEEDS

How can we protect ourselves against interference, discrimination, and threats in our daily lives and at the same time pursue our special interests and goals? We obviously do not confront the same kinds of problems that our Stone Age ancestors did, nor do we need to band together to catch a rabbit, to bring down a fleeing deer, or to fight off marauders. We are usually confident of the protection afforded by our social order in the form of laws, ordinances, and conventions—and the various modes of law enforcement. Further, we rely on the social code of fairness, cooperation, and reciprocity to facilitate the pursuit of our goals in harmony with others. Moreover, religious sanctions against such asocial or antisocial behaviors as avarice, lust, and hostility are designed to discourage overly self-centered behavior. Without these imperatives, social groups would be in a state of chaos. The commitment and obligations necessary for survival and for forming families would be absent, and individuals heeding only their own self-interest would run over one another.

Rules, laws, and sanctions not only serve notice on would-be offenders but serve a cognitive function—insofar as we accept them—in shaping our appraisal of offenses and offenders, crimes and criminals; that is, they influence our behavior and our thoughts and feelings about offenders. The observation of legal and social laws impels us to characterize offenders globally as criminal, sinful, and bad. As we incorporate the rules into our information-processing system, we may have fleeting visual images of the offender as an ugly burglar, smirking lecher, or shifty-eyed confidence man, and we experience corresponding feelings: anxiety, anger, and disgust. Most of the time, however, we do not focus attention on these images (although they become more salient if we pay attention to them), but we are conscious of repellent thoughts about the offenders: bad, revolting, or degenerate.

It is apparent that rules based on expectations and conventions significantly affect our thinking as well as our actions. Our culture and our phylogenetic heritage have apparently molded our receptivity to social regulation. Analogous to societal rules (pressures from outside) are our personal, private rules (the pressures from inside), which govern our behavior and reactions to others. Whatever the source of the rules, however, it is clear that they influence the way we feel about ourselves and others. In fact, the primary impact of the rules—whether societal or private—is on our cognitive systems, and our behavior is derived from these internalized rules.

Although to some degree our cognitive rules, formulas, and labels are replicas of societal rules, we may have an opposite set of rules insofar as we rebel against, devalue, or subvert societal and legal rules. Thus, some of us may regard illicit behavior as "cool" (such as using or selling street drugs) or justified (breaking into affluent homes).

When we examine the imperatives in a variety of contexts, we can see that conventions, mores, and customs play a useful role in influencing, if not controlling, other people as well as ourselves. Social regulations, whether embodied in law or in convention, form the framework for our expectations of others. Many of these expectations are elevated to the realm of demands and obligations: people must be respectful of our interests and goals. Further, our personal wishes become dignified as rights: we have the right to be treated fairly, honestly, and kindly. People *should* be sensitive to our needs and feelings. Viewed as tools for our personal survival, the "shoulds" are levers for control that have power beyond legal injunctions and prohibitions. In fact, our inner dictates cause us to influence or even coerce others to protect us, cooperate with us, nurture us.

By the same token, we are motivated to impose restrictions on others. They *should not* thwart, deceive, or reject us; they *should not* be disrespectful, inconsiderate, irresponsible, or controlling. These practical rules of conduct are frequently glorified in the social canon, and violations produce hurt and anger or rage and a desire for retribution. Thus, social guides to desirable conduct that are potentially useful can become absolute, extreme, and rigid and may paradoxically produce more distress than they are designed to prevent.

AN EVOLUTIONARY THEORY OF THE IMPERATIVES

Why are the "shoulds" and "should-nots" so strong that they often preempt other more adaptive strategies, such as cooperation, negotiation, or gentle persuasion? A perfectionist homemaker, for example, may be driven to maintain perfect order and immaculate cleanliness in her home even though the urgency of her compulsion may interfere with harmonious relations with her spouse and children. We usually find that the "shoulds" in such a case compensate for a fear of being disapproved of or for a self-image of incompetence. A similar exaggerated form of the same process may be noted in psychiatric disorders. Extreme manifestations of these imperatives are seen, for example, in the obsessive-compulsive patient, who washes his hands endlessly lest he be contaminated by invisible germs. The starkness of the "shoulds" is also evident in the depressed patient, apparently immobilized by his psychiatric disorder, who continuously berates himself: "I should be able to do a better job with my office work (or housework)."

To trace the origins of these phenomena, we need to return to an analysis of the fundamental principles that have shaped the thinking, feeling, and behavior of our species over the eons. It appears that the basic programming of our thinking and behavior evolved in order to ensure the fitness and survival of our ancient ancestors.[7] Obviously, if our ancestors had not achieved evolutionary success, our lineage would have died out. Adaptive information processing is the prelude to adaptive behavior. These synchronized patterns of thinking, feeling, impulses, and behavior evolved to enable our ancestors to solve the basic challenges they faced: namely, protecting themselves against various dangers, acquiring resources, and maintaining viable relations with others. The mechanisms they called on to solve these problems were to a large extent automatic and reflexive in character. Thus, an innate program activated by the sudden appearance of a large grizzly bear would evoke, practically simultaneously, the perception of a serious threat (cognitive); the experience of anxiety or panic (affective); the impulse to flee (motivational); and the actual escape (behavioral). These mechanisms, working in synchrony, would be expected to operate automatically today in analogous confrontations.

We can return to the previous example of the angry wife to explore why the conflict was so important that she became enraged. Let us review the incident in a broader evolutionary framework. The division of labor between interdependent individuals is directly relevant to the need to protect crucial resources and to rely on the cooperation of other key individuals. Since performance of crucial activities does not necessarily bring immediate rewards, the anticipation of gratification is insufficient "to make the wheels go around." Consequently, from an evolutionary standpoint, there presumably was survival value in the creation of a *substrate for forming injunctions and prohibitions* to prompt the cooperation of others. The present-day representation of primitive imperatives would be: "I am responsible for managing the household," and, "My spouse *should* attend to his assigned duties."

The presence of an imperative in the expectation of cooperation plays a crucial role in the generation of anger. In this instance, the wife's *intellectual* notion that her husband has a role in helping with home maintenance would not be sufficient to produce such a strong reaction when he failed. It was the pressure of the "shoulds" and their thwarting by the husband that put her into fighting mode. Of course, adhering to a fair division of labor is not a life-or-death matter in present times. Just as with other maximum reactions originally programmed as a response to threat, however, the present derivatives represent overreactions. Just as physical threats can maximally mobilize us, so psychological threats or attacks can produce an overreaction.

7

INTIMATE ENEMIES

The Transformation of Love and Hate

Fred: She wanted me to stay home with her because she had a little head cold. I thought, "If she keeps me home for such a small thing, what will she do when something big happens?"

Laura: Fred wouldn't stay home when I asked him. I thought, "If he would not do such a small favor, what will happen when something really important comes up?"

Fred and Laura had lived together for several years prior to getting married but had no troublesome problems. Just before their marriage, they began to clash over apparently minor differences. These episodes built up until they saw a therapist. They described a typical problem. Laura wasn't feeling well one evening and asked Fred to stay home rather than meet with a professional colleague who was visiting the city. Fred became annoyed and angrily refused. Laura became tearful and withdrew, and Fred stormed out of the house.

This couple's problem can be analyzed at different levels. At first it might appear that this was simply a clash of priorities—that is, both were trying to "get their own way." But it became apparent that the source of their difficulty was the fact that they viewed the same event differently and attached contradictory meanings to it. These contrasting meanings led each to view the other in a profoundly negative way. Laura felt that Fred was running roughshod over her needs. What she didn't fully realize was that her need was based on a childlike fear of being abandoned. Thus, Fred's refusal attained great importance in her mind: "He is deserting me." This conclusion led to a more negative view of Fred as driven, self-centered, and unreliable.

Fred, on the other hand, was invested in his career and wanted to capital-ize on the opportunity to advance himself by meeting with the visitor. He viewed Laura's request as a frivolous resistance to his ability to "do his own thing." He did not realize that he also had a psychological problem: a life-long fear of being restricted, fenced in.

In a sense, the couple's difference in priorities represented a clash of *fears*. These fears controlled their perceptions of each other's behavior and conse-quently shaped their own feelings and actions. The fears were related to underlying personality patterns, which produced a conflict between their respective goals. Fred was predominantly an autonomous person who prized achievement, prestige, independence, and mobility. He felt that Laura's "neediness" was suffocating him. His fears of being obstructed or confined fueled his drive for autonomy and achievement, which in turn made him hypersensitive to any barrier or restriction. Laura was more sociophilic—a "people person." She enjoyed being with other people and felt a void when she was alone. Further, she had a lifelong fear of contracting a serious medical illness and tended to catastrophize when she had a minor ailment. When she was ill, she wanted somebody reliable nearby to help out. As indicated in the earlier dialogue, her fear of abandonment was activated when she felt symp-tomatic. Her desire to have Fred nearby, however, exacerbated his fear of being contained.

Once Laura and Fred became upset with each other, each slid into a hos-tile defensive mode, which interfered with constructive problem solving. As suggested by their therapist, they could have thought of alternatives that would have satisfied them both. For example, Fred could have invited his col-league to the house, postponed the get-together to the next day, or gone to his colleague's hotel for a brief visit and phoned Laura when he got there. Any of these solutions might have indicated to Laura that he really cared and might have reassured her that he would not leave her in the lurch when she "really" needed him. At the same time Fred would not have felt impeded in carrying out his objective. Once they were in the defensive mode, however, each was mobilized to protect himself or herself and to punish the offender. Their thinking was focused on the danger of being abused and on strategies to defend themselves.

Laura and Fred had drifted into a persistent state of conflict based on a continuing series of misunderstandings. As their views of each other polarized—he was the deserter, she the controller—each misunderstanding eased the way to the next. Viewing events simply in terms of their own needs and fears blinded them to the other's needs and wishes. Laura believed that her negative interpretation of Fred's behavior was justified and valid and that Fred's negative interpretation of her behavior was unreasonable and unjustified. Fred's beliefs, of course, were the opposite. They both felt that their strongly held views of reality, as well as their own personal emotional investments, were at stake. The ongoing conflict helped to frame the biased perception that each of them maintained: "I'm right, you're wrong." By perceiving each other as selfish and obstinate, Fred and Laura were able to preserve their own benign view of themselves. Insofar as an individual's overall view of a conflict is one-sided and self-serving, the judgmental rightness involves self-deception. The individual's own thinking, feeling, and behavior appears reasonable and warranted, and the partner's unreasonable and illegitimate.

The behavioral exchange of hurtful remarks and actions leads partners to harden their negative image of each other. This rigidity is often manifested in each partner's complaints: "You are not hearing me," or, "You ignore everything I say," or, "You turn everything I say against me." Only rarely is a complaint in a heated disagreement such as this accepted as valid by the other person. Even when they are not involved in overt conflict, each spouse is placed in a malevolent category devoid of any positive redeeming qualities.

AMBIVALENCE

Ambivalence, the alternation or simultaneous occurrence of positive and negative images, beliefs, feelings, and desires, is common—perhaps universal—in close relationships. The capacity to view the same person in opposite ways at different times is an expression of the dualistic organization of the primal information-processing systems. At the most obvious level, ambivalence becomes apparent when partners are loving one moment and in the next moment feel like tearing each other apart. It is often difficult to explain how or why people switch suddenly from affection to dislike, or why the previous warm feelings do not form a cushion against the experience of angry feelings.

This kind of reversal of feelings in the relationship may not be apparent during the halcyon days of the romance, when all differences are submerged or sweetened by affection. It becomes more obvious, however, as the partners become more involved in their own goals and desires. Potholes and rifts appear in the previously smooth landscape of their romance. The balance between reciprocal interests and self-interest gradually shifts: "What's best for you is best for me" becomes "What's best for me is best for you." Actually, both orientations may persist throughout a marriage, but the self-centered beliefs become more active than the altruistic beliefs in distressed relationships.

The kind of ambivalence that undermines a relationship and may lead to its ultimate demise often involves more than simple differences in preferences or styles, or the normal ebb and flow found in any relationship. Most people have contradictory constellations of goals, beliefs, and fears at the core of their own personality. At one time sociophilic goals—the desire for intimacy, sharing, and reciprocal helping—may be dominant. At another time more autonomous goals of independence, achievement, and mobility may be prominent. A person may alternate between the magnetic pull for closeness and the craving for independence. These modes are reflected in how people feel toward each other. In the autonomous mode, they are likely to feel distant, but they swing into a more affectionate phase when their sociophilic mode becomes dominant. When the partners' modes are out of synchrony with each other, they are more likely to have problems in the relationship.

While the affiliative mode promotes bonding and dissolution of boundaries, the autonomous mode maintains boundaries, erects barriers against incursions by others, and facilitates freedom of action. The self-protective aspect of this mode is expressed in drawing away from or repelling demands, manipulations, or seductions. It functions to protect the individual self-interest and "space." At a deeper level, the strategy of distancing may serve to compensate for a fear of being enveloped (as in Fred's case), while the strategy of clinging may compensate for a fear of abandonment (as in Laura's case.)

When accentuated, ambivalence in everyday life may be expressed by alternate liking and disliking, attracting and repelling, and, at the extreme poles, loving and hating.

Some people clearly oscillate between the autonomous and sociophilic concerns: the fear of being enveloped or the fear of being deserted. Consider the following exchange between a husband and wife who have a moderately distressed marriage.

> *Alva*: You don't really care about me. You are so preoccupied with yourself.
>
> *Bud*: I do care about you. You are everything to me. I want to take care of you, to help you—and I do try to provide for you.
>
> *Alva (angrily)*: I don't need your help, I'm not an idiot child. I can very well take care of myself.
>
> *Bud (hurt)*: Okay, so I won't help you.
>
> *Alva (hurt)*: There you go again—rejecting me.

Alva alternated between a dependent and an independent mode. Each aspect of her personality was exposed in her mixed signals: "Help me. . . . Leave me alone." Whichever tactic Bud took ran afoul of one of her ambivalent goals (or fears): "Now he's belittling me. . . . Now he's abandoning me."

Alva's sociophilic mode incorporated not only the craving for closeness and dependency but also the fear of rejection; her autonomous constellation consisted of a sense of pride in self-sufficiency and a sensitivity to being indulged. Actions by Bud that impinged on either of these modes triggered the relevant fear incorporated in the mode: indulgence of her dependent wishes hurt her pride; yielding to her desire for independence represented distancing and withdrawal of support. Also invoked were the contrasting imperatives: "Bud should be more helpful" versus "Bud should let me be." Any violation of an imperative distressed her and led to an angry outburst.

Bud, a social worker, was more independent than Alva but had strong doubts about his competency. He bolstered his own sense of mastery through helping people. Alva's rejection of his desire to help her injured his self-esteem and reduced his sense of competency. To "protect" himself he would withdraw from Alva, who would then berate him for being cold and uncaring.

THE BIG SWITCH—FROM LOVE TO HATE

One of the puzzles in contemporary life is that love that feels so right, that is so gratifying, even exhilarating, can vaporize and leave behind a trail of resentment and hostility—even hate. A clue to the reversal may be found by examining the accentuation of ambivalence in a relationship. As various frictions develop, a structure of negative attitudes begins to crystallize. Not infrequently, this new constellation is derived from previously formed negative attitudes toward key figures in the past, perhaps a sibling, parent, or former lover. These negative attitudes become fortified over time and may take the form of a negative image of the partner. They experience a change—usually gradual but sometimes relatively sudden and dramatic—in how they perceive each other. This is illustrated in the following case of a highly distressed relationship.

Ted and Karen met at a street corner during a rainstorm while both were waiting for a bus. Since the bus was delayed and they were obviously going to become soaked, Ted suggested that they get a cup of coffee at a nearby café. Karen agreed readily, and they had a nice time. Ted's thoughts about Karen were, "She's very agreeable and pleasant." He was very much taken by her spontaneity and joie de vivre. Karen was delighted to meet a man who was mature, decisive, and well organized. Eventually this casual meeting progressed to a courtship and then to marriage. Karen admired Ted's intellect and his ability to discourse at length on literature, world affairs, and history. Ted enjoyed Karen's charming small talk, her penchant for describing people and their experiences.[1]

Although they were initially happy with each other, their personalities, which seemed to blend so well at first, began to grate against each other. The specific beliefs each held about the relationship, the arrangement of domestic responsibilities, and social activities—or, in general, the way the marriage should go—clashed. The very qualities that each had initially prized in the other were now obscured or tainted. What seemed good and desirable at first now appeared bad and undesirable.

The change in their view of each other is illustrated in the change in their evaluation of each other's characteristics.

TED'S VIEW OF KAREN

Before	After
Carefree	Flighty
Spontaneous	Impulsive
Lighthearted	Empty-headed
Charming	Superficial
Lively	Emotional

KAREN'S VIEW OF TED

Before	After
Steady	Rigid
Organized	Compulsive
Decisive	Controlling
Intellectual	Stuffy
Objective	Cold

The switch to the negative beliefs about each other resulted in a biased interpretation of whatever the partner said or did. The strongly held views of each of the distressed partners enclosed the other in a category (or frame). As with any strongly held negative evaluative belief, the polarized images led to polarized conclusions. This kind of framing leads to the characteristic thinking errors of selective abstraction, overgeneralization, and arbitrary inference.

It is interesting to note how the switch in the balance of a particular trait can change it from endearing to unbearable. Ted, for example, initially enjoyed Karen's easygoing, spontaneous, impulsive qualities because they complemented and enriched his own more serious style. Later, as he tried to impose his own mode of operation on her—organized, planful, serious—she rebelled. Hurt by her rejection, he began to view her playfulness as frivolous and childish. By the same token, Karen began to view him as heavyhanded, joyless, and rigid.

Several lines of research bear out these clinical observations.[2] One study, for example, showed that distressed couples made negative characterological explanations of their mate's behavior ("She's late because she's irresponsible"), but more specific, situational explanations for the identical behavior by someone other than the mate ("She's late because she was probably tied up in traffic").

The positive evaluation of the mate at the height of the infatuation is the mirror reversal of the evaluation during the depths of the deterioration. The positive frame at the peak of the relationship usually involves some thinking errors, such as selective abstraction and exaggeration of the positive traits and minimization or disqualification of the negative. Further, any favorable behaviors of the lover are overgeneralized to extend, in principle, to all behaviors. The lovers see each other as uniformly wonderful and are oblivious to or dismiss any flaws or unpleasant traits. Finally, characterological explanations are made for pleasant behaviors: "He is kind, sensitive, loving," and so on, amid positive explanations for problematic behavior: "I'm sure he tried to get here on time." These inflated attributions set a standard to which the mate will be held when the relationship starts to flounder. The difference in appraisal between the positive and negative extremes is epitomized in an observation by Ted: "When we were in love, I could do no wrong. Now I can do no right."

EXPECTATIONS AND RULES

Many people entering into marriage have a glorified set of expectations and rules regarding their partners. These expectations may not be explicit or even conscious, but they become apparent when problems begin to accumulate and are used as yardsticks to measure the partner's value. These imperatives—the "shoulds" and "should-nots"—also lead to strategies designed to coerce compliance with the standards and to designate transgressions where the rules are violated. The imperatives, while possibly latent during the more carefree period of the courtship, become more obvious when the mates assume responsibilities for each other and for their joint obligations. Among these imperatives, as illustrated below, are the usual ones that emphasize the caring behaviors.

MY SPOUSE SHOULD BE . . .	MY SPOUSE SHOULD NOT BE . . .
Sensitive	Insensitive
Caring	Uncaring
Considerate	Inconsiderate
Accepting	Rejecting
Uncritical	Critical
Responsible	Irresponsible

It is interesting that during the harmonious period the spouse is regarded as fulfilling more of the positive criteria (on the left), and in the distressed period the spouse seems to fulfill only the criteria in the right column. Although it is likely that the spouse does behave better in the happy period, the partners are inclined to exaggerate the degree and significance of the undesired behavior during the distressed period.

Ironically, instead of being more secure, partners often become more vulnerable to each other after marriage. The increased responsibility for each other's welfare and dependence on each other for carrying out the multiple roles of home maintenance and parenting set the stage for fears of being let down. There are more opportunities for being hurt, frustrated, and disappointed. The intimacy itself intensifies their need for affection and support and, at the same time, creates the potential for the fear of loss of the support. To protect themselves they may therefore feel impelled to set up more protections and rules. Thus, qualities such as consideration, sensitivity, and empathy, which may be taken for granted during the romantic period, become more vital when clashes about household responsibilities, economics, parenting, and leisure time activities arise. These situations evoke an increased desire for reciprocity, reasonableness, and acceptance but, by their very nature, activate self-protective tendencies that interfere with these qualities.

EVOLUTIONARY ASPECTS OF DISCORD

The deadly significance of the close relationship is epitomized in a conventional marriage ceremony by the phrase "until death do us part." The life-and-death aspects are dramatized in the genetically determined production of a new life through the partners' physical, if not emotional, intimacy. Conversely, the death-dealing impact of the disruption of the close relations may lead to homicide or suicide or both. Nowhere within normal society are the unleashed furies as dramatized as in the violence within a family.

In order to understand more fully the intensity and death-dealing significance of these reactions, it is valuable to look for clues in the embedded mechanisms that presumably evolved to fulfill the evolutionary imperatives of reproductive success. It seems clear that innate mechanisms play a role in the initiation and maintenance of an intimate relationship. With the exception of arranged or forced marriages, individuals, drawn to each other for a variety of reasons, are rewarded with pleasure from the rapport and support, as well as from the sexual relationship.

It seems inescapable that this reward mechanism cued to intimate relations is part of our biological heritage. The feelings of gratification play a role in reinforcing various activities designed to provide for individual survival, but also for parenting the next generation. The gratification from sexual activity, of course, promotes activities that, willing or not, lead to propagation.

Hostility as a defense against the noxious behavior of an antagonist is also favored by natural selection. Defensive hostility is engendered by the way the opposing spouses frame each other: their noxious representations. One story of a custody battle illustrates this phenomenon quite well.

Both marital partners were in a hostile mood. As part of the divorce settlement, the judge awarded custody of the children to the mother. The father had the right to see the children two weekends a month. The wife contested this on the basis that the husband was a bad influence on the children.

The mates saw each other as the Enemy, and a battle raged, just as much to defeat the "Enemy" as to provide parenting for the children.

THE HUSBAND'S REPRESENTATION OF THE WIFE

- A "shrew"

- Manipulative

- Controlling

- Power-driven

- Dangerous to children

THE WIFE'S CHARACTERIZATION OF THE HUSBAND

- A pig

- Self-indulgent

- Irresponsible

- Infantile

- Devious

- Harmful to children

When the judge ruled in favor of the husband's continued visitation rights, the wife felt depressed and anxious. She felt that he had defeated her, and she worried about the impact of his behavior on the children, who would now be "unprotected."

The mutual negative images were the essence of the problem. These highly charged images were exaggerated, based on the hostile behavior that they displayed toward each other, and were not representative of their behavior with the children and with other people. In distressed marriages, the mates' negative views of each other become increasingly exaggerated as their negative actions escalate. We can see the continuous cycle of the negative frame leading to negative behavior, leading to a more negative frame, and so on.

Consider the following scenario: Karen and her husband Ted were describing to some friends a reception that they attended. Karen said to the group, "There were a *ton* of people there." Ted interrupted with the statement, "There weren't all that many. In fact, it wasn't much of a crowd." Karen became so enraged that she could hardly speak, and when they got home that night, she screamed at Ted for "putting her down" by claiming she exaggerated. Her thought was, "He makes me look foolish to the others. He likes to humiliate me."

Karen's strong emotional reaction was dependent on many more factors than simply the implication that she exaggerated. After all, people commonly exaggerate shamelessly and are not necessarily corrected. In essence, the way she construed his remark determined whether she took it as an insult. Although we acknowledge that the meaning attached to another person's behavior influences how we feel, we may not recognize that the theory we apply to explain a person's actions is responsible for the degree of upset we experience. *What* is said is less decisive than *why* we believe it was said. Karen inferred that the causal factor in Ted's correction was his wish to demean her in public; in her mind, Ted felt the need to correct her because of his devalued image of her. His presumed image of her upset her.

The impact of this image was magnified by its communication by Ted to their friends. Karen had a strong conviction that from then on they would regard her as an exaggerator and would doubt her credibility. In her mind, her social image would suffer. In actuality, Ted's intention was not to demean her but to "educate" her to be more accurate so as to *increase* her credibility. However misguided his correction, his intent was not malevolent, as Karen had assumed. Nonetheless, his public comment severely damaged an already shaky relationship.

Even though many people would agree that Ted's remarks could be hurtful, we can observe that such remarks do not inevitably evoke the degree of hurt and rage experienced by Karen. To fully understand her reactions—and to cast further light on the intense reactions in couples' conflicts—we must consider the development of her self-image and her social image. Karen had a lifelong view of herself as socially awkward and believed that she presented this image to other people. However, she developed a compensation for this negative image by painstakingly cultivating her conversational skills and having a flair for telling interesting stories.

Ted's supposed disqualification of her storytelling undermined this compensation and exposed her (in her mind) as a socially inept girl. Ted's penetration of her social shield made her feel vulnerable. Karen's desire to punish Ted derived from the "marital imperatives," the standard of behavior she had implicitly imposed on Ted: he must be her champion and supporter and must not do or say anything that could expose her to criticism or disdain. Although Karen had never articulated this rule to Ted, she assumed that he knew it and expected him to abide by it.

Ted, thus, had violated an important marital rule: "Do not sully your partner's social image." In Karen's eyes, he had violated her trust and exposed her. When the social facade is cracked, the "victim" feels vulnerable to a variety of social abuses, such as loss of status, ridicule, and rejection. Ted's violation of the rule intensified Karen's reactions because of the long-range consequences: his action would leave Karen permanently defenseless and vulnerable. Karen saw herself faced with two possibilities: punish Ted so severely that he would never break the rule again or terminate the relationship.

The decision to punish an offender is generally not the result of reflective reasoning. Rather, it is built into the entire pattern consisting of damage to the self-image or social image followed by hurt, fixation of responsibility on the rule-breaker, and attempts to restore the status quo by retaliating.

The presumed value of punishment, even excessive punishment, is that it reestablishes the correctness of the broken law, establishes that the law is enforceable, compensates for feelings of powerlessness, and restores some of the lost self-esteem. Inflicting pain (retribution) also serves to direct attention away from the victim's pain to that of the victimizer.

MORTAL MESSAGES

Consider a couple in a shouting match. Their fists are clenched, teeth bared, spittle on the corners of their mouths, bodies poised for attack. All systems are on "go." Although they are not clutching at each other's throats, it is easy to see from the tension in their muscles that their bodies are mobilized as though for a struggle to the death. Though these adversaries exchange no blows, they "attack" each other with their eyes, facial expressions, and tone of voice, as well as with their angry words. A stony glare, curled lips, and snarls of contempt—these are all weapons in their arsenal, ready for the fight. In the heat of battle, spouses may hiss like snakes, roar like lions, and scream like birds.

Stares, growls, and snorts are signals of attack, even when the antagonists exchange apparently innocuous words—or no words at all. These signals are designed to warn the opponent to back off or compel him or her to capitulate. The "sharp edge," the threatening tone of voice, the volume and speed of speech can be more provocative or hurtful than the literal meaning of the words spoken. It is no surprise that people often respond more strongly to the tone of voice than they do to the words themselves. Nonverbal messages expressed through the eyes, face, and body represent a more primitive—and usually a more persuasive—form of communication than words. Consider the following conversation between a couple:

Tom: Dear, will you remember to call the electrician?
Sally: I will if you ask me in a nice tone of voice.
Tom: I did ask you in a nice way!
Sally: You always whine when you want me to do something.
Tom: If you don't want to do it, why don't you say so!

Tom *intended* to make the request in a civil way. But he had some resentment over Sally's past intransigence, so his request was tinged with the tone of a reprimand. Although his words were civil, they were fused with a decidedly negative message transmitted by his tone of voice.

When there is a double message such as this, the recipient is likely to respond to the *nonverbal* signals as the significant message and to ignore the words, just as Sally responded to Tom's tone of voice. Tom, not realizing that his tone was provocative, interpreted her reproach as a refusal of his request and retaliated. Sally would probably have agreed to make the call if Tom had not edged his words with a reprimand. But they both got caught up in scolding and retaliating, and so they never got around to addressing the practical problem, namely, calling the electrician.

When some expression of hostility is justified, we may become so angry that we actually could fight to the death—even though we limit ourselves to scolding or name-calling. Such total mobilization in a marital fight so far exceeds what is called for that it prompts the calmer partner to discredit the other as "hysterical" or "irrational," or to retreat in fear.

A more serious problem occurs when total mobilization for attack breaks through inhibitions and leads to physical abuse. Several years ago, I was consulted by a couple who complained that although they loved each other, they were always fighting. On several occasions the husband was so abusive physically that his wife had called the police. They described the following incident.

Two days previously, as Gary was leaving the house, Beverly said, "By the way, I called Girardo's (a private trash collector), and they will remove all the trash from the garage." Gary did not say anything but became increasingly angry as he thought about her statement. He ended up punching her in the mouth. Beverly ran to the phone and started to call the police until Gary restrained her. After much struggling, followed by a heated discussion, they agreed to see me in consultation.

Gary's reaction seemed inexplicable on the basis of the story that they told me at first. As the story unfolded, however, the incident became more understandable. When asked why he had struck her, Gary said, "Beverly really made me mad"—as though the provocation was self-evident. As far as he was concerned, she was at fault for his hitting her because she made him angry by talking to him the way she did. Since Beverly made him angry, Gary believed that he was justified in striking her. His unstated assumption was that, despite her apparently innocent statement, she was in effect saying that he was irresponsible, and that she was morally superior.

Beverly, on the other hand, maintained that she was merely "giving him information," not accusing him. She had been asking him to clean out the garage for some time, and since he had not attended to it, she decided to take care of it herself by calling the trash collectors.

In order to obtain "hard data" about what really had gone on, I decided to ask the couple to re-create the incident in my office. I asked Beverly to give the background and then to repeat her statement to Gary. As he heard her words, his face flushed, he began to breathe heavily, and he clenched his fists. He looked as though he were going to hit her again. At this point, I intervened and asked him the fundamental question of cognitive therapy: "What is going through your mind *right now?*" Still shaking with fury, he responded, "She's always needling me. She's trying to show me up. She knows she drives me up the wall. Why doesn't she just come out and say what she's thinking—that she's such a saint and I'm no good?"

I suspected that his very first reaction to her statement—which he clearly took as a put-down—was the thought that he was a failure as a husband. He had quickly wiped out this painful thought, however, by focusing on her "offensive statement." Although she had repeated this to him in a measured way during the role-play in the office, I suspected that in real life she may have spoken in a cavalier or critical tone.

She did acknowledge that when she spoke she had the demeaning thought, "See, I can't count on you for anything; I have to do everything myself." Although she did not express this thought at the time, it evidently came through in her tone of voice. Also he was sensitized to this kind of message from past experience. A provocation can be concealed in an apparently innocent message. But how can we understand the intensity of Gary's reaction? The explanation lies in facets of his personality, as well as in their marital history of accusations and retaliations.

Before marrying, Gary had been self-sufficient and considered himself successful. Raised in a poor family, he worked his way through college and became an engineer. He opened his own consulting firm and prospered from the very beginning. He thought of himself as a successful, rugged individualist. Beverly was attracted to Gary because of his good looks and his uninhibited, independent manner. She had been raised in a "proper household" in which the emphasis was on good manners and fitting in socially. Somewhat inhibited herself, she was attracted to a man who did not seem bound by social conventions, was an independent thinker, and, above all, appeared *strong*. She admired him for his successful career and had the fantasy of a knight in shining armor who would always take care of her. Indeed, during their courtship, he assumed responsibility for all the plans for the time they spent together. Because she regarded him as superior, she felt very comfortable with this arrangement. Gary was attracted to Beverly because she was pretty, dependent on him, and admired him. She was submissive and accommodated her own wishes to his.

After they were married, Beverly was, at first, intimidated by Gary. But she gradually discovered his human foibles—for instance, he procrastinated over household chores and could not relate to the children. With the passage of time, she became more mature and self-confident and no longer regarded herself as inferior to him. In fact, from time to time she got satisfaction from demonstrating that, far from being a "perfect doll," she was more mature than he in many ways. She attended to details better, was a more conscientious parent, and managed their social life more deftly than he could.

At the same time, Gary had brief episodes of mild depression, during which he thought that he was an inadequate father and husband. At such times, he accepted Beverly's implied criticisms as valid. He felt hurt by them but did not fight back. When he was no longer depressed, however, he refused to tolerate her criticisms and lashed out at her.

But why did his retaliation progress to physical abuse instead of being limited to the kind of verbal attacks that other people in his social circle engaged in? First of all, he had been raised in a "rough" neighborhood where conflicts were frequently settled by physical fights. Moreover, Gary described his father as a "violent man." When angered, his father had hit Gary's mother, as well as Gary and his siblings. Apparently, Gary learned early in life: "When you're angry, you should let the other person have it."

Gary never had an alternative model from whom to learn nonviolent ways to settle problems. He did not have much control in dealing with any of the people in his life, including his employees and clients. If he felt provoked by an employee, he would fire her—and then try to hire her back later. If he was in conflict with a client over plans or fees, he would break off negotiations. This lack of control gave him the reputation of being tyrannical—but, oddly, instead of deterring clients, it attracted them. He conveyed the image of the ultimate authority—superbly self-confident, decisive, and intolerant of opposition. In short, a strong man.

Although his authoritarian style was successful in his line of business, it was obviously ill suited to married life. At first, when Beverly attempted to stand up to him, he yelled at her. As she began to fight back verbally, he gradually became physically abusive. Eventually, if he detected a note of derision or deprecation in her tone of voice, he was moved to react with a physical attack.

In working with this couple, it emerged that self-esteem issues were paramount. Beverly was continuously trying to protect her self-esteem by not yielding to Gary when he told her what to do. To Gary, her opposition meant, symbolically, that she had little regard for him. After all, he *knew* the correct course of action—his employees and clients listened to him and did what he told them to do. Thus, her resistance had a deeper meaning to him—that perhaps he was not really as competent as he liked to believe. This notion was painful; his angry attacks served, in part, to dispel this idea.

It further transpired that while Gary was growing up, his older brother used to torment and tease him, calling him "weakling." Despite his successful career, he was never able to shed this image of himself as a weakling. He was only rarely haunted by the notion of being a "pushover," however, because in most of his dealings with people he had the upper hand.

With Beverly, though, things were different—he felt vulnerable. By attacking her, he tried to stave off the pain of having his "weak" side exposed. If she were to get the upper hand, it would confirm in his mind that he was indeed a "weakling"—a very painful idea. In fact, at times when she was critical of him, he had the thought, "If she really respected me, she would not talk to me that way—she thinks I'm weak." Thus, both partners were in a sense trying to equalize the relationship by putting each other down. Gary wanted to maintain his self-esteem, which was based on exercising control over other people. His polarized thinking—"If I'm not on top, I'm a flop"—reflected his hidden fear of being revealed as weak.

When we consider the pervasiveness of the kinds of cognitive difficulties and interpersonal clashes experienced by Gary and Beverly as well as the other couples described in this chapter, we can get a sense of why there are so many unhappy marriages and the divorce rate is so high. However, problems such as disillusionment, retreat into self-absorption, the increasing sense of vulnerability, and the cumulative misunderstandings might seem to be intractable. However, identification of the self-defeating beliefs can help couples to undercut their misinterpretations and lay the groundwork for a more rewarding relationship, companionship, mutual enjoyment, and satisfaction in creating a family together.

PART 2

VIOLENCE

Individuals and Groups

8

INDIVIDUAL VIOLENCE

The Psychology of the Offender

Destructive individuals are commonly viewed as having a "violent streak," a pattern of willfully harming other people as a deliberate strategy for getting what they want, or as an expression of uncontrolled rage. Given this profile of the violent offender, it seems plausible for an authority to punish him severely enough that he learns that "crime does not pay." Thus, the management of the offender generally focuses on control and deterrence. However, this group is highly variable; understanding the workings of the mind of the specific offender is crucial for appropriate intervention as well as for prevention.

Despite differences between individual offenders and variations in their typical violent behavior, certain common psychological factors can be identified across various forms of antisocial behavior, such as delinquency, child abuse, spouse battering, criminal assault, and rape. The common psychological problem lies in the offender's perception—or misperception—of himself and other people. Simply "beating the devil out of him" may have a temporary deterrent effect on the juvenile offender, for example, but does not change—in fact, reinforces—his view of himself as vulnerable and other people as hostile toward him. Therefore, it can be argued that such "deterrents" can actually perpetuate the violent behaviors stemming from his interrelated maladaptive beliefs.

Let me clarify: as a result of the interaction between his personality and his social environment, an individual may develop a cluster of antisocial concepts and beliefs. This cluster shapes his interpretation of other people's words and actions. The offender's sense of personal vulnerability is reflected in a hypersensitivity to specific kinds of social confrontations, such as domination or disparagement. He reacts to such perceived assaults by fighting back or by attacking a weaker, more accessible adversary. Whether a juvenile or an adult, the violent offender sees himself as the victim and the others as victimizers.[1]

The offender's thinking is shaped by rigid beliefs such as:

- Authorities are controlling, disparaging, and punitive.

- Spouses are manipulative, deceitful, and rejecting.

- Outsiders are treacherous, self-serving, and hostile.

- Nobody can be trusted.

Because of these beliefs and a shaky self-esteem, the potential offender often misinterprets the behavior of other people as antagonistic. Moreover, he operates on the belief that any degree of control or disparagement by another person makes him vulnerable. Hence, he develops a set of interrelated beliefs designed to protect himself from other people. These beliefs set the stage for violent counterattacks against suspected aggressors:

- To maintain my freedom/pride/security, I need to fight back.

- Physical force is the only way to get people to respect you.

- If you don't get even, people will run all over you.

The beliefs and perceptions of the violence-prone individual are analogous to those of a boxer in a fight. As he steps forward into the ring, he focuses all of his attention on the behavior of his opponent. Each movement of his opponent becomes a threat, which must be matched by a countermove. If he lets down his guard for an instant, he leaves himself open to a decisive punch. In a fight to the finish, he must knock out his opponent or be knocked out. As the match progresses, he alternates between a sense of vulnerability and one of mastery. The satisfaction of landing a punch outweighs the pain of receiving one.

Similarly, the violence-prone individual regards his entire life as a battle. As he defends himself against perceived physical and psychological threats, he alternates between feeling vulnerable and feeling secure. He is continually mobilized to fight because of his never-ending pattern of perceiving belligerence in other people's behavior. In a typical sequence, a confrontation triggers a general defensive *belief* of the offender, which shapes the meaning of the situation; for example, "They are trapping (or demeaning or dominating) me." This negative *interpretation* of the event initially stimulates feelings of distress, then anger and a desire to restore his sense of autonomy and efficacy. He believes he can accomplish this by physically assaulting the threatening individual, or some other available target, and experiences a *desire* to do so. Since the offense against him (for example, a punishment) seems unfair, the offender feels entitled to do something violent to repair the wound to his psyche. Consequently, he gives himself *permission* to consummate his desire. If he does not recognize deterrents in the immediate situation, he mounts an attack.

A crucial element in this process is the activation of hostile beliefs when an event strikes at the person's specific vulnerability (for example, rejection or disparagement). When these beliefs are primed, the offender's information processing shifts to the primal mode. His thinking about the incident is biased and highly exaggerated, often exhibiting the following features:

- Personalizing: He interprets other people's actions as directed specifically against him.
- Selectivity: He focuses only on those aspects of the situation that are consistent with his biased beliefs, blocking out contradictory information.
- Misinterpretation of motive: The offender interprets neutral or even positive intentions as manipulative or malevolent.
- Overgeneralization: He views a single adverse encounter as the rule and not the exception; for example, "Everybody is against me."
- Denial: He automatically holds others responsible for any violence, while he himself is wholly blameless. His denial may be so complete that he totally forgets his role in a violent interchange. When confronted by authorities with allegations regarding his role in an altercation, he minimizes any provocation on his part.

The generic cognitive model, in addition to describing the psychology of the reactive offender, can be applied more specifically to a range of violent behaviors, such as spouse abuse, juvenile delinquency, child abuse, and individual violence, such as mugging and robbery.

SPOUSE ABUSE

The popular image of an abusive husband depicts a man who gains pleasure from knocking his wife around and flies into insane rages with lightning speed.[2] This image, however, does not accurately portray most men involved in domestic violence. Despite individual differences, certain threads run through the provocation and execution of the behavior of most spouse abusers. Domestic violence does not occur in a vacuum; it is frequently the culmination of a conflict in which husband and wife use whatever resources they have to attack or defend against each other. The wife may resort to needling, name-calling, hitting, and throwing objects, while her husband swears, threatens, and finally beats her.[3] Since he is stronger, physical assault is his ultimate weapon.[4]

Although the most striking feature of domestic abuse is a violent assault on the physically vulnerable wife, it is crucial to remember that the husband perceives *himself* as psychologically vulnerable to her words and actions. In his eyes, she has wronged *him*, and he must use force to reduce the perceived threat and restore the proper balance to the relationship. In reality, his skewed beliefs magnify the damage to his psyche and channel his thoughts toward violence as the only solution. Thus, his wife's disagreement and resistance represent attacks on his authority; nagging and criticism are signs of disrespect; sexual refusal and emotional withdrawal mean total rejection. His wife's most devastating offense, however, is action or words that he perceives as a threat of infidelity.

The same types of cognitive distortions observed in nonviolent disruptions of relationships, such as being diminished or wronged, lead to violent clashes in cases of spousal abuse. Feelings of emotional distress precede the husband's hostile response, which is produced by his negative interpretations. He tends to overinterpret his wife's behavior as a devaluation, based on what he assumes is her image of him: "She thinks I'm a jerk, a weakling, a nobody," and so on. When this projected image becomes fixed, it automatically shapes his interpretation of her every action or utterance that could conceivably be a negative evaluation of him. Even when a derogatory remark, for example, truly represents his wife's view of him, he exaggerates its seriousness.

The hurt feelings resulting from a real or fantasized slight quickly lead him to label her action a flagrant offense: "She has no right to treat me this way." This sense of an unjustified violation produces anger and an urge to punish her. Any inhibitions against hurting her are lifted by the belief that she deserves it. This sequence can occur very rapidly. It is also representative of other kinds of violent behavior, such as child abuse, bar fights, and rape.

A number of systematic studies have tracked the cognitive processes of men prone to hostile reactions, as well as those who have assaulted their wives. Anger-prone men, for example, are especially attentive to hostile cues in their environment.[5] They are also less sensitive to friendly or empathic statements than controls. Those who have physically abused their wives are particularly likely to attribute negative intentions, selfish motivation, and blame to the wife.[6] Sexist attitudes—such as, "Women like to sleep around," "She will do it to get back at me," and, "I have to control her because I can't trust her"—feed into overrestrictiveness, jealousy, and suspiciousness. Further, these husbands have dysfunctional beliefs and biased thinking relevant to marital violence.[7]

An additional catalyst to physical aggression is the husband's hypersensitivity to threats to the balance of power in the marital relationship. According to his dichotomous way of thinking, if he is not clearly dominant, he is submissive; if not in total control, he is helpless; if not in power, powerless. Appearing to be submissive, weak, or powerless lowers his self-esteem and makes him feel vulnerable to further violation. Much of his violent behavior serves as a form of self-protection, a defense against his wife's oppositional or disrespectful behavior, and a reinvigoration of his shaky self-esteem.

Beliefs about the desirability and acceptability of violence as a strategy in marital conflicts differentiate violent from nonviolent husbands. When deadlocked in a conflict, the violence-prone husband believes:

- Physical force is the only language that my wife understands.

- Only by inflicting pain can I get her to change her abusive behavior.

- When she "asks for it" (physical abuse), I should respond and give it to her.

- Hitting is the only way to get her to shut up.

His self-esteem acts as a barometer, not only of how he values himself but also of how much he thinks his wife values (or devalues) him. If he assumes, "She thinks I'm a piece of dirt," he feels he has to knock that idea out of her head; if she thinks she can get away with being attracted to other men, he is impelled to make the attraction too painful for her to maintain. The impulse to use physical force becomes a strong imperative, practically as much a reflex as defending oneself from physical attack. The wife's aversive behavior must be neutralized at all costs, he believes.

Although these pro-violence beliefs can be modified, they exert a profound influence on the husband's behavior as long as he does not examine them. They stem in part from the archaic cultural and legal principle of treating a wife as chattel. Many violence-prone men assume that they have proprietary rights to their wives. Noting the tendency of other primates to exact total obedience from their mates and punish any approach to other males, some writers assert that male sexual possessiveness and jealousy have evolutionary roots.[8]

When primal thinking is triggered and the husband's impulse to hurt his wife becomes intense, pathological processes come into play. The abusive husband experiences tunnel vision, especially under the influence of alcohol.[9] His attention span is narrow and focused on the wife as an antagonist, even the Enemy. He sees only the image he has projected onto her: a bitch, harpy, or slut.

Not all husbands who feel like hitting their wives yield to this impulse, of course. Most husbands in distressed marriages control their hostile impulses and utilize nonviolent strategies to deal with marital conflicts. They may damp down a vicious cycle with self-statements such as, "Maybe we should talk this over before it gets out of hand," or, "It's time to cool it."[10] Violent husbands, in contrast, are deficient in self-control strategies and social skills, such as communicating their problems to their wives, engaging in mutual problem solving, and asserting themselves constructively, especially in situations involving potential rejection or abandonment.[11] Lacking such interpersonal skills, they may believe that violence is the only way to resolve a troublesome conflict. Moreover, such husbands show an inordinate frequency of clinical depression, severe personality problems, and alcoholism.

The connection between men's sexual possessiveness and their recourse to violence has been observed in all cultures in which it has been studied. The expectation of absolute fidelity and the consequent jealousy when this expectation appears to be violated is indicated not only in husbands who abuse their wives but also those who murder their wives.[12] In the majority of the cases, battered women attribute jealousy as the motive for their husband's assaults. Violent husbands also acknowledge that jealousy is the most common motive in beating their wives.

The abusive husband's exquisite sensitivity to the possibility of being betrayed by his wife leads him to adopt a number of coercive strategies to fence her in. According to the battered wives' reports, their husbands try to limit their contact with family and friends, continually interrogate them about where they were and who they were with, and limit their access to the family income. This curb on the wife's freedom of action frequently leads to her defiance and a rebound assertion of her autonomy, which may further threaten the husband.

Significant numbers of the severe assaulters are clinically depressed. Their insistence on absolute control of their wife, their hypervigilance, and their monitoring of their wife's activities represent an attempt to fend off the sense of uncontrollability and helplessness that epitomizes depression. When their controlling strategies fail, and the wives do consort with other men, the husbands often experience a wave of intensified hopelessness. When they see no possible solution to their problems and misery, their thoughts may turn to homicide and suicide. They no longer see any point in living but want to destroy the supposed cause of their misery—the wayward wife—before they kill themselves. Their entire focus, their only reason for continuing to live, is to inflict the ultimate punishment on the unfaithful spouse: "If I go, she's going with me." The husband's malevolent perspective on his wife occurs automatically and is accepted as reality.

Consider the case of Raymond, a typical wife abuser. Abused by his parents and tormented by his peer group as a child, Raymond developed a worldview filled with antagonistic people lying in wait for an opportunity to pounce on him. Although he had a warm relationship with his wife most of the time, his image of her soured when she pressured him to attend to household chores or criticized him for stopping off for a drink with "the guys" after work.

Pressure and criticism were flash points; he would then see his wife as a reincarnation of the critical, abusive, coercive people in his past. She became an Enemy who had to be controlled. He did not realize that he was threatened by the image he projected onto her, not by the real person. His defensive strategy was to attack immediately. Later, when his rage subsided, he was puzzled and troubled by the intensity of his outburst. His guilty feelings, however, did not prevent the next outburst, because he never examined or attempted to modify the beliefs that led to violence in the first place. In therapy, he was able to recognize his dysfunctional beliefs: mild coercion by his wife meant total domination and made him feel helpless; criticisms meant rejection and made him feel abandoned.

Raymond was typical of the largest group of individuals involved in violent behavior: the "reactive offenders." Some confine their violence to spouse or child abuse, while others may resort to violence in a broad range of interpersonal conflicts. They have in common a hypersensitivity to disparagement and rejection and a tendency to react with violence—hence the term "reactive offenders." When not upset, however, they are capable of positive feelings of caring and concern, and of experiencing shame and guilt for past transgressions.

ABUSIVE PARENT AND DELINQUENT CHILD

Terry, an eight-year-old boy, was referred to the clinic because of a history of demandingness, disobedience, disruptive behavior at school, continuous fights with his younger siblings, and rebellion against his parents and teachers. He had shown evidence of low frustration tolerance from the age of three. His parents complained that he was difficult to discipline. Spanking and slapping by his father in order to curb his attacks on his younger brother were largely unsuccessful. Terry would become further enraged and start to hit his father. His father also punished him, often unpredictably, for minor misdemeanors, such as being noisy.

Terry always seemed to irritate his father. By the time Terry was six, his father would slam him against the wall, wrestle with him, or drag him to his room and lock the door. Terry's father justified the punishment on the basis that his son needed more control, but it reflected his disappointment at having an "undisciplined" child. His mother played a passive role and was disposed to be lenient when his father was absent.

In his various hostile encounters at school and home, Terry felt misunderstood, mistreated, and rejected. One of his most common complaints was,

"Everybody is against me." This belief shaped his interpretation of other boys' behavior. If a fellow student walked by without making a sign of noticing him, he took this as a deliberate attempt to put him down. His interpretation was, "He's trying to show that I'm a nobody, not worth noticing." He regarded the "insult" as deliberate and unfair. After his initial hurt feeling, he felt a craving to salve his injured self-esteem by yelling at the other student and precipitating a fistfight. He regarded his "counterattack" as defensive and justified. He would not consider the possibility that he might have played a role in provoking the brawl. Characteristically, he justified his actions on the basis: "The other kid started it." According to his distorted recollection of the event, he was completely innocent.

Although Terry's behavior at home was attributed to a "bad temper," the core of his problem resided less in weak control or impulsivity as in his negative beliefs about other people. He continuously pictured himself as the innocent victim. He assumed that other children and teachers picked on him because they enjoyed it. When asked about his feelings about the other students, he said, "They are my *enemies*." Sometimes he would react to discipline at school by seeking a weaker target, usually one of the girls, whom he would tease, pinch, or shove. When admonished by the teachers, again his explanation was, "She started it." When his parents attempted to confront him with a specific report of misconduct, he would deny it ever happened: "It's not true. . . . She made it up."

This youngster showed the typical features of the cognitive model of hostile aggression. He perceived himself as vulnerable to the malevolent actions of others, whom he regarded as antagonists. He felt diminished by their actions, which he interpreted as biased against him. He was impelled to restore his self-esteem by counterattacks. In addition, his belief that he was innocent of initiating fights distorted his recollections of the events and made corrective feedback difficult.

Many studies have shown that a significant proportion of delinquent children emerge from families in which they have been subjected to harsh physical punishment such as that suffered by Terry. These parents use coercive strategies, punishment, and force rather than more adaptive techniques such as reasoning, explanation, reward, and humor. In a typical delinquent child case, the household consists of a single parent, generally the mother, who is under severe social and economic stress. Under these circumstances, the parent has a low threshold for frustration and overinterprets the child's acting-up as a personal affront.[13] In addition, she often has age-inappropriate developmental expectations for her children's behavior.

Children are most likely to become delinquent when the punishment is extreme (punching, beating, burning) and capricious. A warm relationship with the parent can, on the other hand, buffer to some extent the effects of periodic beatings and temper the child's misconduct. Also, physical discipline in keeping with the norms of a particular cultural or ethnic group is less likely to produce delinquency, unless it is extreme. Harsh punishment meted out by a father to his son or by a mother to her daughter is more likely to produce delinquency than when parent and child are of the opposite sex.[14]

Kenneth Dodge and his group at Vanderbilt University have shown that the psychological impact of harsh parental discipline can be ascertained as early as the age of four. Testing of children who later become delinquent shows that they were more likely than other children to attribute hostile intent to another child in an ambiguous situation. For example, if someone spills milk on another person or bumps into him, they explain the actions as deliberate rather than accidental. Overinterpretation of other people's behavior as hostile became a consistent cognitive pattern as the child progressed from preadolescence to adolescence to adulthood.[15]

Harsh parenting shapes the child's inimical views of others and his view of himself as vulnerable to the hostile actions of others. Even though the child may dislike or even hate his parents, he often mimics their behavior and incorporates their attitudes. They fail to present constructive role models and do not give the child the kind of guidance, support, and understanding that he requires. The parent also may directly affect the child's notion of the best way to influence others: by intimidation, domination, and force. The notion that other people "have it in for him" may lead the child to engage in a variety of "unsocialized" behaviors: lying, cheating, bullying, cruelty, disobedience, and destruction of property.

His early experience structures his perception into the beliefs that other people treat him unfairly, that they "have it in for him." Of course, part of this belief has a basis in reality. Other children—and adults as well—tend to shun difficult children. The delinquent child may not fully realize that his aversive behavior makes people wary and antagonistic toward him.

Hence, he is all the more likely to feel that they are treating him unfairly.

Given the harsh treatment and lack of guidance and support at home, the child is drawn to other delinquent youngsters in the community. Ultimately, they consolidate their interests in the formation of gangs. The gang members have the same disposition to see themselves as right and other people as their opponents or enemies that they possessed prior to joining the gang. Hence, the gang reinforces a sense of victimhood and, at the same time, provides moral support and justification for fighting the perceived enemies.

Parents of children with conduct disorders frequently manifest the same kind of thinking disorder as their children. They tend to interpret normal developmental behaviors, such as impulse activity or crying, as attempts to annoy or manipulate them. Further, they view certain difficult behaviors, such as sloppiness or defiance, as a reflection of bad character traits. They also regard behavioral problems at school as indicators that "he's a rotten kid." It is not surprising that, given these hostile beliefs and interactions, the child's misbehavior gets worse over time.

A child's behavior, furthermore, often strikes at the heart of a parent's sensitivities. For example, a mother who fears being unlovable may interpret her child's lack of obedience as a personal rejection. A father who values order and control may perceive the same disobedience as a blow to his macho self-image. In both cases, the parent may be hurt, and consequently overly angry with the child.

REACTIVE ASSAULTERS AND PSYCHOPATHS

Many different kinds of people engage in violent acts; though their overt behavior seems similar, their psychological makeup may be at polar extremes. The disparity is particularly obvious when comparing individuals who react with violence only in specific provocative situations to those who adopt deliberate violence as a way of life. The former group may be labeled "reactive offenders" (or sociopaths) and the latter primary psychopaths (or hardened antisocial characters).

Let us examine the case of a serious "reactive offender" named Billy. He received his original prison sentence after he assaulted one of his drinking buddies at a bar. What began as a typical political disagreement escalated into an exchange of insults and blows. Billy got the worse of the interchange and went home in a high state of fury, obsessing about the incident. He located a pistol, returned to the bar, and shot, but did not kill, his adversary. After serving six years in prison, Billy was let out on good behavior, with instructions to report to his parole officer on a regular basis.

Billy's shaky self-image and violent beliefs can help explain his violent act. When subjected to the perceived insults at the bar, he experienced a catastrophic drop in self-esteem, almost to the point of clinical depression. Since brute force would not subdue his adversary at the bar, he decided that he could best punish him and salve his injured pride by shooting him. For Billy, violence carried a powerful message: he was not a weakling and would not tolerate disrespect.

An even clearer picture of Billy's psychological problem emerged when he was released from prison and had to meet with his parole officer. Prior to the appointment, he was stirred up and felt ready for a fight. The anticipation of meeting with the parole officer accentuated his specific sensitivities: he would be controlled by the officer's power; he would be belittled and disparaged by his superior attitude; and he would be threatened by the prospect of being returned to prison. Indeed, after his meeting with the officer, he felt weakened, trapped, and subordinated. His response to these emotions was an urge to fight back, an act that would restore his sense of power and equalize the relationship in his mind. But he restrained himself in consideration of the overwhelmingly negative consequences.

When Billy left the parole officer, he felt very tense and hostile and decided he needed a few drinks to "chill out." He had continuing obsessive thoughts about how everybody was putting him down. Once again, he got into a brawl at a bar when he felt that one of the customers addressed him in a disparaging way. He felt that he had to punish anyone who insulted him. He punched the other man in the mouth, whereupon the opponent backed off and moved away. Billy felt triumphant. By this effective act, he was able to counteract his unpleasant feelings. He no longer felt tense, weak, powerless, or inferior. The violent act was an efficient way to restore his self-esteem and self-efficacy (temporarily, of course) and neutralize his feelings of distress.

It is useful to analyze Billy's problem by viewing the world from his perspective: he saw himself as an innocent victim and others (society, officials, his peers) as victimizers. He therefore felt justified in assaulting "the Enemy." The progression may be summarized as follows.

VISIT TO THE PAROLE OFFICER → FEELING WEAK, POWERLESS, TRAPPED →
URGE TO RESTORE SELF-ESTEEM → MOBILIZATION FOR VIOLENCE →
JUSTIFICATION THAT HE HAD BEEN MISTREATED → VIOLENT ACT

Billy represents an unsocialized offender. These individuals are capable of experiencing the usual human emotions, such as shame, guilt, and empathy, but lack the inhibition, control, and reflectiveness that can interrupt an escalating urge to attack ("response modulation").[16] They are characteristically poor at problem solving and assertive social skills, often feeling weak and inadequate. As a result, the reactive offender feels vulnerable in interpersonal conflicts and is disposed to use the only problem solving that is familiar and comfortable to him: violence. Unfortunately, violence often works for the offender in the short run, and he is consequently positively reinforced each time he punishes an adversary and restores his self-esteem.

PRIMARY PSYCHOPATHS

Reactive offenders may be contrasted with primary psychopaths, who, while constituting a minority of the prison population, account for a much larger proportion of violent crimes, and especially the most violent. This group, originally described by Cleckley as grandiose, lacking in empathy or guilt, impulsive, sensation-seeking, and unconcerned about punishment, has been studied extensively in recent years.[17] While many offenders have, to varying degrees, characteristics of both the reactive offender and the psychopath, primary psychopaths constitute a group of criminals with a well-defined cluster of beliefs and behaviors.

Professionals who have worked with psychopaths have been struck by their extreme egocentricity. They are totally self-serving, feel that they are superior to others, and, above all, think that they have innate rights and prerogatives that transcend or preempt those of other people.[18] They rise to the challenge when other people oppose them and generally resort to antisocial strategies to remove the opposition: lying, deception, intimidation, or actual force. These manipulations are rewarded with feelings of pleasure when they work and do not produce shame when they are unmasked.

Careful examination of the psychopath's response in experimental situations has demonstrated an information-processing deficit. Joseph Newman and his associates have demonstrated that when the psychopath is engaged in a plan of action, he is relatively impervious to cues that would prompt other people to stop and reflect. This insensitivity, lack of reflectiveness, and defective response modulation explain, in part, the psychopath's impulsiveness and apparent lack of inhibition. Since he does not automatically anticipate the consequences of his actions, they convey a sense of fearlessness, regarded by David Lykken as a central component of the psychopath's antisocial behavior.[19]

Their lack of empathy for the people they hurt is a major component of their congeniality with violent crimes. Although they may be quite adept at reading other people's minds, they use this skill only to dominate and control others and not to identify with the people they hurt. They have not incorporated the socialized rules that prompt people to feel shame when they make a social transgression or guilt when they hurt other people. They know the rules very well but simply don't apply them to themselves.

The primary psychopath and the reactive offender can best be illustrated in terms of how they contrast.

	PSYCHOPATH	REACTIVE OFFENDER
Self-concept	Invulnerable	Vulnerable
	Superior	Fluctuates
	Preemptive rights	Fragile rights
Concept of Others	Dupes	Hostile
	Inferior	Oppositional
	Weak	The Enemy
Strategies	Manipulative	Inadequate problem solving
	Violence	"Defensive violence"

These two groups have some features in common, but they serve different purposes. They both assert their prerogatives—for example, "You have no right to treat me that way"—but for different reasons. The psychopath takes for granted that his rights are supreme and confidently imposes them on other people. The reactive offender feels that nobody recognizes his rights and reacts with anger and sometimes violence when others reject him or do not show him respect. Both have a low tolerance for frustration and punish their frustrator. The reactor, however, may feel shame or guilt afterward, while the psychopath feels only triumphant.

The clinical approach to these two types of offenders differs, of course. The clinical strategies for the reactor are directed toward helping him with his sense of inadequacy and training him to assert himself constructively and to solve problems. Initially, it is important to educate him in anger control. He is coached in trying to defuse conflicts with self-statements such as, "It's not such a big deal," or, "It's not worth getting excited about." The approach to the criminal psychopath is much more difficult but utilizes empathy training, increasing sensitivity to feedback, and incorporating more reflectiveness about the long-range effects of his antisocial behavior. The biggest task is modifying his egocentricity and grandiosity.

SEDUCTION AND RAPE

Sexual violence against women can be understood, in part, within the framework of the masculine, and especially macho, mythology endorsed by a subgroup of the male population. This web of distorted images, concepts, and expectations is based on the typical ingroup-outgroup mentality. The masculine stereotype incorporates the characteristics highly valued by this subculture: toughness, superiority, competence, and boldness. The feminine stereotype is epitomized by weakness, inferiority, incompetence, and fearfulness. The male, by virtue of his self-arrogated superiority, assumes rights and privileges that supersede any claims of women. "Men are born to dominate, women to submit."[20]

This cultural mythology is reflected in macho attitudes toward sex. Women are perceived as sex slaves or playthings, whose role is solely to provide pleasure to their male masters. Women's resistance to male advances is seen as a part of the game leading to ultimate capitulation. The male's goal is to manipulate, deceive, trick, and ultimately seduce. Behind their beliefs is the sexist doctrine that men and women are adversaries; each sex is out to get what it can from the other—through exploitation, deception, cheating. Beyond seduction is rape, the supreme expression of masculine power, dominance, and ownership.

Coercive sexuality brings additional pleasure because it helps to reinforce the aggressor's masculine self-image and boosts his self-esteem.

Martha Burt compiled a list of "rape myths" that are endorsed much more frequently by sexually coercive men than by other violent male offenders.[21] These myth-beliefs, which provide a justification for the coercive sex offenders, include:

- The rape victim is usually promiscuous and has a bad reputation.

- Women hitchhikers who are raped get what they deserve.

- Stuck-up women who think they're too good to talk to the guys on the street deserve to be taught a lesson.

- Women who wear tight tops and short skirts are asking for it.

- A healthy woman can resist being raped if she wants to.

Enhancing the power of these beliefs is the attitude that force and coercion are legitimate ways to get another person to comply with one's demands. This attitude is carried over to intimate and sexual relations.

Of course, this cultural prejudice varies in intensity, but it provides clues to the workings of the minds of the men who have raped. When we examine individual cases, we can see that a wide variety of factors lead to a depersonalized image of women. Often the woman is perceived not as a real person but as a body or an object. These debased images of women and distorted beliefs about them shape the way these men interpret women's behavior and how they behave toward them. Polaschek, Ward, and Hudson present a strong case for biased information processing by rapists.[22] They point out that the offenders not only interpret the way the victim dresses as a "come-on" but also interpret her frightened compliance and passivity as a sign of enjoyment. In addition, rapists are more sexually aroused by rape scenes than by scenes of consensual sex. These men believe that women enjoy being dominated, and indeed, they assume dominant roles in heterosexual relationships. In fact, the position of power appears to prime the rape beliefs and the desire for a sexual liaison. The aggressor is often oblivious to signs of the victim's resistance or aversion, or he misinterprets them as part of the feminine game. Experimental studies, in fact, have demonstrated a cognitive deficit in rapists: they do not read women's cues accurately. Other authors have found that sexually assaultive men have "suspicious schemas" about their victims: they think that the women actually mean the opposite of what they say. Thus, a display of anger means, "She protests too much."[23]

Some adolescent or adult males, obsessed with memories of past rejections or humiliations by women, experience rape as a kind of vindication or revenge. Sometimes the ideology of a gang sways a new member into adopting the more extreme sexist views of other gang members. He may, for example, participate in a gang rape as much out of allegiance to the group as out of a desire to engage in a violent assault. Adolescents and adults may be drawn to coercive sex as a compensation for their own deficient self-image: weak, unmanly, unattractive.

A subgroup of men, particularly loners, use forcible sex as a kind of self-medication for unpleasant feelings, analogous to the mechanisms in some substance abusers. It is not simply the gratification from sex itself that drives them, but the experience of power and dominance over another person that neutralizes their own feelings of helplessness. In addition, the concept of dominance is closely tied to the concept of sex in the rapist's mind.[24] The progression from helplessness to dominance to sex is reminiscent of the psychology of the "reactive offender," whose violent interpersonal behavior is driven by dysphoric or depressive feelings.[25]

Some rapists, on the other hand, are similar to the supreme egoists, the psychopaths, who make a life goal of reinforcing their narcissism.[26] This group engages in a wide variety of antisocial behaviors, of which rape constitutes only one example.

In general, the men who rape, whether solitary or in a gang, out of psychopathic tendencies or reactive dysphoria, have a number of common characteristics. During the pursuit and forcible sexual act, they are lacking in empathy for the victim and interpret ambiguous female behavior as sexual invitation. Their tunnel vision precludes recognition of or concern about the obvious pain and humiliation of the woman. They are either oblivious or indifferent to her cries. They underestimate both the physical and psychic pain ("It's just sex"). The usual inhibitions against harming others are lifted by the notion that the aggression is justified, that the victim "deserves it," or that she will enjoy being raped. The fear of future punishment is doused by the immediacy of the experience. Afterward they minimize the amount of physical or psychological trauma and are likely, if possible, to blame the victim. They believe that the laws are wrong, not their behavior.

Various statistical surveys add weight to the psychological formulation. The majority of rapists blame the victim, and many believe she benefited from the experience. About 60 percent of rapists, however, acknowledge that the motive is to humiliate and degrade the victim.[27] Many acknowledge past experiences of being humiliated or degraded by a woman.[28]

People who engage in violent acts against others in the form of child abuse, spouse battering, or juvenile or adult assault or rape, can be seen to share certain common psychological factors. Their negative beliefs prompt negative interpretations of the victim's behavior. They are deficient in social skills and in reading feedback from other people; they justify their assaultive behavior on the basis that *they* are victims. Finally, they are frequently emotionally stressed or depressed and resort to violence as a strategy to neutralize their distress and restore their self-esteem.

9

COLLECTIVE ILLUSIONS

Group Prejudice and Violence

> A crowd scarcely distinguishes between the subjective and
> the objective. It accepts as real the images evoked in its
> mind, though they most often have only a very distant rela-
> tion with the observed fact. . . . Whoever can supply them
> with illusions is easily their master; whoever attempts to
> destroy illusions is always their victim.
>
> *Gustav LeBon,*
> **The Crowd: A Study of the Popular Mind** *(1896)*

Picture a football game: the crowd on one side of the stadium cheers as its
team scores, and groans as it is pushed back. The fans in the opposite stands
reciprocate with their own boos and cheers. The synchrony on each side gives
the impression of an indivisible unit, like members of a chorus, controlled by
the same script. The warmth and empathy of the members of the crowd for
each other and their disdain, even hostility, for the other side are striking.
The polarized thinking of a team on the field and of their supporters in the
stands has much in common with the more extreme thinking involved in
prejudice, race riots, and persecution.

Now imagine a parade of jackbooted storm troopers goose-stepping in
unison with the rousing music of a military band to the cheers of adoring
crowds. This scene resembles the spectacle of the uniformed football players
and their cheering partisans at the stadium. The enthusiasm of the supporters

generates contagious fervor, exalting their champions (and by association, themselves) and denigrating the opposition.

The hostility in the sports arena is generally contained and time-limited, while military ferocity can be extensive and all-consuming—yet the dichotomy between "us" and "them" exists in both arenas. In fact, the line between sports competition and violent attacks on the adversary is not infrequently crossed. Witness the "soccer wars": rampaging fans of the losing team assault the supporters of the winners and even the members of their own team by whom they feel betrayed.

The most notorious of the sports wars occurred in Constantinople in the sixth century A.D. The Hippodrome in that city was the scene of the competition between rival chariots, distinguished by the green or light blue colors of their drivers' liveries. The city was divided into two factions, green and blue. Catalyzed by an admixture of religion and politics, the chariot races exploded into riots and later massacres. A fight between the greens and blues in the Hippodrome in 532 A.D. escalated into a war between rival factions and culminated in the burning of much of the city and, ultimately, the massacre of thousands of greens.[1]

The thinking, feeling, and behavior of a person in a group can be only partially explained in terms of how that person might interact with another individual. Although dualistic thinking and bias occur in both individual and group interactions, certain phenomena such as camaraderie, commitment to a leader or cause, and collective illusions need to be understood in the context of the group. "Groupism" is the collective counterpart of egoism. The person in the group transfers his own self-centered perspective to a group-centered frame of reference. He interprets events in terms of the group's interests and beliefs. Ordinary selfishness is converted into "groupishness." He not only subordinates his personal interests to those of the group but opposes the interests of outgroup members unless they are compatible with the interests of his group.[2]

The group-centered member may promote the enhancement of the image of his comrades (and consequently himself) and the demeaning of outsiders. Confrontations with other groups accentuate the positive bias toward his own group and the negative bias toward the adversarial group. There is a reciprocal relationship between his evaluation of the ingroup members versus the outgroup members. The more he perceives opposition from the outsider group, the more he elevates his own. His fellows become more worthy, noble, and moral as the others become increasingly unworthy, ignoble, and immoral.

Much group behavior rests on the communication, often subtle, of beliefs, images, and interpretations across the group. Group members are tuned in to special meanings assigned to events affecting the group, and they readily accept opinions and policies advocated by the leader. Despite their extremeness, these beliefs are relatively plastic (in contrast to those of psychiatric patients). Beliefs in which they have a strong investment can be reversed on signals from the group leader. For example, Hitler could call for the destruction of the Soviet Union amid roars of approval in one speech, and at a later date arouse enthusiastic cheers for his announcement of a non-aggression pact with the same country. The multiple interactions among his followers served to sway mass opinion in the chosen direction. And the same group contagion spread when he reversed himself again and marched against the Soviet Union.

The tendency of people to adjust their report of their perceptions to conform to the evaluations of other group members has been demonstrated in experiments by Solomon Asch.[3] He showed that experimental subjects would change their stated observations of a stimulus under the influence of other people. The subject would assume, for example, that her initial pinpointing of an object in space was wrong and change her judgment to conform to the judgment of the others. Such collective thinking, often leading to clear cognitive distortions, helps to bind a group together.

This cohesiveness, in turn, induces members to submerge their own thinking into a collective mentality. The unanimity of opinions and the resulting gratification from sharing goals with their comrades accentuates their commitment to the group. The sacrifices they make and the risks they take for other group members further intensify their commitment and cohesion. This allegiance facilitates their willingness, even eagerness, to abandon the usual ethical and moral norms, even to the point of participating in torture and wanton killing.

The biases leading to cognitive distortions such as arbitrary inference and overgeneralization are similar whether a person is engaged in conflict with another individual or with a member of another group. The passionate hatred of a husband and wife toward each other during a turbulent divorce action is based on some of the same psychological processes as the rage of uniformed thugs pillaging the homes and stores of a defenseless minority. In group actions, however, people are moved by the collective biases and the "contagious" sweep of feelings. An individual substitutes his group's values and restrictions for his own as the group establishes boundaries between "us" and "them."[4]

A fire in a theater, the loss of a football game, or the news of a military victory—each concentrates people's attention on a theme, whether danger, defeat, or victory. The shared meaning of the event leads to shared feelings—panic, distress, or euphoria—and incites the same type of behavior—stampede, riot, celebration. Rioters in a lynch mob or a pogrom are driven by the same kind of diabolical image of their innocent victims; victorious celebrants are elevated by the same glorious image of their group or nation. The meshing of the individual's beliefs with those of the group energizes ethnic conflict and acts of prejudice, persecution, and war. The subordination of personal interests to those of the group is played out in self-sacrifice and most dramatically in suicide bombings.

The individual's yearning for personal success, combined with his or her hunger for attachment and bonding, is satisfied by identification with the success and intimacy of a dedicated group. The psychological mechanism that produces the subjective feeling of pleasure following a personal achievement also operates in a group triumph. But group allegiances bring rewards beyond those of exclusively personal experiences. Since members of a group interact, their joy in victory is amplified as it reverberates across the group.

This "emotional contagion," or synchrony of group reaction, may be observed at the very earliest stages of human development. Anyone who has watched newborn infants in a nursery can testify that when one baby starts to cry, the other babies howl.[5] The same ripple effect occurs later in life: when a single person in an audience starts to yawn, one by one the rest of the audience starts to yawn. Similarly, laughter, once started, can cascade uncontrolled through an entire group.

Since empathy (or at least mimicry) is present so early in life, it seems probable that receptivity and responsiveness to other group members' emotional expressions are "hard-wired" into the mental apparatus.[6] Cheers or groans, smiles or grimaces, all are perceived, rapidly processed, and reproduced by other people in a group. Voices, facial expressions, and body language trigger feelings of excitement, joy, or anguish. To the extent that they attach the same meaning to their perceptions, group members experience a similar emotion.

A considerable amount of research substantiates the crucial influence of interpersonal "displays" in human interactions.[7] People continuously monitor and mimic one another's emotional reactions without realizing they are doing so. Elaine Hatfield and her colleagues have documented automatic "mimicry" of subjects' facial expressions in their reactions when viewing a videotape of an actor's happy or sad expression. In a review of the research in this area, Janet Bavelas and her coworkers proposed that such synchrony of expressions represents a kind of innate communication system that promotes group solidarity and involvement.[8]

An intercommunication network is obvious in many species, especially in social insects, which work in unison to accomplish enormous feats because they can signal "instructions" to other colony members. An organized crowd of humans may also be likened to a network of transmitters and receivers. Nonverbal signals like screams, laughs, and flag waving produce a kind of reflex reaction in the receivers. The simple concrete stimuli moving in waves across the group are transformed into complex meanings. As a leader harangues and exhorts his audience, his verbal messages trigger a chain of nonverbal responses—cheers, head nodding, and foot stamping—that cycle back and forth among the receptors.

The signals from the group leader and among group members do not arrive at empty receptors. They are picked up by cognitive structures, schemas, which consist of specialized algorithms and gross images for converting signals into meaningful constructs. Since the context of the schemas is consistent across the group, the collective meaning is relatively uniform.

Intergroup conflict may be viewed in terms of the network construct. Group members generally have stereotyped images about members of an outside group. Negative messages about the other group's actions generally trigger the stereotypes that help to mold a biased interpretation of the other group's behavior. These stereotypes are often robust. When encased in a rigid schema (or frame), they do not permit any modification of the prejudiced belief and bear the earmarks of what we call a "closed mind."

Stereotypes that are embedded in the matrix of a militant ideology may shape the image of an adversarial group into the Enemy. If this ideology includes a moral code based on the principle that the end justifies the means, persecution and murder of the stereotyped others may occur.

IMAGINATION AND GROUP HYSTERIA

The power of imagination and the enormous impact of its spread from person to person have been observed in practically all societies. The capacity to conjure up images of unspeakable, even supernatural, acts perpetuated by stigmatized others is practically unlimited. When the rumors pertain to inflammatory issues, they create vivid imagery consonant with the message—for example, robed figures sacrificing an infant on an altar. Although the image is generally pure fantasy, the group process enhances its credibility and members experience it as though it were real.

Even highly educated, intelligent people have succumbed to lurid tales of ritual child murder and cannibalism—without a shred of corroborating evidence. As recently as 1997, uncorroborated testimonials by adults that as children they had witnessed and participated in ritual sacrifices of babies were accepted as fact by many psychotherapists and evangelical religionists.[9] In previous eras, the mere allegation of child torture and pacts with the devil was sufficient to stir the imagination of listeners and incite them to torture, burn, or hang the supposed malefactors. The credibility of recent tales of ritual child torture has been based alternatively on the believers' notions of the infiltration of Satan into society, and their belief that child abusers in a male-dominated society are capable of engaging in the most fantastic acts.

To be accepted as fact, it is often sufficient that the concocted tale of horrors by a stigmatized group be compatible with the belief system, the ideology, of the listener. A frightening story of alleged misdeeds by stigmatized others (who are suspect on the basis of the believers' preconceptions) stirs up distressing images, which are taken at their face value as facts. The formation of a persecutory perspective is facilitated by the tendency of an excited imagination to displace reason—particularly when the imagination is primed by the group leader and other members of the group.

Whatever the nature of the alleged crime, the group's lust for revenge springs from its sympathy with the supposed victims (such as babies) and its framing of the stigmatized others as villains. The tales do not rest on any concrete, visible evidence but on the mental images of wrongdoing, often deliberately planted. These images inspire conviction because of the credibility of the rumormongers and the propensity to believe the worst about a suspect group. In the mind's eye, the imagined event is just as real as if personally witnessed. In fact, contemporary legends of ritual murders have stimulated individuals to "recall" vivid memories of events in their childhood.[10]

Unlike the actual observation of a real event, the imaginary tale is not subjected to rational reflection or consideration of evidence. Further confirmation is provided by other believers, and objections by skeptics are discounted. The images stirred up by outrageous stories have a powerful impact not only because of their horrific nature but also because they make the believers feel more vulnerable. In the danger mode, people imagine the worst. After all, if members of a diabolical group can torture innocent babies, there is no limit to the evil things they can do.

If a person's cultural heritage is embellished with notions of demons, evil spirits, and diabolical possession, then his imagination is especially susceptible to fantasies of magic spells, witchcraft, and ritual sacrifice. The destructive power of the imagination is illustrated by historic persecutions, including burning at the stake of innocents believed to be witches, sorcerers, or warlocks.

The well-known Salem witch trials of 1692 show the influence of imagination on the victims as well as on their persecutors. In Salem voodoo tales by a West Indian slave induced a wave of mass hysteria in suggestible adolescents. The signs and symptoms of epileptic-like seizures, unnatural positions, trances, and other strange behavior were similar to those readily induced by hypnosis. Since no medical explanation was forthcoming, the village physician expressed the opinion that the affected children were bewitched.[11] As the hysterical fear of witchery spread through the community, scores of citizens were accused of witchcraft. Ultimately 19 persons were hung as witches and 150 more were imprisoned.

Since notions of the devil and witches were endemic in the Salem community, the diabolical accusations fit into the preexisting religious ideology. Interestingly, there was considerable racial, political, and economic flux in the region, and it made the villagers more susceptible to supernatural explanations. Historically, times of unrest sensitize people to images of conspiracy. Epidemics of witch burning occurred during periods of social upheaval in the

Middle Ages. An estimated 500,000 innocent people were convicted of witchcraft and burned at the stake in Europe between the fifteenth and seventeenth centuries.[12]

The mythology of blood libel provides a good illustration of the history of an illusion. Accusations that Roman children were kidnapped for ritual sacrifice were leveled against Christians in the early centuries A.D. The imaginary sacrifices were a symbol of the absolute evil of Christians. The story resurfaced in the Middle Ages, but this time with Jews targeted by the Christians as kidnappers of Christian children. The blood libel myth against the Jews has recurred right up to the present time. Threats to the established religion by a competing religious group have prompted persecution of the supposed heretics. Throughout history, upstart religious groups have been charged with being in league with the devil.[13] The fantasy of a reviled group engaging in abominable rituals such as child sacrifice is an expression of belief in the eternal war of evil against good. Even people who do not believe in the existence of the devil can enjoy the purifying experience of unmasking and denouncing the stigmatized others alleged to be engaged in these abhorrent activities.

During periods of social change and economic upheaval, people are more amenable to adopting a paranoid perspective, if that is communicated by an authority.[14] The condemnation of groups of individuals as witches throughout history has provided the oppressed population with a convenient explanation for impoverishment, plagues, and famines. Church and state have collaborated in stimulating and perpetuating the witch mania to deflect blame from themselves and preserve their own status and power.[15] The enemies of the people thus were not the princes and popes, but the witches. Framing certain groups of people, placing them into a category, is an expression of the universal tendency to stereotype others.

STEREOTYPING AND PREJUDICE

Walter Lippmann, the noted political commentator, is credited with originating the popular meaning of the word *stereotype* in 1922.[16] According to Lippmann, we create stereotypes—simplifications—to guide our perceptions of people and aid us in interpreting their behavior.

The psychologist Gordon Allport suggested in 1954 that placing people in categories has an adaptive function: "The human mind must think with the aid of categories. . . . Once formed, categories are the basis for normal prejudgment. We cannot possibly avoid the process. Orderly living depends on it." Allport pointed out the necessity of reducing the vast complexity of our social world to manageable dimensions. By placing people in categories, we help to make our adjustment to life "speedy, smooth, and consistent."[17] Some categories, of course, are rational. It is likely (thought not inevitable) that somebody of Mediterranean stock has darker hair or skin than a person from Scandinavia. But other characteristics attributed to a category of people are unreasonable—for example, that Scots are stingy or Asians are crafty. Stereotyping erases the unique characteristics of outgroup members. As soon as boundaries are drawn around an outgroup on the basis of religion, race, or creed, its individual members are perceived as interchangeable. In particular, uniform images of people of competing classes, political or economic organizations, or ethnic groups (the political left versus the right, labor versus management) are constructed. This kind of ingroup-outgroup division provides the matrix for biased thinking and prejudice.[18]

The tendency to "think in categories," the prototype of prejudice, has been subjected to intensive study by social psychologists. Simplifying through categorization readily leads to oversimplifying and consequently to distortion. Because groups are biased in their perception of themselves and others, their people are likely to attribute better motives and more sterling character to members of their own group than to those of an outgroup. When things go wrong, people are likely to assign more blame to a member of an outgroup than to a member of their own group.[19]

Biases have been observed even when individuals are assigned randomly to arbitrary groups. Experiments demonstrate that people in an artificial group evaluate others in their group as more friendly and cooperative, and their physical and personal attributes as more desirable, than those who have been arbitrarily assigned to another group.[20] This perceived superiority of the ingroup is reminiscent of the attitudes of summer campers assigned to competing teams in the traditional "color wars." Members of one arbitrary group draw closer to one another and distance themselves from members of the other group; they tend to overestimate the similarities within their group and their differences from the other artificial group. The greater the competition between the groups, the more these similarities and differences are accentuated.

In one experiment at a summer camp, boys were divided into two groups that engaged in competitive activities with each other. Boys in each group developed antagonistic attitudes toward those in the other groups. The group-engendered hostility culminated in raids on each other's belongings and other destructive behavior.[21]

Even the words used to designate different affiliations, specifically *we* and *they*, can distort perception. The mere designation of the personal pronoun *we* to an arbitrary group induced a "we-group" to make more favorable ratings for their group members than they did for those designated as *they*.[22] The motive to boost one's own self-esteem can lead people to view their own group more positively. People tend to search for differences that are favorable to their own group and unfavorable to the outgroup and to minimize differences that favor the outgroup. Group experiences also affect how people think and feel about themselves. Tafjel demonstrated that people show an increase in personal self-esteem following a group success.[23] Moreover, people whose self-esteem has been diminished by a group failure show an increase in prejudice.[24]

It is important to recognize that people may be prejudiced against different races or ethnic groups without realizing it.[25] Most subjects who view a picture of a person of a different race, for example, will subsequently show faster reaction time to unpleasant words and more prolonged reaction time to pleasant words. The reverse is true when they see a photograph of a person of their own race.[26] The rapid reaction time indicates that a person is keyed to make an automatic evaluation. The negative bias toward another race facilitates automatic negative labeling and a positive bias toward one's own race favors positive labeling.

The extravagantly destructive behavior of some people, whether rioting after a racial incident or massacring innocent villagers in civil strife, may be traced to their beliefs and thought processes. They show the same kind of conceptions and misconceptions, interpretations and misinterpretations in conflict with other groups as they do in a dramatic personal conflict with another individual. They are likely to attribute the cause of friction or adverse interaction to indelible character defects in the members of the opposing group rather than to the relevant circumstance or situation.[27]

In intergroup conflicts, the crisscrossed interactions within each group are multiplied. They fortify the members' resolve, authenticate their biases and misconceptions, and give license to acting out their destructive impulses. Hostility toward another group merges the group-serving biases with the self-serving biases of the individual members.

People form the same kind of negative attributions and overgeneralizations in power struggles with their social, ethnic adversaries as they form with a parent, sibling, or mate. They also frame an adversarial group with the same type of negative global evaluations that they use during a "hot war" with a personal antagonist.[28] They blur the distinctions between individuals in the outgroup, force them into the same unattractive or malevolent frame, and view them as psychological and moral (or immoral) clones of each other. As their hostility intensifies, they can no longer see the human qualities of the outgroupers.

Automatic negative appraisals may be attached to outsiders who are not even members of an oppositional group—simply because they do not belong.[29] In many cases, exclusion from the ingroup is based on a devaluation of all outgroupers: they do not belong because they have unacceptable values or beliefs, lack the required virtue or purity, or possess "obnoxious" traits.[30]

The tendency to place people in either a favorable or unfavorable category has been observed in all cultures: we or they, friend or foe, good or evil, honest or dishonest.[31] Some authors regard this dualistic thinking as an expression of a basic principle of mental functioning.[32] It seems likely that under stress people revert to this primal dichotomous thinking. Certainly, clinical observations bear this out. Patients in the grip of depression, anxiety, or paranoia organize their experiences in terms of opposite attributes: worthy or unworthy self (depression); safe or dangerous situation (anxiety); benign or malicious others (paranoia).

Adjectives or nouns describing personal characteristics or traits are almost always evaluative, either complimentary or derogatory, desirable or undesirable. Whether applied to individuals or group members, these terms reflect esteem (honorable, dynamic, brilliant) or disparagement (dishonorable, manipulative, devious). The same invectives used by a marital partner in a heated conflict with her mate may be used by members of one group toward an antagonistic group (treacherous, manipulative, hostile, dangerous). As soon as somebody stigmatizes another person or a group with a characterological label, he or she uses this assigned trait to explain any "undesirable" behavior of the other person. Thus, an outsider who seems to oppose the interests of the ingroup is judged to be motivated by his inherently unworthy impulses, his intrinsic "badness."

The very act of condensing the diverse characteristics of an outgroup into a homogeneous image stamps all of its members with the same pejorative label (inferior, aggressive, immoral) and distorts perception of the individuals and interpretation of their actions. This kind of reduction to a few arbitrary undesirable characteristics inevitably deletes the positive qualities. The more extreme the undesirable derogatory adjectives, the less human the outgrouper appears and the easier it is to aggress against him or her with impunity.

THE CLOSED MIND

A complete picture of the nature of prejudice requires an understanding not only of *what* prejudiced people think, but also of *how* they think. A clue to this understanding is found in studies of intolerance, specifically of the "closed mind." As described by Milton Rokeach, people who receive high scores on tests of ethnic prejudice are rigid in their problem-solving behavior, show concrete thinking, and are narrow in their understanding of subjects of vital interest to them. They are also prone to make snap judgments, to dislike ambiguous situations, and to show distortion in their recall of significant events. Of paramount importance is their active resistance to any change of belief. Rokeach points out, "Acceptance of those who agree (opinionated acceptance) is as much a manifestation of intolerance as is rejection of those who disagree (opinionated rejection)." Opinionated acceptance of the group's view of the opposition forms the basis for prejudice. In contrast, tolerance is the acceptance of other people, regardless of whether they agree or disagree with us.[33]

The closed mind is impermeable to information that is contradictory to the highly charged beliefs enclosed within its rigid frame. As pointed out by Rokeach, certain conditions seem to contribute to the mind's closure: feeling helpless and miserable, living in a lonesome place, fearing the future, and looking for someone to solve one's problems. These findings suggest that the degree of rigidity in people's thinking may be in part a function of the stress they have experienced.

External pressures that heighten the desire for approval by the group or authority figures tend to freeze the beliefs of the closed-minded person and lead her to reject those with different beliefs. External threats in particular tend to make one's thinking more rigid and categorical; the person is even less likely to make judgments independent of the expectations of the group and authority.[34] In contrast, the open mind is characterized by the ability to evaluate information on its own merits, unencumbered by one's affiliations and beliefs.

A relationship exists between rigid thinking, ideology, and prejudice. A person with an extreme commitment to a religious belief, for example, tends to "sacrilize" the difference between the believer and nonbeliever and to dismiss all those who do not fit into his hallowed world. On the basis of interviews with Fundamentalist Protestants in New York City, Charles Stozier concluded that the "absolutism" of the highly committed predisposes them to intolerance.[35] Other studies have shown a correlation between fanatic religious belief and prejudice. An interesting study in Germany showed that both personality type and economic position biased the subjects' thinking; the investigators found that a low threshold for anger and marginal economic status correlated with prejudiced attitudes.[36]

Under pressure from the group or a leader to accept the prevailing attitudes and values, people can slip into the rigid thinking style. "Groupthink" is a term derived by Janis from the "newspeak" in George Orwell's futuristic novel 1984. As described by Janis, groupthink is the product of the "deterioration of mental efficacy, reliability testing, and moral judgment that results from in-group pressures."

After observing some of the policy making during the Vietnam War, Janis concluded that the high degree of amiability and esprit de corps among members of the policy-making group, coupled with the seriousness of the military threat, militated against independent critical thinking. Although Janis and more recent writers, such as McCauley, have applied the term "groupthink" to decision makers,[37] it is applicable as well to the more or less uniform thinking of any cohesive group engaged in intergroup conflict with other groups. Groupthink can result in irrational and dehumanizing actions directed against the opposition, based on such implicit assumptions as: we are a good group, so any deceitful actions that we perpetrate are fully justified. Further, anyone who is unwilling to go along with our version of the truth is disloyal.

In addition to closed-mindedness, groupthink embraces such patterns as the illusion of invulnerability, collective rationalizations for destructive acts, and stereotypes of outgroup members. The group often has self-appointed "mindguards," members whose role is to exclude outside information that might undermine confidence in the group's beliefs and decisions. The reliance on mindguards suggests that in some instances group decisions may not be totally incorporated by some group members, who consequently require pressure to conform.

Janis points out the various losses that ensue from this kind of group process. Decision making may be defective, since there is an incomplete survey of alternative courses of action. The group may also fail to appreciate the risks involved in a chosen plan and to reappraise the plan after setbacks. Groupthink also limits exploration of all channels of information and produces biased judgments of the information that is obtained. McCauley points out that despite the obvious hazards associated with groupthink, it may sometimes be an efficient mode of solving problems. When there is ample available information and when the circumstances are clear-cut, groupthink has proven to be successful.[38]

In viewing the phenomena of the closed mind and groupthink, we can see how prejudice and enmity not only hurt the targeted group but can also impair the judgment of members of the aggressive group. These characteristics provide the matrix for more malevolent group attitudes and behavior, such as hatred for the stigmatized Others and violence against them. The same kind of psychological processes are evident when the Enemy is not a stigmatized group but the government. In this instance, the political leaders and bureaucracy are perceived as the Enemy, and their actions are subjected to distorted interpretations. The ideology driving this kind of biased thinking leads inevitably to violent strategies, since these seem to be the only means to defeat the powerful force of the tyrannical government.[39]

GROUP HATRED AND TERRORISM

On April 19, 1995, Timothy McVeigh and Terry Nichols engineered the bombing of the Alfred P. Murrah Federal Building in Oklahoma City. Killed were 167 people, including 19 children. Subsequently, McVeigh was found guilty of the crime and, on August 15, 1997, was sentenced to die by lethal injection. Terry Nichols acknowledged his guilt, entered into plea bargaining with the authorities, and received life in prison for conspiracy, although he was acquitted of the actual bombing and murder charges. The bombing at Oklahoma City can serve as a paradigm for the violent, destructive acts of terrorists on the left as well as the right, and particularly for the antigovernment fervor in various sectors of society.

Violent actions against the government by extremist groups of the right and left can be traced to the convergence of several interpersonal and psychological processes: the contagion of highly charged antagonistic ideas, the stereotyping and ultimately the framing of the opposition as the Enemy, and the lifting of inhibitions against murder. Indeed, the preemption of traditional notions of the value of human life by a higher, sacred cause that sanctions terrorist activities removes the psychological block to committing violent acts, including murder.

The sociopolitical ideology of the domestic extremist groups in the United States and elsewhere is centered on a powerful aversion for the government and other groups that they see as opposing and undermining fundamental rights. Whether their ideology originates in a far right or far left orientation, their image of the government is of a monolithic organization dedicated to violating their basic rights. To extremists on both sides, the government is coercive, corrupt, and conspiratorial.

American groups of the far right, such as the militiamen and skinheads, view the government as the tool of ideologues promoting the interests of various ethnic groups as well as international financiers. They visualize international bodies such as the United Nations as composed of conspirators plotting to establish a "New World Order." The far left groups of the 1960s and 1970s, such as the Weathermen and Black Panthers in the United States, the Red Army Faction in Germany, and the Red Brigade in Italy, had a corresponding image of the government as the puppet of corporate capitalists and the military-industrial establishment. Another group preaching and practicing terrorism, Aum Shinrikyo in Japan, rails against the impure world and foresees ultimate redemption in the death and rebirth of the world.

The collective self-images of the extremist groups at both ends of the political spectrum also have much in common. They view themselves as righteous and committed to a noble cause. They and their protégés and sympathizers are regarded as victims of the power structures, including the government, the media, and big business.

The Order, a clandestine offshoot of the Aryan Nations, had its origins in the book *The Turner Diaries,* which inspired Timothy McVeigh and Terry Nichols to perpetrate the Oklahoma City bombing. The book, written in 1978 by William Pierce, the leader of a neo-Nazi group, presents a plan of action that was duplicated in the actual bombing of the Oklahoma City Federal Building. The mission of the organization was to wage war to overthrow the U.S. government, to kill Jews and other nationalities, and to make America into an all-white fascist society. Guerrilla warfare tactics, including robberies, bombings, and other terrorist acts, would be used to consummate the revolution.

The ideology of extreme right-wing groups in the United States is essentially reactionary, directed toward returning the government to the principles espoused by the Framers of the Constitution: liberty, freedom, and patriotism. They claim to be infused with the spirit of the American revolutionaries who fought for their freedom and civil liberties. One of the groups, indeed, took the name of the Minutemen, who defied the British at Bunker Hill. For many of these militant groups, the doctrine of racial purity mandates the clearing out of those "alien" elements in the population (blacks, Jews, Hispanics, and Native Americans) that sully the image of a pure white Christian America.[40]

THE PSYCHOLOGY OF THE MILITIAMEN

The paramilitary units of militiamen in the United States are scattered among a large number of states, primarily in the West, and are loosely connected with each other. They exchange information and propaganda with other units through the Internet, radio, pamphlets, and books. Individuals attracted to these paramilitary organizations share a special personal sensitivity to being controlled by others and a macho orientation to life. Their collective sensitivities, rugged individualism, and ultrapatriotism are expressed in an aversion to restrictions and regulations and a yearning to return to the alleged ideals of the Founding Fathers. The supposed violation of their collective principles by the government or by unwanted elements of society is taken as a personal injury by each of the members.

The militiamen have essentially a frontier state of mind. They prefer to live in relatively small communities. They are fiercely independent and value their mobility and the freedom to maintain their lifestyle without interference by official authorities. They accept no authority higher than the county sheriff, who is responsible for local law and order. In their eyes, the government and its agencies are hell-bent on disrupting their way of life: imposing taxes, passing gun control laws, and setting up multiple layers of bureaucrats and enforcement agents. They resent the diversion of their taxed income to "favored" minorities of alien race and nationality. These restrictions and intrusions by law enforcement agents produce a kind of claustrophobic reaction: they feel hemmed in and threatened. Their fear of being controlled—and contaminated—extends beyond the national boundaries: witness their sensitivity to the possibility of a multinational or world government.

The militiamen try to fulfill their social and political agenda primarily through organizing paramilitary units and building up arsenals to protect themselves. When confronted by government agents and then armed representatives serving arrest warrants for technical violations, enforcing court judgments, or seizing arms from their compounds, they are determined to stand their ground.

They proclaim the constitutional provision for establishing a militia and the right to bear arms as symbols of liberty and freedom. They engage in paramilitary training, organize autonomous militia groups, and build up a cache of armaments in order to protect their rights and foil intervention by "illegitimate" government agents.

Randy Weaver, who was involved in a lethal shootout with federal agents in 1991, typifies the personal philosophy of these individuals. Weaver, like many other extreme rightists, cherished the open country, away from the stifling atmosphere of the city and the decadent East. Under threat from government agents, he retreated to a practically inaccessible cabin in a remote area of northern Idaho to make his last stand against "illegal" authorities.

Several events in recent American history have served as catalysts for the growth of a violent animus against the government. These events assumed enormous symbolic significance for the militants, analogous to the "Boston Massacre," the annihilation at the Alamo, and the sinking of the battleship *Maine* in Havana Harbor. The 1991 siege by federal agents in an attempt to arrest Randy Weaver at Ruby Ridge, and the subsequent fatal shooting of his pregnant wife and his son, crystallized the image of the government as a ruthless agent of destruction. Cognitive errors made by federal agents led to some of these deaths. The final catalyst, the disaster at Waco, Texas, on February 28, 1993, enraged the far right. The messianic David Koresh and seventy-nine of his Branch Davidian followers, including eighteen children, perished in flames after an assault by federal agents. These events branded the fiery memories into the minds of the militants and became a rallying cry. Even though several federal agents were killed in both confrontations, the message to the militants was clear: the government was committed to destroy any resistance to its illegal activities.

The fifty-one-day standoff at the Branch Davidian compound in Waco was the second of the three events that catalyzed the growth of militia groups across America. The Branch Davidians, an offshoot of the Seventh-Day Adventists, had been operating in Texas since 1935. They had a history of preaching doom. The leader, David Koresh, had a specific message based on unusual religious beliefs. He focused on the decoding of cryptic apocalyptic passages (such as the Seven Seals of the Book of Revelations). He believed that a cataclysmic event, a cosmic struggle between good and evil, was about to occur. The forces of evil, concentrated in the government of the United States, would be engaged in the Battle of Armageddon. Exceptionally poor mind-reading by the agents, who mistook a passionate religious fervor for a simple hostage situation, confirmed the paranoid belief of the cult and contributed to the ultimate catastrophe.

The third event that elevated the antigovernment pitch was the passage of the Brady gun control bill. This legislation was widely interpreted by rightist groups as an attempt to suppress their right to bear arms to defend themselves against a monstrous government. It also demonstrated to them once again that their government was attempting to interfere with their most basic rights.

McVeigh and Nichols were influenced by the passionate antigovernment feelings. The fact that the Oklahoma City bombing was scheduled to take place on the second anniversary of the Waco disaster indicates the powerful symbolic meaning of what appeared to be the actions of a government that was out of control. It was clear from statements by McVeigh that he considered a "dramatic" counterattack necessary to punish the government, even at the cost of innocent lives.

TERRORISM FROM THE LEFT

The development of extreme left terrorism in the United States emerged from the increasing radicalization of student groups during the Vietnam

COLLECTIVE ILLUSIONS   161

War. The Black Panther Party was formed in California in 1966 by Huey Newton and Bobby Seal. The political philosophy of this group was inspired by a variety of radical heroes, such as Che Guevara, Malcolm X, Ho Chi Minh, and Mao Tse-tung. This group initially emphasized cultural nationalism but was converted to a terrorist philosophy after one of its members was killed in 1971 while trying to escape from jail. The Black Panthers were involved in several shootouts with the police and launched several bomb attacks. Concurrently, the Symbionese Liberation Army, formed in Berkeley, committed a number of bank holdups and several murders. This group achieved large-scale notoriety following the kidnapping of the heiress Patty Hearst and her subsequent conversion to an "urban guerrilla."

The extreme wing of the Students for a Democratic Society, the Weathermen, went underground at the end of 1969. The Weathermen formed an image of corporate capitalism as "an incredibly brutal and dehumanizing system whether at home or abroad." Their ideology incorporated the notion that modern society had created a new proletariat of middle-class and professional workers who presumably were oppressed by social conditions. These social conditions also deprived minorities.[41]

In contrast to the reactionary philosophy of radical rightists, the New Left had a revolutionary utopian view of the future based on commitment to the liberation of the oppressed segments of society and also to the exploited states of the Third World. As radicalization increased, the individuals merged their own identity into that of the group until, at the terrorist stage, the group identity reached its peak.[42]

THE PARANOID PERSPECTIVE

In a comprehensive treatise on extremist groups in the United States, Hofstadter applied the term "paranoid style" to characterize their thinking and behavior.[43] The term "paranoid perspective," however, seems to capture their worldview more completely. The development of a paranoid perspective would seem to be almost inevitable in a group that has a collective self-image of vulnerability to a controlling, intrusive government. The paranoid perspective leads to interpretations and expectations of malicious behavior that go far beyond the objective evidence. This perspective imposes hidden malevolent meanings and motives on relatively innocuous events. Because of their paranoid perspective, for example, the militiamen suspected their enemy of using secret methods to further their malevolent aims. In this case, the government was believed to have formed a conspiracy to subordinate the United States to a world government.

The militiamen have circulated stories that markings on federal highway signs are actually secret codes inserted by the government to guide UN armor when it moves in to take over the United States. Photographs of the Russian tanks in Michigan have been interpreted as signs of a Russian military presence in the country. And the sight of black helicopters hovering overhead has led the militants to conclude that the government was monitoring their movements.[44]

Deliberate misrepresentations and falsehoods on talk shows, Internet websites, and videotapes have facilitated the formulation of the paranoid perspective. One widely circulating videotape presents a highly distorted view of the catastrophic events at Waco, Texas. The tape was deliberately edited to make it appear that federal agents started the fire in the compound. The unedited version clearly demonstrates that the fatal fire was started within the compound.

Media clips, Internet postings, and militia literature convey the persecutorial beliefs and fears of the extremist groups. Among the various "exposés" circulated by these media are reports that the government plans to put dissidents into forty-three concentration camps; that Hong Kong policemen and Gurkha troops are training in the Montana wilderness for the purpose of taking guns away from Americans; that the government plans to give the North Cascades Range in Washington State to the United Nations and the CIA; and that an international group that wants to take over the world is changing the weather. These stories presumably expose a global conspiracy to create a New World order.[45] The opposition of the government to their ideals has made the militiamen feel vulnerable and, consequently, enraged at the threat to their values. Since the militants cannot match the military or police power of the government, they have turned to active resistance and acts of sabotage, which they hope will rally other sympathizers to their side.

Although there are decided differences between people who are members of an extremist group and those who are psychologically disturbed, it is illuminating to examine the similarities in their beliefs and thinking. The comparison between militant groupthink and paranoid delusions is useful for the light it shines on the nature of the human mind and its tendency to create fantastic explanations for distressing circumstances.

In common with the paranoid delusion, the paranoid perspective centers on the Enemy and its "plot." The escalating conflict with the persecutor exacerbates the paranoid perspective. Just as the aggressive, paranoid patient will lash out at his supposed persecutors, the militiamen who perceive themselves as oppressed by tyrannical government agencies will retaliate against supposed enemies, to wit, the 1995 bombing of the Oklahoma City Federal Office Building. Both patients and extremist group members have a huge psychological investment in their grandiose, as well as their persecutory, beliefs: "We can overthrow the tyrannical government," or, "We can save the world." Their presumed Enemy draws on covert or hidden powers to threaten their security and goals. The opponent is envisioned as carrying out its baleful activities deviously; it has no moral code or standard to restrain it. The members of the group see themselves not only as right but as endowed with a messianic purpose: to restore the purity of their nation and to save kindred persons from the hegemony of the Enemy.

Both the delusional and paranoid perspective have the earmarks of the "closed mind." Their beliefs are impermeable to evidence that contradicts their mythology (group) or delusion (patient). In fact, since they view their Enemy as using all available tools of deception, any disconfirming evidence is interpreted as proof of the Enemy's deceptions. Consequently, the group employs counterstrategies of stealth and subversion to counteract the invisible as well as the overt manipulations of the Enemy.

For both the members of the extremist groups and the paranoid patient, the outward display of hate and hostility obscures an underlying problem, their sense of vulnerability. Since their own ideology involves a malevolent view of government and the goal to reform or overthrow it, they are prone to resist governmental intrusions. As government agencies make them conform to the expectations of the citizenry, they feel more and more vulnerable and driven to "counterattack."

Lest one fall into the temptation to label militant group members mentally ill, it is important to underscore the ways in which the militiamen differ from the delusional patient. The militants, first of all, confine their conspiratorial beliefs to a relatively circumscribed domain: their relation with the government and their group. They have normal relations with members of their family and friends, carry on normal business transactions, and appear rational when testifying in court. The paranoid patient, in contrast, generally shows disturbed thinking in his relations with other people and may be in a continuous state of agitation. In contrast to groupthink, he does not receive consensual validation of his beliefs from other people in a group, and if his beliefs "normalize" under medication, that is evidence of a mental disorder. In contrast, members of the militant group can modify their beliefs when circumstances change or their leaders change their theories.

CULTURAL RULES:
THE SOUTHERN CODE OF HONOR

Many people have a particular set of sensitivities or vulnerabilities based in part on their personal experience or on the notions they absorbed from other people. The importance of receiving respect is evidenced in the specific rules incorporated into the codes of prescribed behavior of various cultures and subcultures. Attaining objectivity toward one's own perspective and consequent harmful behavior is far more complicated when the rules for defining a transgression and reacting to it are imbedded in one's own culture. Generally accepted rules in cultures and subcultures dictate appropriate rules of conduct, the specification of what constitutes a transgression, and the appropriate remedy. Examples of such culture-dictated rules are the codes of honor prevalent in the southern United States and the Mediterranean countries and the code of the streets in urban United States.

In *Culture of Honor: The Psychology of Violence in the South,* Nisbett and Cohen describe the overriding importance to southern men of maintaining a public image of strength and power.[46] A man's assumed reputation, how he believes he is viewed by others, depends on being sensitive to affronts and retaliating forcefully when such a provocation occurs. The values of the southern subculture become embedded in the dualistic belief system of the individual. In his interpersonal encounters, he appears either vulnerable or invulnerable, weak or omnipotent. The beliefs of the individual man can be summarized as follows:

- Any negative action or statement directed at me diminishes me not only in my own eyes but also in the eyes of my fellows.

- If I do not retaliate, my status (honor) will deteriorate.

- If I don't retaliate, I will lose the respect of my fellows and will be vulnerable to attacks from others as well.

- The retaliation must be violent even though the affront may be purely verbal (slight or insult).

- Successful retaliation will promote my image as a person with honor, somebody who is entitled to respect.

- A real man would fight someone who insulted his wife or girlfriend and would be justified in shooting somebody who has taken her away from him.

The authors point out that this value system accounts for the relatively high homicide rate among southern whites as compared to their demographic counterparts in the North.[47] The system of values had its origin in the distant past when it might have had some functional value. The Scotch-Irish colonists in the South, descended from Gaelic herdsmen, carried their values to their new country, irrespective of whether they engaged in herding themselves. Herdsmen are universally associated with a readiness to retaliate violently for affronts to their reputation. The source of this propensity was originally economic: herdsmen have historically been vulnerable to the theft of their animals, which would be an economic disaster. Consequently, it was important for them to establish a social image of toughness and a reputation for violent retaliation against rustlers. Moreover, they had a low threshold for provocation, evidently based on their sense of vulnerability to possible offense.

Since this economic rationale is no longer present, the keepers of the code can justify its value on the basis that non-adherence would mean that they were not only regarded as weak but were indeed weak. Whether they would be bullied if they did not retaliate, they believe they would be. The belief becomes self-sustaining. Children, particularly little boys, are coached by their mothers as well as their fathers to fight to defend their rights. The southern whites are no more violent in other situations than their counterparts elsewhere in the country. They tend to be religious and generally law-abiding. The laws have supported the use of force in protection of home, family, and property.

The power of the belief in retaliatory violence is evident not only in the high homicide rate among southern whites but also in their strong physiological and behavioral reactions when insulted. In laboratory studies they experience more stress, as indicated by increase in their cortical levels, and are more charged up for aggression, as indicated by increases in their testosterone level. They are also more likely to consider violent solutions to situations involving an affront and generally display more anger in these situations.

Despite the violence of their reactions to arguments and slurs, there are many features of the southerners' belief system that could provide opportunities for amelioration of their behavior. Their retaliatory beliefs apply only to certain circumstances, namely, personal insults and threats to their property or the integrity of their marriage. Nisbett and Cohen emphasize the problem in trying to change deeply rooted violence. They recognize the romance and allure of the Masai warrior, the Druze tribesman, the Sioux Indian. Nonetheless, they propose that intervention programs that prompt individuals to examine their beliefs about loss of reputation if they don't retaliate or that teach ways of obtaining community respect without engaging in violence may have a limited chance for success.

Some improvement in the culture of violence could be brought about by the repeal of existing laws and thus could weaken the legal justification for violence. Educational programs directed at bellicose child-rearing attitudes might be helpful in modifying the admonitions to boys regarding insults and fighting to maintain respect. Other programs in religious or educational institutions could explore the moral justification for violent behavior. A distinction could be made between obligation to the community and obligations to the self. Men are expected to sacrifice their lives in war because "that is the right thing to do." The community expects good, honorable men to comply. But community standards also decree that they engage in retaliatory behaviors because "those are just things you have to do." Change the community expectations, and you can change the behavior.

THE NORTHERN CODE OF THE STREETS

The system of beliefs underlying the problems of violence that beset the poor, inner-city, black community is similar in many ways to the system contributing to the culture of honor among southern whites. Violence in the form of muggings, burglaries, hijackings, and drug-related shootings in the urban North has been attributed to a kind of street culture labeled by Elijah Anderson the "Code of the Streets."[48] The rules incorporated in the street culture stipulate both a proper comportment and the proper way to respond when challenged.

Like the southern code of honor, the issue of respect is at the heart of the street code, with its emphasis on being treated properly or granted the deference that one is entitled to. If an adolescent presents a strong social image and receives the right amount of respect, he can avoid "being bothered" (hit, shoved, robbed) in public. As with whites in the South, if he is "dissed" (disrespected), he is disgraced.

A well-known example of being dissed occurs when another person maintains eye contact for too long. The rationale for this designated offense appears to be that prolonged eye contact may be an indication of the other person's hostile intentions. Similar to the situation in the South, being dissed leads to a loss of status in the group—which in both subcultures is remedied only by violent retaliation. These northern urban youths are said to suffer from chronic low self-esteem, for which they compensate by demonstrations of physical prowess and fierceness and by dressing with expensive jackets, sneakers, and gold jewelry (which they generally have robbed from another vulnerable youth). To maintain his honor, or "juice," he must communicate that he is capable of violence and mayhem if the situation warrants.

The specific belief system centers on the messages they receive from street-oriented adults:

- If someone messes with you, you have to straighten them out.

- Beat somebody up to get more "juice."

- You have to take risks (for example, the risk of getting killed) to show your "nerve."

- Watch your back and don't "punk out."

The subculture provides more than just the rules of conduct: how to get by on the streets, what goes and what doesn't. It also provides a cognitive framework according to which individuals attach meaning to their own actions and those of other people.

Without knowledge of the interpretative rules, an outsider would be baffled by the hostile, sometimes violent reactions as well as the impression management strategies of the youths. The rules regarding disrespect, for example, are built into their information processing and lead to taking offense automatically over what an outsider would consider a neutral or trivial statement. The rules regarding self-esteem, similarly, lead to self-inflating explanations, such as, "They resent me because I wear an athletic suit and sneakers every day." Justification for mayhem and killing are built into the code. A murder may be excused on the grounds that the victim should have known the code: "Too bad, but it's his fault. He should have known better." The ideology also seems to focus on anti-establishment values, which can flourish when enforcement or respect for the law is weak or nonexistent. The lack of faith in the police or judicial system, which are representative of the dominant white society, is profound. Thus, the code of the streets substitutes for established laws and justice.

Unlike the culture of honor of the South, which originated in the distant past and whose initial rationale has long since disappeared, the Code of the Streets is of recent origin and is supported by current environmental conditions. As pointed out by Anderson, chronic unemployment, the drug culture, and continuous conflict with the authorities generate and maintain the ideology. Since there does not seem much likelihood for adequate alleviation of the socioeconomic conditions in the predictable future, it is necessary to look elsewhere for other remedies. Reeducation, religion, and recreational sports do not seem to have a sufficient impact to change the ideology. Even families that have a "decency orientation" and are opposed to the values of the code often encourage their children's familiarity with it—albeit reluctantly—to enable the children to negotiate the environment in the inner city.

One of the factors that elevates the importance of the streets and underpins its ideology is the deficient child rearing provided for many of these youths. Many of the delinquent youths have been reared by a single parent—usually a single mother who is under severe economic and social stress. This parent typically lashes out unpredictably at the child, who develops a worldview of omnipresent hostility and learns that violence is presumably the most effective way to influence people as well as to survive. The streets consequently represent the best way to gain respect, obtain a sense of power, and build self-esteem.

Kenneth Dodge and his group at Vanderbilt University have engaged in a promising project to address the problems of delinquent youths.[49] One aspect of the project is helping the parents to develop better child-rearing skills and thus undercut the child's hostile mindset and low self-esteem. Although it is too early as yet to determine the efficacy or generalizability of their intervention, their rationale appears sound.

PERSECUTION AND GENOCIDE

Creating Monsters and Demons

To Create an Enemy

Start with an empty canvas
Sketch in broad outline the forms of
men, women, and children.
Obscure the sweet individuality of each face.
Erase all hints of the myriad loves, hopes,
fears that play through the kaleidoscope of
every finite heart.
Twist the smile until it forms the downward
arc of cruelty.
Exaggerate each feature until man is
metamorphasized into beast, vermin, insect.
Fill in the background with malignant
figures from ancient nightmares—devils,
demons, myrmidons of evil.
When your icon of the enemy is complete
you will be able to kill without guilt,
slaughter without shame.

Sam Keen, Faces of the Enemy *(1986)*

The killing of an entire tribe or ethnic group has been noted throughout recorded history. The Book of Samuel (15:3) in the Old Testament records the command to smite the tribe of Amalek and destroy every living creature in the tribe. Genghis Khan and Tamerlane were notorious for their mass murders. The Crusades generally started with the slaughter of Jews and ended, when "successful," with massacres of the Muslims.[1] During the Thirty Years War (1618–48) much of the population of Germany was annihilated.

Various episodes of mass murder in the twentieth century have much in common with each other, but also with episodes of individual violence. Whether acting in a group or as individuals, people have built-in mental categories that identify good and bad, right and wrong. These are generally embellished by memories of wrongs inflicted on them. When they experience serious damage or a severe threat—whether real or imaginary—the mental categories are primed and transform the "noxious" entity into an image of *the Enemy*. The priming may also be conducted by other people or by national leaders. They become mobilized to correct the situation: expel, punish, or eliminate the noxious agent.

Humans overcome the natural restraint against hurting or killing other humans by drawing on permissive beliefs that provide justification for violent behavior. The people then use their available instruments—knives, guns, bombs—to consummate their goals.

Persecution and mass killings follow a similar path. The persecutors have acquired a *belief system,* founded in primitive notions of good and bad, that isolates stigmatized subgroups as alien. Input from their cultures produces memories of past misdeeds, real or imaginary, by members of the vulnerable subgroup. Although the negative image may persist for long periods of time in a latent or mildly active form, it can become fully activated by outside influences.

A variety of external situations—economic distress, war, political propaganda—can *activate* these beliefs and transform the image of the vulnerable subgroup into that of the Enemy. Political leaders impose a political, social, or racial ideology that reinforces the image of the Enemy.

A kaleidoscopic array of associations to the Enemy that have been implanted are stirred up: conspiratorial, deceptive, manipulative. When the stigmatized subgroup has become prominent in the economic and cultural life of the country, it is accused of trying to usurp traditions, political authority, or economic power.

Political leaders explain economic distress and social turbulence as due to the treachery of this isolated subgroup. As the negative image of members of the subgroup intensifies and hardens, they are seen increasingly as dangerous, malevolent, and evil. By depicting the stigmatized subgroup as traitors, revolutionaries, or counterrevolutionaries, the political leadership exploits the negative image of the subgroup to advance their own political agenda.

At some point, metaphors about the stigmatized subgroup are reified, and its members are seen as monsters, demons, or parasites. The dominant group mobilizes its forces to wall off, expel, or exterminate the noxious entity. Psychological and moral restraints against killing are removed by ideological justifications: the end justifies the means; it is necessary to amputate a desired limb to save a life; subhumans do not deserve to live.

Although there is an innate reflex horror at killing, people do become desensitized and indeed are rewarded by their own group for their murderous acts. The stigmatized subgroups end up in front of firing squads, in lethal gas chambers, or in deadly forced-labor camps. Wartime conditions, especially, facilitate such killings through the utilization of an efficient bureaucracy and the armed forces.

This sequence can be usefully applied to the ideological or political massacres in Turkey, Nazi-occupied Europe, Germany, the Soviet Union, Cambodia, Bosnia, Indochina, and Rwanda.[2] The Turks labeled Armenians as traitors and decimated them (1915–18). Stalin labeled any political opposition as counterrevolutionaries in league with imperialist powers and starved them to death or killed them outright in 1932–33. Hitler used the anti-Semitic platform to gain power and then, having no place to which he could exile the Jews in wartime, had them killed outright (1942–45).

The Indonesian government accused ethnic Chinese of being in cahoots with the Communists and slaughtered hundreds of thousands in 1966. In 1975 and 1976, Pol Pot and his Khmer Rouge labeled Cambodia's professional class and intellectual elites as exploiters of the peasants and tools of the American armed forces and forced them into the farms where most of them perished.

Transcendental genocide—war against certain factions to achieve political goals, such as clearing out the indigenous inhabitants of a land to make room for colonists[3]—is based on an ideology of hatred toward a stigmatized group within the country; this kind of mass slaughter is exemplified by the destruction of vulnerable groups in Turkey, Germany, Russia, and Cambodia.[4] In

Turkey, Germany, the Soviet Union, and Cambodia, those in power exacerbated the prejudice of a favored group (the Turanese, the *Volk*, workers, peasants) against a vulnerable subgroup (the Armenians, the Jews, *kulaks*, the bourgeoisie). They mobilized and manipulated the bias against the stigmatized group by accusing it of having exploited the favored group. By identifying the subgroup as the Enemy, they raised the collective self-esteem of the idealized group. Violent action then satisfied the thirst for revenge for the alleged abuses attributed to the despised group.

The Turks, supposedly threatened by the possible "treachery" of their Armenian minority, began a systematic campaign of extermination in 1915 on the basis of national security. Hitler, Stalin, and Pol Pot did not need detailed positive programs to consolidate their power. Each painted a dark image of the Enemy, remarkably similar in each instance: decadent, corrupt, conspiratorial, exploitative. The German Jews were accused of being in cahoots with foreign powers: the Soviet Union, France, England, the United States. In the Soviet Union the opposition was denounced as a tool of Western imperialists. The intellectuals and bourgeoisie in Cambodia were depicted as agents of the Vietnamese and the Americans.

Lavish praise was heaped on the "chosen people" in all instances: the workers in the Soviet Union, the *Volk* in Germany, the peasants in Cambodia. They were portrayed as noble, pure, and virtuous. In each of these states, aggression against the stigmatized group was more appealing than commitment to a positive political agenda. It is easier for the populace to blame and attack an alien group than to understand the intricacies of economic and political problems and the complexities of a positive political and economic program. Class war is more appealing and achievable than class harmony, and the appetite for purging the recalcitrant peasants, intellectuals, or ethnic minorities is stronger than the motivation to work toward a real solution.

The breakup of the Yugoslav government provided the impetus for Serbian nationalists and old guard Communists to expand Serbian territory. The Serbian government then created memories of ancient injustices by previous generations of Muslims as though the present Bosnians were reincarnations of their ancestors and therefore warranted punishment. The Indonesian government, beset by numerous political and economic problems, identified the Chinese as Communist revolutionaries geared up to overthrow the government and thus requiring extermination. The final example of state-sponsored mass killing was demonstrated in Rwanda, where a political elite attempted to consolidate its power by designating the Tutsis as the Enemy and stirring up the Hutus to eliminate them.

An obvious requisite for genocide is control of the police and military power of the government. It is easier to operate the machinery of genocide during the exigencies of war, when there is total mobilization of resources and a defined external enemy. The German genocide of the Jews was easier to implement during World War II, as were the killings in Cambodia by the Pol Pot regime during the Vietnam War.

CAUSALITY AND CONSPIRACY

The disposition to ascribe unfortunate happenings to alien groups has its roots in ancient notions of causality, which ascribed natural disasters such as floods, droughts, famine, and epidemics to the malevolent intervention of supernatural forces. The early superstitions were full of angry gods, duplicitous demons, and evil spirits. Prophecies of the battle between the forces of darkness and light were woven into religious beliefs. Eventually the malefactors in these legends influenced the behavior or assumed the form of human beings, conspiring to destroy what was most sacred.

Certain categories of people were readily identified as having acquired secret malevolent powers. The Jews and other heretics, for example, were regarded as representatives of Satan seeking to overthrow Christendom. Consistent with this belief was the popular assumption that diseases and disasters were produced by Satan working through his agents, the Jews.

Conspiracy theories like this are an elaboration of the belief that evil groups have designs on their innocent neighbors. Usually members of the dominant group believe that the stigmatized subgroup has been plotting to trap and control them. In modern times conspiracies attributed to Jews and Armenians were proclaimed as the justification for their persecution.

Lumping together all members of a suspect minority (*overgeneralization*) can further the notion of an internal conspiracy. Members of a subgroup who become conspicuous because of their economic or political success are suspected of conspiring with their brethren to advance their own interests at the expense of the unsuspecting majority. Their success diminishes the self-esteem of the insiders, who frequently conclude that the subgroup resorted to shady tactics and schemes to gain an unfair advantage. As the more successful subgroup is alleged to be exploiting the larger dominant group, it is supposed that its members are working in concert, according to a secret plan, to usurp economic and political power.

This disposition to suspect plots and hidden influence is an extension of the sensitivity of individuals to deception by other group members or by outsiders. The flip side of this sensitivity is the universal tendency to deceive others, expressed in such relatively benign activities as joking, play-acting or bluffing, as well as cheating, lying, and plotting.

Since members of the subordinate group have already been tagged as both bad and powerful, they can provide a convenient explanation for unfortunate political or economic happenings. Instead of seeing economic and political reverses as due to the inefficiencies of the systems, the dominant group attributes them to sabotage by the target group. As the image of the minority becomes distorted, the ruling group takes control and converts the subgroup members into hostages of the state.

Ideas of control by malevolent groups are also found in paranoid delusional patients, of course. Although one cannot link the hostile outpourings of a political group to the pathological creations of disturbed patients, their similarity suggests the human disposition to see conspiratorial patterns and intrigue when they do not exist. Such observations form the basis for the concept of the "paranoid style,"[5] or paranoid perspective, in political groups or nations.

Obviously the accusation of malicious behavior by members of a vulnerable group does not in itself lead to mass murder. People do not engage in organized killing, no matter how strong the urge, unless they feel justified in doing so at the time. Ordinarily, homicidal behavior is kept in check by the moral code, empathy for the intended victim, and the fear of punishment.

The suspension of the moral deterrent to killing can be clarified by examining the belief system of street gangs. The "moral disengagement" of juvenile offenders and terrorists, studied extensively by Bandura and others, rests in part on the offenders' ability to regard their destructive acts as justified and to view their victims as the villains.[6] They are thus able to diminish their own personal responsibility by displacing it onto the group or the leader. Finally, by dehumanizing the victims, they can extinguish any empathy they may feel. Similarly, the ideological mandates, justifications, displacement of responsibility, and dehumanization of victims during acts of terrorism and persecution suggests that a person can suspend or reconstrue his moral code, whether his actions are mandated or prohibited by the state.

THE HOLOCAUST

The Holocaust has been the most widely analyzed transcendental, or ideological, genocide. Although many aspects of this disaster are unique, the psychology of the perpetrators, the onlookers, and the leaders can clarify essential characteristics of mass murders in general. Many authors have described the Holocaust as "the ultimate evil" and have pondered what kinds of people could have engaged in its crimes against humanity.[7] The assignment of the label "evil" to explain the actions of the Nazis and their supporters does little to further the understanding of their thinking and behavior. In the minds of the perpetrators and passive participants, the Jews were the evil ones and had to be liquidated.

The personnel who moved the victims along the deadly assembly line from home to gas chamber did not consider themselves evil. Many believed that they were doing the right thing. This self-righteousness was true of the police who rounded up the Jews, the engineers who transported them on the trains, and the guards who herded them into the concentration camps. The Jews were viewed as morally degenerate, lusting for world domination and polluting the culture. The perpetrators believed that the "evil nature" of the Jews was pervasive, and that to protect themselves and their civilization it was necessary for them to annihilate all Jews—man, woman, and child. As long as a single member of the diabolical race survived there would be danger.[8]

Their image of the Enemy packed enough force to motivate them to commit murders, whether or not they experienced sadistic pleasure in doing so. Given the strength of irrational assumptions, even a delusion can follow a logical, rational progression to destructive behavior. The genocidal ideology acquired credibility in Germany because it was endorsed by scientists, academicians, and professionals. It was taught in the classrooms and propagated by the leaders of the nation.

Formation of the Diabolical Image

The development of anti-Semitism and its progression to genocide may be analyzed from the perspective of the shifting images of the Jews and the

Germans' national self-image. The historical background of the diabolical image of the Jew can be traced back to the teachings of early Christianity. The curse of deicide (Christ-killing) directed at the Jews has followed them up to the present time. During the Middle Ages, Jews were accused of poisoning wells and conducting ritual ceremonies in which they sacrificed Christian children. In the religious plays, ballads, and folktales of Europe, the Jews were cast as evildoers. The image of the vile Jew, reinforced by the teachings of Martin Luther, was woven into the fabric of German folklore. From the earliest Crusades, when the Jews were among the first to be massacred, to the Inquisition and the expulsion from Spain, France, and England, to Martin Luther's excoriation of the Jews, they were portrayed as murderers, an image that impelled their enemies to murder them preemptively.

The Enlightenment and the Napoleonic era, with their emphasis on human rights, had a paradoxical effect on the fate of the Jews. Napoleon's liberation of the Jews from their social and economic confinement, culminating in the granting of essentially equal rights by Bismarck, led to a burst of active participation by the Jews in most aspects of the secular life in Germany and Austria. Only positions in the civil service, judiciary, and officer corps of the armed forces were explicitly closed to them. Within a relatively short period, Jews became conspicuous in business, politics, and the press.

The image of the Jews did not improve, however, in proportion to the elevation in their social, political, and economic status. In many sectors of German society, the centuries-old image of the diabolical Jew remained a chronic irritant. In others, it persisted in a latent form until it was fully activated by the Nazis. The success of the Jews provoked a new version of the ancient story of their conspiracy to dominate and subvert their Christian neighbors. Fears of Jewish manipulation, greed, and materialism were embedded in the writings of philosophers like Fichte, Hegel, and Kant. Despite the changes brought about by the Enlightenment, the German *Volk* remained essentially conservative, if not reactionary, and regarded the progress of the Jews as an encroachment on and a threat to their institutions.

This reaction to the sudden prominence of the Jews and the perceived threat to fundamental values produced a spurt of anti-Semitism. The mass political movements led by the German conservatives and the Austrian clergy in the last part of the nineteenth century were fueled by fear of the social, economic, and political change that was threatening the established order. As active participants in the liberal movements, and as representatives of a rapidly evolving capitalism, Jews were identified as a threat to the established order. The new secular, political anti-Semitism was superimposed on the religious mythology of the Jew as the killer of Christ and the agent of the devil. Anti-Semitic theology continued to spread at church, school, and home, and in the passion plays, right through the Hitler era. The political ideology, however, undermined the traditional notion that the Jews had to survive so that they could continue to serve as "witnesses" to the creation of Christ's kingdom on earth.

Symbolic of the notion of the Jew as a danger to society was the credibility attached to a forged document, *The Protocols of the Learned Elders of Zion*.[9] The origin of the legend of an organized Jewish plot to take control of the world can be traced back to Napoleon's convocation in 1806 of an innocuous advisory group of prominent French Jews—predominantly scholars and rabbis—which he labeled the "great Sanhedrin," after the high court in ancient Israel. The convening of this assembly gave rise to the idea that a secret group of Jewish elders had been in existence since ancient times and would, with Napoleon's backing and an alliance with the Masons, seek to overthrow the church of Christ.

The theme of a Jewish conspiracy to take over the world resurfaced in a novel, published in Germany in 1868, in which representatives of the twelve tribes of Israel meet to discuss their strategy of domination of Europe. By 1872 the fictional episode was circulating in Saint Petersburg as a pamphlet, darkly suggesting a factual basis for the story. Later permutations of this myth were incorporated into the broader, forged document, the *Protocols*. The publication of this novel was the forerunner of a stream of propaganda in Germany. Starting in the 1880s, Germany became the leading producer of anti-Semitic tracts, and the anti-Semitic platforms of the nation's political parties accentuated the fear and hatred of the Jews. As the perception of a danger to the moral and physical health of the community grew, the idea of an enemy that had to be destroyed crystallized. Described by Goldhagen as the "eliminationist ideology" and by Weiss as the "ideology of death," the mandate for the annihilation of the Jew was already in place decades before Hitler.[10] The mythology was propagated through the very institutions that the Jews were alleged to have targeted for corruption: the educational system, politics, and the economic structure. The *Volk* feared that the twin demons of bolshevism and capitalism supposedly unleashed by the Jews would culminate in the complete demise of German civilization.

The German *Volk* looked back to the romanticized past, with its glories, mythic experiences, and legendary heroes. The titanic struggles against the enemies that had striven to subjugate or annihilate them, to extinguish the German spirit, molded their current perspective. The historic vulnerability of Germany to invasion and the residual memories of the disastrous Thirty Years War and the calamitous First World War helped to create and maintain a paranoid view. The idealized image of strength, beauty, and purity contrasted with their post–World War I self-image of a persecuted people surrounded by enemies, defeated through treachery, and humiliated by the callous peace terms at Versailles.

Despite the malicious ferment, the Jews received a modicum of social acceptance among the more tolerant Germans. Many people believed Jews were enhancing German culture, science, and medicine, a benevolent belief fostered by the political platforms of the more liberal political parties, which attracted a disproportionate number of Jews. These political parties, in turn, came under fire by their opponents for advocating equal rights for the Jews.

People are naturally more susceptible to conspiracy theories during periods of malaise. The illusion of the treacherous Jew, previously dormant, became more salient after the First World War. Primed and reinforced by the leaders and media and cycled and recycled in conversation, the belief led to the intensification of the age-old anti-Semitism. The diabolical image of the Jew as a central element in the Germans' information processing provided simple, acceptable explanations for adverse circumstances. Unfavorable events were perceived as results of a Jewish conspiracy. Unfriendly diplomatic moves by England or France were seen as the work of Jewish politicians, economic crises as the manipulations of Jewish bankers, and rumblings from the Soviet Union as inspired by Jewish Bolsheviks.

The demand for a face-saving explanation for the German humiliation was satisfied by attributing the calamities to the machinations of the Jews. They were charged with having sabotaged Germany on the home front during the war ("a stab in the back"), conspiring with the Allies, and contributing to the economic hardships following the war. Above all, they were held responsible for the rise of the Communists and for the weak Weimar Republic. The

Red Scare produced by the emergence of the Communist Party and the short-lived revolution in Bavaria accentuated the image of the dangerous Jew who attacked the fundamental institutions and values of the German people.[11]

With the advent of the Nazi era, the negative image of the Jews was further reinforced as a matter of state policy. Courses in racial hygiene and biology in German primary and secondary schools in the 1930s attempted to demonstrate scientifically the defectiveness of the Jews and other marginal groups. The textbooks portrayed Jews as diseased; Nazi posters equated the Jews with typhus, disease, and death. This preoccupation with disease harked back to the Middle Ages, when Jews were believed to be the agents of the Black Death. The metaphors of the rodent, the snake, and the bacterium were reified by the Nazis into the image of the repulsive, diseased Jew.

Robert Jay Lifton describes the "imagery of sickness and purity" and of the "deadly poison" envisioned as threatening the German culture.[12] This picture of the infected, poisonous bodies of the Jews not only denied them their humanity but also affirmed the necessity of their elimination. James Glass has proposed that this sick image produced a phobic reaction, if not a paranoia, among the Germans.[13] Such portrayals of the Jew culminated in the image of absolute evil, the solution to which was by necessity absolute: total annihilation. As stated by Himmler, "Having exterminated a germ, we do not want, in the end, to be infected by the germ and die of it."[14]

Priming the Diabolical Image

The deadly image of a stigmatized group generally arises in the context of the perpetrators' ideology and national self-image. The doctrine, legends, and memories are interwoven with past injustices and present enemies and generally specify the internal as well as the external adversaries. In the modern era, the nationalist ideology in Germany had evolved a doctrine of racial purity, a collective self-image of power and superiority, and a conspiratorial view of the power of others. In accord with Nazi propaganda, the Aryan "race" viewed "aliens" in German society as a kind of rot that had to be burned out. Nazi ideology explicitly regarded Jews, Gypsies, homosexuals, and the mentally abnormal as corrupting the Aryan race. Jews were considered a particular threat, since they presumably had forged the weapons of capitalism to squeeze the state from above and bolshevism to undermine it from below. The discrepancy between how the Germans saw themselves in the present and what they had been in the past fired their determination to regain the lost paradise. This called for liquidating the enemy within as well as conquering the enemy without; an expansive ideology was tightly bound to fears of the treachery of the other powers.

Individual Germans were exposed to anti-Jewish rhetoric in their social environment: conversations, speeches, writings. The informal indoctrination at home was reinforced and embellished at school and church. University students received a further dose of nationalist—if not Nazi—philosophy from their professors, a surprisingly large percentage of whom were members of the Nazi Party. These academics espoused the theory of social Darwinism, which reached its ultimate perverseness in the notion of racial supremacy. The rage in intellectual circles, this doctrine proclaimed the superiority of Aryan blood based on distorted notions of "survival of the fittest." Not only the Jews but also the "Mongol hordes" from the Slavic nations were proscribed from mixing their blood with that of Germans.[15] Racial supremacy justified the dream of world domination (a fantasy they also projected onto the Jews). The Nazis aroused popular support by reinstating an idealized image of a Thousand Year Reich dominating the world.

Two negative images predominated. As the Jews gradually moved into most aspects of German society, there was a growing fear that they were "taking over" business, the professions, and the arts. This is the fear that led to charges that they intended to dominate Germany, if not the world. In this image, the Jews were seen as superhuman and diabolical. The other image—of the subhuman Jews—was derived from the notion that the Jews were polluting the pure racial stock of the Germans. Presumably, through assimilation and intermarriage, the Jews would mix their inferior blood with that of the Christians. Both in cartoons and verbal descriptions, the Jews were portrayed as monsters, rats, or vermin.[16]

The myths of past crimes, present wrongs, and future catastrophes thus produced anxiety and hate. That there was no evidence to support these accusations was interpreted as proof of the Jews' talent for deception and covering up their wrongdoing. The more distressing the events attributed to the Jews, the more diabolical their image became.

Anti-Semitic beliefs were by no means uniform throughout the German population. In addition to variation in the extremity and toxicity of the beliefs, there were changes in the intensity over time. In good times, a

Prussian landowner might subscribe to the belief that Jews were a menace but would disregard them. In times of war, however, military defeat and fears of betrayal would prompt the conclusion that the Jews had been traitors.

It seems likely that a sense of humanity and morality held the destructive attitudes in check, but as hostile attitudes toward the Jews were accentuated by Nazi propaganda, the negative aspect of the ambivalence became stronger and the protective, humanistic one weaker. The negative beliefs were continually being primed by government-instigated anti-Semitic slogans, news items, and posters. When certain national misfortunes could be attributed to the Jews, the hostile attitudes deepened.

The prominence of Jews in the government of the Soviet Union promoted the equation of Jews with bolshevism. As the fears induced by Nazi propaganda helped to stereotype the Jews as enemies—even though the majority of German Jews were not Communists—the propaganda not only solidified the hateful image of the Jews but provided a remedy: eliminate them.

At first, only a minority of the German population adhered to the idea of eliminating the Jews, but as the distortions circulated by the fanatics worked their way into the belief systems of the non-Nazis, the fear of Jews set the foundation for the strategy of genocide. A sequence of events exacerbated the fears aroused by the distortions, and the hatred of the Jews spread like panic in a theater when somebody yells "Fire!"

The Mind of the Perpetrators

The images of a father playing with his children and of a camp guard coolly shooting a feeble prisoner seem incongruous and raise the question of how one person can be both a ruthless killer and a kind parent. Even more paradoxical are the pictures of a physician ministering to his patients and then deciding who should live and who should be sent to the gas chamber.

The description of the Nazi physicians by the psychiatrist and author Robert Jay Lifton casts light on the psychology of perpetrators in general. Lifton proposes in his study of five Nazi doctors that the dual roles were possible because of the psychological process of compartmentalization. The guard and the physician could assume different selves in their different roles, a phenomenon Lifton labels "doubling." The apparently contradictory behaviors shared a theme: the men were doing good and were worthwhile members of society.[17]

The Nazi doctors were committed to, and believed strongly in, the bio-medical model, which combines the certainty of science and the humanity of medicine. Their application of this model was integrated into the Nazi racial theories that considered the *Volk* a sacred organic unity, susceptible to pollution by alien blood. An Auschwitz physician, contending that his extermination of the Jews was a consummation of his Hippocratic oath, said: "Of course I am a doctor and I want to preserve life. And out of respect for human life, I would remove a gangrenous appendix from a diseased body. The Jew is the gangrenous appendix in the body of mankind." The unifying theme of the different roles was the belief that he was serving mankind. The killing mode incorporated the biomedical model: "A perception of collective illness, a vision of cure, a set of motivations to discover and apply that cure."[18] The Nazi physician switched from healing mode to killing mode depending on the circumstances, which activated the relevant "compartment" of the personality. The healing mode and its beliefs, motives, and procedures were primed when the physician stepped into his consulting office, and the killing mode when he entered the death camp.

A currently popular view of the personality of the perpetrators stems from Hannah Arendt's notion of "the banality of evil." Taking Adolf Eichmann as the prototype of a political murderer, Arendt suggests that the bureaucrats and technocrats involved in the processes of genocide were ordinary people, creatures of their time. Practically anyone assigned the same role might have followed orders in the same way.[19] Yet even if the psychological imperative to kill the Jews (that is, to stamp out evil) were activated, how could it override ordinary moral restraints, the prodding of conscience, the empathy for innocent, helpless humans?

Although this question may never be completely answered, it is clear that many factors facilitated implementation of the homicidal plan. The desire for revenge for the Jews' alleged role in the defeat in World War I, in the postwar Communist revolutions in Germany and elsewhere, and in economic privations was strong. The prevalent stereotype of the Jews working with foreigners to corrupt German society provided convenient support for Nazi propaganda. By eliminating the Jews, the Nazis could unify and express their own power. Wartime conditions, a live-or-die situation in World War II, provided a warrant for genocide.[20]

In trying to understand the mind of the perpetrators, it is important to emphasize that not all Germans or Austrians shared the same image of the Jews or the same plan for dealing with the "Jewish problem."[21] Although people find it convenient to lump together all Germans, or Americans, or British, as though the populace of each nation has a homogeneous set of beliefs, there is often a considerable variation in the public at large, even in a dictatorship. In pre-Hitler Germany, the spectrum of public opinion ranged from the extreme anti-Semitism of the nationalist right-wing parties to the philo-Semitism of the liberal left.

The statistical spread of public attitudes can be represented graphically by a figure similar to the well-known bell-shaped curve. The predominant opinion is represented by the highest density near the center and peak of the curve. The distribution of opinions slopes to the ends (or "tails"), which represent the less popular views, presumably the fanatically homicidal at one end and the more benign or benevolent at the other. Although statistics are not available for the distribution of beliefs about the Jews in World War II Germany and Austria, it seems likely that there was a considerable shift of public opinion in the more negative direction. Judging from the active participation of a significant segment of the German population in genocidal activities, it can be inferred that the anti-Jewish beliefs became more extreme and more intense. Nonetheless, it is apparent from the fact that at least fifty thousand rescuers have been identified that a significant number of Germans either did not endorse or actively opposed the genocidal policy.[22]

The Image of the Führer

The leaders of the embittered majority or embattled minority are frequently charismatic and can inspire and mold their followers with their words or even by their presence. Followers not only see their group as superior to other groups but see their leader as supreme. These idealizations boost their personal self-esteem and sense of power. The glorified image of the leader of the nation or state—especially intense when the group is under duress or attack—often leads the followers to do things that would be unimaginable to them at other times.

The framework for the creation of the national savior in Germany was formed long before Hitler appeared on the political scene. The vision of the Führer of the people had been heralded in the nineteenth century. The mythical image of the Germanic leader was embodied in the cult of the nation.

The romantic strains of folk thought revolved around the themes of valor, victory, and heroism, expressed in celebrations of early nineteenth-century victories. Social festivals of fire and light were accompanied by Germanic pagan and Christian symbolism and ritual. The fantasized leader was an expression of this mythic symbolism. The future leader, "the bearer of godly power and grace," would be tough, straightforward, and ruthless.[23]

This heroic image was formed and ready to be projected onto an individual whose beliefs matched the ideology that produced the image. Although accepted initially by only a small group of dedicated followers, Hitler gradually was identified with the image by the nation at large. His charismatic qualities and simplistic program represented the kind of strong figure that would lead the state to the greatness that it desired, crush the internal and external enemies, and expand the German empire. To promote the projection of the heroic image onto Hitler required a continuous flow of adulatory information. Goebbels, a genius at providing the appropriate propaganda, was able to pump up the fantasized image of Hitler. Once the Nazis came into power, their control of the mass media ensured that only flattering material about Hitler, consistent with the heroic image, ever reached the people.

As pointed out by Stern, Hitler's speeches involved "layer upon layer of invectives in which a whole network of accusations, recitals of injustices, threats, real and imaginary fears is personalized as existential attacks against the Führer of the German nation and thus against each single member of that nation."[24] Hitler's polemics followed a logical course in triggering and then accentuating the paranoid perspective of his followers. His speeches, which often lasted for hours, started by playing on the people's fear of the Jews, Communists, and other unfriendly countries. The litany of wrongs was designed not only to revive the pains over past humiliations but also to arouse fears of future abuse. After upsetting his audience with tales of past persecutions and the diabolical portraits of the enemy, he empowered them by providing the solution: wreak revenge on this accursed people. The switch from the image of the Germans as innocent victims to the image of the avenger infused the followers with a sense of omnipotence and feelings of exhilaration. The national pride and honor would be restored and the enemies would be eliminated. In this way, he successfully marketed the elimination ideology described by Goldhagen.[25]

Hitler reduced all complex issues to a few simplified formulas and gave his listeners very little factual information aside from the diatribes. He provided minimal rationale for his accusations, nor did he explain the paradox of Jews being simultaneously the Bolsheviks and the capitalists, joined in a conspiracy to destroy German culture. Further, he did not clarify how a tiny minority of the population could wield such enormous power. He made his impact through his rhetorical skills, gaining the people's awe and molding their thinking to conform to his. He had an uncanny talent for reading the minds of his various audiences and adapting his message to their particular views and biases. His "hypnotic" power apparently emanated from his ability to form a powerful rescue fantasy and dreams of glory in the minds of the audience. He became the personification of the Germany they yearned for.

Hitler's early international and domestic victories primed the national self-image of the "master race." These events also energized the despised image of the Jews as a corruption, politically subverting or poisoning the pure Aryans. This kind of antithesis appears to be characteristic of the way national leaders and their followers represent themselves: their glorified image is so pure that the image of the opposition is vile by comparison. Through his public appearances and portrayal in the media, Hitler came to represent ideals that the people cherished. According to Kershaw, he projected an image of authority, power, and decisiveness, qualities that demonstrated his ability to lead them successfully to their national goals.[26] He was also seen as reasonable, moderate, virtuous, and sincere, even saintly, qualities that helped to create trust. The German people viewed him as the defender of morality and racial purity with a fierce dedication to their cause.

Hitler's image was apparently more important in rallying the people than the ideology that he preached.[27] Obviously both were important. His packaging of the ideology was inspiring, but what he had to say was attuned to the people's own goals, illusions, and biases. This personal image was bolstered by his spectacular successes in foreign relations: the bloodless conquests and easy victories at the beginning of the war. He was generally regarded by Germans as a statesman par excellence and as a superb military commander. The considerable impact of his personality, rhetorical skills, and program made the people feel more powerful. In addition, he succeeded in establishing a bulwark against the dangerous Bolsheviks and in first creating and then neutralizing the specter of the dangerous Jew.

Hitler evidently had a continuing personal obsession with annihilating the Jews.[28] In his final testament prior to his suicide he called for—and predicted—the eventual demise of the Jews and reiterated his claim that they had caused the war.[29] This obsession is analogous to the obsessive beliefs of a patient with an obsessive-compulsive disorder who thinks that he has lethal germs on his hands or body and must wash continually to ensure that every microbe has been removed. If even one germ survives, it could multiply and destroy the host. By analogy, the Jews would have to be *completely* eliminated.

The intensity and extremeness of Hitler's belief about the Jews embodied in his final testament borders on the delusional.[30] In fact, there is some evidence that Hitler had become increasingly dysfunctional during the last year or two of his life. It would be a mistake, however, to attribute the genocidal mentality of the Nazis to mental illness. In fact, a study of psychological records by Eric Zillmer and his coauthors could not attribute the violent behavior of the Nuremberg defendants to gross psychopathology. Nor did they find any consistent personality profile.[31] Browning and Goldhagen similarly found no special abnormality in the men who participated actively in killing Jews.[32]

The kind of ideology expressed in Hitler's speeches and in the media propaganda was similar to that displayed by other tyrants. Stalin, for example, had declared war on capitalism—more specifically on the bourgeoisie—while extolling the superb virtues of the proletariat. He also was able to portray opposition to him or his plans as an incipient counterrevolution. Prior to World War II he ruthlessly exterminated farm owners, Ukrainian peasants, intellectuals in his own party, and military officers. Similarly, Mao Tse-tung in China and Pol Pot in Cambodia deified the peasant and persecuted intellectuals, professionals, and urban dwellers.

Progression to the Holocaust

Radicalization of beliefs and malevolent images of a stigmatized group, coupled with the desire to eliminate that group, are not sufficient to mobilize firing squads or to train people to herd others into gas chambers. Even an intense desire to kill a group of people can be deterred by the moral code against murder. The image of the ruthless enemy may be balanced by a complementary image of the helpless victim. To engage in an inhuman act requires giving oneself permission to kill. Inhibitions against killing are often lifted when a person is in a rage and sees an offender as all bad or experiences a continuous cold hatred toward an "evil" group.

When such malevolent beliefs become charged, they may preempt more humane beliefs and press the perpetrator to act. A state that is intent on exterminating a hostage group not only provides the instrument for this but also gives permission for the homicidal activity. It introduces a form of morality, loyalty, and patriotism that replaces the conventional morality embedded in the social order and religious canons.

Involvement in war tends to accentuate beliefs about the evils of the enemy. In a true life-and-death struggle, it may be life-preserving to view the enemy in an absolute and distorted way. When attention shifts to the stigmatized group, the internal enemy, the same "kill or be killed" beliefs become charged. The implementing beliefs are also energized: "When in doubt, wipe them out." In addition, the portrayal of the enemy as subhuman vermin not only prompts the perpetrator to get rid of them but also so dehumanizes the victim that the perpetrator feels little remorse or guilt for eliminating a subhuman individual.

During the successive phases of the Holocaust, Jews were first considered undeserving of the social, political, and economic rights enjoyed by pure Germans; then, since they were held responsible for all of Germany's ills, they deserved to be punished; finally, deemed threats to the human race, they had to be eliminated like any epidemic. These themes were sounded repeatedly through Hitler's writings and speeches in the 1920s and early 1930s and during the Second World War.

The Holocaust started with social, political, and economic restrictions. The socialization of state policy progressed to the extermination of the mentally and physically handicapped, the persecution of homosexuals, and the murder of political opponents. Following the start of the Second World War, Jews were rounded up for slave labor and were incarcerated in concentration camps, in preparation for the Final Solution.

The wartime dichotomies of friend and foe, loyalist and traitor accelerated the process of lifting inhibitions against killing civilians. Studies of documents relating to the behavior of people involved in the killings indicate that they were carried out not only by Nazis and their sympathizers but also by "ordinary Germans" and citizens in occupied Poland and the Soviet Union.

During the invasion of the Soviet Union in 1941, Hitler used death squads and police battalions, often composed of recruits from local ethnic groups, to shoot Communists and Jews. Eventually Jews in Germany and in the conquered lands were shipped to the death camps, which were equipped to administer lethal gas. In his studies of the German police battalions, Browning demonstrated that much of the torture and systematic slaughter was carried out by individuals with no special allegiance to the Nazi party, as did Goldhagen in his broader survey of perpetrators in Holocaust activities.[33]

The anti-Jewish measures intensified with the spread of the diabolical image of the Jew, permissive attitudes toward killing, a concomitant relaxation of the moral code, and a loss of empathy for the victims. The acceptance by the public of the increase in severity of anti-Jewish measures followed an established principle of social psychology. When people overcome internal resistance to a harmful policy (for example, disenfranchisement), they frequently experience a shift in their attitude toward hurting the stigmatized group. Conventional rules against hurting or killing other people are replaced by the belief that "under certain circumstances, it is permissible to harm or kill others." Once this attitude is assimilated, people are likely to accept progressively more destructive activity and so continue to the ultimate: genocide. As their image of the potential victim becomes dehumanized, they are more likely to endorse inhuman policies.

OTHER GENOCIDES:
CAMBODIA, TURKEY, AND THE SOVIET UNION

The forced death of perhaps three million people in Cambodia between 1975 and 1979 provides another example of ideological genocide. The Cambodian revolution was planned by a group of Communist intellectuals who had studied in Paris, where they learned revolutionary strategies. The Khmer Rouge, the Communist group organized in Cambodia and headed by Pol Pot, plotted to seize the reins of power during the Vietnam War. This group was able to capitalize on a series of events that stirred up people's hatred for the United States and for their own government, widely viewed as the puppet of the United States.

The influx of American aid had produced a service industry of restaurateurs, waitresses, maids, taxi-cycle drivers, and civil servants. Further, the Cambodian army had become increasingly corrupt with the influx of American dollars. To add to the negative image of the United States, a coup engineered by the army and the CIA replaced Prince Sihanouk as president. The new government was subsequently undermined by the American and South Vietnam invasion intended to interdict supply routes through Cambodia used by North Vietnam. The secret American bombing of Cambodia in 1973 in another fruitless attempt to destroy North Vietnamese bases added to the growing antagonism of the populace toward the United States and its "stooges" in the government. Relentless bombing was interpreted as an unwarranted destruction by an imperialistic, racist, capitalist superpower against an innocent, helpless population. These actions played into the hands of Pol Pot, who then gathered enough popular support to overthrow the government.

The revolutionary strategy of the Pol Pot regime followed the outline created by the regimes in the Soviet Union and Nazi Germany. The Enemy was identified: the army of the corrupt regime and the intellectual, commercial, and professional class in the cities. The image of these individuals as stooges of the United States and parasites was paralleled by the image of the Khmer Rouge as pure, sincere, and cooperative. Class warfare was stirred up between the predators, the city dwellers, and their prey, the farmers. In order to pacify the farmers, the Pol Pot government exiled the city dwellers to work on the farms. The Enemy also included ethnic groups, such as the Chan Muslims, Buddhists, Chinese, and Vietnamese.

The ideology of the Khmer Rouge called for a complete transformation of society, wiping out any Western influences and converting the country to the purest form of socialism. Individuals were expected to surrender their own freedom of will to that of the collective. The revolutionaries stood for a complete annihilation of modern values, practices, and customs. They sought a revival of ancient glories and purification of the present by emptying out the cities, forcing the corrupt parents to obey their uncorrupted children, and turning their education over to the peasants and workers, who had not been corrupted. Everything inconsistent with the goal was to be wiped out: individualism, personal possessions, families.

As in other revolutions, conspiracy theories were rampant. After the revolutionaries had wiped out the hated bourgeoisie, they turned on members of their own group, who were accused of being secret agents of Vietnam. Possibly as many as one million transplanted city dwellers either died from starvation while working as forced labor or were killed outright. The "purification" continued until the 1979 invasion by the Vietnamese, who drove out the Khmer Rouge and restored order and some semblance of sanity.

The origins of the Turkish genocide of 1915–18 can be traced back to the aftermath of the Russian-Turkish War (1877–78), when Armenia was divided between Russia and Turkey. Following the Russian victory, the Armenians had appealed to the commander of the Russian army for protection under the terms of the peace treaty. Although it seems reasonable to expect an unprotected minority ethnic group to seek help from the victor in a war, this action greatly antagonized the Turks. The image of the Armenians was transformed from simply unwelcome subjects to traitors to Turkey. This malevolent image was subsequently aggravated by the pressures by Armenians to be treated in the same way as the other nationalities in the Ottoman Empire. A series of massacres followed in 1894–95 during which 100,000–200,000 Armenians were killed.

The subsequent decline of the Ottoman Empire, coupled with successive military defeats, culminated in a 1908 coup by the "Young Turks," an ultra-nationalist group of young officers. The subsequent murders were the result of their policy. The Pan-Turamian ideology behind the massacres was the "Sacred Cause." The essence was to purify the nation by eliminating the traitors in its midst. The Young Turks pursued an extremely virulent form of nationalism that often arises after a defeat and disintegration of an empire. (The Austrians manifested the same kind of intolerant nationalism and anti-Semitism after the fall of the Austro-Hungarian Empire.)

The Young Turks proclaimed the mystic unity of the Turkish people. They had visions of a united Turkish state that would extend through Eastern Europe, including Turks in the area of Russia and into Central Asia. They imposed a stringent definition of who belonged in Turkey. They tightened the definition of acceptable individuals to those who spoke Turkish. Any outsiders within the borders of the country were labeled as alien and were made suspect. The Young Turks then repudiated minority rights and the whole doctrine of multinationalism and pluralism.

A precipitating factor in the progression of the anti-Armenian action was the reluctance of Armenians in Turkey to follow the requests of the Turkish government to incite Armenians living in Russia (the enemy of Turkey during World War I) to support the Turkish army. In fact, many Armenians did join Russian troops and volunteer in fighting Turkey. The dénouement came with the savage defeat of the Turkish army by the Russians in the winter of 1914. The subsequent genocide in Turkey consisted of the execution of able-bodied males in Armenian families followed by the forced march of women and children to detention camps, which in itself was deadly. It has been estimated that as many as one million Armenians died.[34]

The forced collectivization of agriculture in the Soviet Union resulted in the deaths of countless numbers of independent farmers (*kulaks*). This group was demonized for purportedly opposing the virtuous working class and the state as a whole. The politicide in the Soviet Union was also ideologically driven; the victims were debased as alleged counterrevolutionaries, enemies of the people, agents of foreign powers. However, the perpetrators from the top down recognized that they used "bad means" to accomplish "worthwhile ends." Paul Hollander quotes George Lukacs:

The highest duty for communist ethics is to accept the necessity of acting immorally. This is the greatest sacrifice that the revolution demands of us. The conviction of the true communist is that evil forms itself into bliss through the dialectic of historical evolution.[35]

The Communist elite accepted the notion that they were doing evil, but they were committed to the principle that the ends justifies the means. This redefinition of morality was warranted by the belief that there was a sharp separation between the unholy present and the glorious future, presumably uncontaminated by the nasty practices of the present. What drove the perpetrators to engage in a sequence of homicidal or genocidal actions against their own people? The engine of destruction was fueled by the ever-changing needs of the revolution as formulated by the Party. The Party was the beacon that illuminated in succession the various groups that were to be purged— the capitalists, the peasantry, the military officers of the Red Army, suspect Communist officials, and ordinary members of the Party who had fallen out of favor.

Various nationalities in the Soviet Union were also subjected to terror. For several generations, the revolutionaries and then the Soviet specialists in violence displayed their dedication to the Party by engaging in horrors that they recognized as immoral. The moral responsibility was shifted totally to the Party. Political violence was justified as a defensive strategy to protect the system from its internal enemies.

The concept of the glorious Party was an abstraction just as the Enemy was an abstraction. In actual practice, the Party was synonymous with the Party leadership, made up of ever-changing members who moved in and out of favor. The Party leaders, especially Stalin, were fallible people subject to all the human frailties, including a paranoid perspective. By dumping the repudiated individuals into the Enemy category, the perpetrators could deny the humanity of the victims, who generally were innocent of the crimes for which they were condemned.

Many of the specialists in terror and violence were evidently driven by the fascination of power. Paul Hollander sees continuity between the sense of power derived from the physical force applied to large groups of people and the power of the torturer. Many of those engaged in inflicting violence "found it a congenial rather than a painful exercise of duty."[36]

PROPAGANDA AND THE IMAGE OF THE ENEMY

Propaganda used by the politically elite in totalitarian regimes is designed to play on the people's vital concerns and to stir up grandiose dreams. When Hitler asserted that the Jews stabbed Germany in the back during the First World War, the people felt wounded and burned with fury. When he promised them the Thousand Year Reich, they exploded with excitement. The choice of words and phrases, the images he created, whipped up either their primal fears or their desire for grandiosity.

When people are aroused in this way, they revert from their usual more open-minded, relativistic, pragmatic line of reasoning to closed-minded, extreme categorical thinking. Their beliefs are compressed into absolute categories, such as, "The Jews (or *kulaks,* or capitalists, or intellectuals) are our enemies." Such primal thinking automatically pigeonholes people: friendly or unfriendly, good or bad, pure or evil. When Stalin raised the specter of the oppositional *kulak,* or Pol Pot defined urban residents as parasites, the offenders automatically fell into the evil category, while the Party, of course, continued to occupy the righteous category.

Stalin and other Communist leaders drew on a Manichean lexicon to designate their supporters and opponents. People (or states) were either cooperative or obstructionistic, peace-loving or vengeful, progressive or reactionary. The party line contrived to show the falsity of the "pretensions" of the democratic capitalist states by making a distinction between true democracy (communism) and formal (that is, in appearance only) democracy, genuine humanism and false humanism.[37] Totalitarian speech was caricatured in Orwell's description of newspeak in his novel *Nineteen Eighty-four*.[38] Particularly characteristic of this kind of thought control is the denial of fact and logic and the abolition of independent thinking. Hannah Arendt argues that the intrusion of factual reality can attenuate the impact of propaganda, undermining the myths and falsehoods.[39] Realizing this, the Communists jammed the broadcasts from the "free world," banned books, and muzzled dissidents.

One of the most effective antidotes to the falsehoods and deceptions in the Communist satellites was said to be smuggled copies of *Nineteen Eighty-four*, which caricatures the regimentation of thinking and the freezing of logic and rational thought in a totalitarian country. People in the satellite countries told me that the book produced a shift in their perspective of their own governments during the cold war. They began to examine their own assumptions and beliefs, consider alternative explanations for the actions of the capitalist states, and view what they read and heard with skepticism, if not disbelief.

When successful, propaganda unites the populace behind the leader and focuses their energies on defeating the enemy. The usurpation and exercise of power has a particular fascination for the leader and his elite cadre. From the beginning of the movement until its culmination in taking over the reins of government, each increment in the size and influence of the movement brings pleasure and encouragement to keep them progressing. The success reverberates across the party. Because of the nature of group dynamics, the enthusiasm and increased self-esteem circulate continuously through the group.

The belief of the leader and his mass followers that they have an ideology superior to that of other groups feeds the sense of power and solidarity. The framing and degradation of stigmatized groups further enhances the collective image and power. In contrast to the traditional power grabs by religious movements that depended on faith and divine inspiration to deliver their message to their followers, the more modern political movements promoted their theories on the basis of science: the science of Marxism or of racism, endorsed by leading intellectuals and scientists in their countries.[40]

The power of the movement increases as the nation gains increasing success in foreign relations. Hitler's rapid series of successes bred more success, eventually leading him to risk war to achieve his goal of total domination of Europe. His exercise of absolute authority over the German people and their obviously enthusiastic submission to regimentation further strengthened the sense of power of Hitler and his governing elite.

Some writers believe that people have specific innate mental modules that facilitate interpersonal adjustment, including the detection of deception.[41] We learn at an early age to read another person's facial expressions, tone of voice, and behavior to decide whether he is kidding, teasing, or manipulating us. This sensitivity to deception is widespread through the animal kingdom within and between groups and species. Because the concealment of hostile intentions by others is potentially harmful or life-threatening, people develop "counterintelligence" strategies, such as suspiciousness and hypervigilance, to cope with such problems. When alerted to such a potential danger, an individual looks for disguised patterns and hidden meanings in others' behavior.

Like other survival strategies, sensitivity to deception can become exaggerated. It is better to misclassify benign behavior as deception than to miss actual deception. It is possible to correct a mistaken inference of concealed hostility, but there may be no second chance if one is oblivious to an actual plot.

Recognizing insidious manipulations by others is related to the survival strategy of distinguishing friend from foe, amity from enmity. Similar strategies are employed by national leaders to determine the intentions of leaders of other nations: whether their signs of friendship are genuine, whether they bargain in good faith, whether they are honest in their disclosures. Secret coalitions of other states pose a particular danger: governments must be alert to the possibility of foreign conspiracies because secret alliances and treachery may lead to war.

When threatened by ethnic neighbors, states are inclined to exercise strict vigilance toward internal enemies, "the fifth column." In times of national danger the vigilance is transformed into an obsession, and "alien" groups within the border are subjected to stringent regulation. During World War II, Japanese in the United States and American citizens of Japanese descent were subjected to severe restriction although there was no evidence of their collaboration with the enemy. Their possessions were stripped from them, and they were sent to concentration camps.

Although different pathways have been taken to mass murders in general, and to genocide in particular, there are successive stages discernible as the attitudes of the ruling elite progress from prejudice to group annihilation. In the earliest stage, a vulnerable minority is stigmatized as alien to the body politic of the nation and acquires the image of being dirty, unprincipled, and malodorous. The ghetto Jews bore this stigma through much of their history in Europe. Governmental policy was directed toward containing them as much as possible. In the next stage, associated with partial emancipation, the stigmatized group moves more into the mainstream of the political, cultural, and economic life of the state. This increased prominence and success evokes a more malevolent image: polluters of the culture, usurpers of political power, and overlords of the economic life of the country.

The ruling class paints a picture of the stigmatized group as exploitative, conspiratorial, and treacherous. The favored group (the *volk*, the workers, the peasants) are idealized and endowed with virtue, purity, and righteousness. In times of distress, the majority perceives the stigmatized subgroup as having caused the problem. The state removes its protection from the vulnerable group and engages in active persecution.

In times of war, the image of the vulnerable minority is transformed into the Enemy of the State. The image becomes increasingly diabolical as the threat to the survival of the nation from external military attacks becomes more real. Although much of the literature on genocide has focused on the motives, character, and behavior of the perpetrators, the driving force behind their actions is their negative image of the victim.

The progression of the pernicious image can be seen in terms of the perceived threats. Initially, the threat of aliens arouses contempt and disgust. Then the fear of being controlled, dominated, and having cherished values undermined becomes dominant. Finally, the threat to survival and the fear of internal treachery permeates the thinking.[42]

With the progression through the stages, the image becomes not only more malevolent but more accentuated. The image of the malicious Jew became an obsession for Hitler, for example. At each stage, the image and the associated negative beliefs lead to hostile action by the government. In the first stage, the minority is walled off from the rest. In the second stage, after emancipation, measures are introduced to set them back. In the final stage, ideology not only promotes but mandates their destruction.

IMAGES AND MISPERCEPTIONS IN WAR

The Deadly Construction of the ENEMY

Is war inevitable? This crucial question for governments and their citizens was posed by Hinde and Watson in a 1995 volume.[1] The popular stereotype and the historic persistence of war seem to suggest the answer is yes. Who can resist the glamour of military parades, smartly uniformed soldiers marching with their regimental colors waving in the breeze, bugles blaring, and drums beating? Who is not thrilled by the appearance of the country's flag and the cheering crowds?

War encourages people to subordinate their personal interests to the greater good, so that they are willing to make extreme sacrifices for a higher cause, cooperating with others in a selfless enterprise. Their esprit de corps may surpass that of any previous collective enterprise. Proud citizens willingly accept the call to arms and prepare to submit to their commander's orders. Soldiers pride themselves on their unit's citations for courage under fire. War engages all of the energy, skill, and motivation of the community.

As with any institution, all the participants fit into their roles in the various sectors: factories, transportation, and the war zone. The civilian population works overtime to produce equipment and supplies, often showing prodigious industry. And the ecstasy when victory is achieved is overwhelming. Victorious commanders, war heroes, and wounded veterans receive their decorations and may become prime candidates for high political office when the war is over.[2]

When people crowd into the streets, excited by the anticipation of war, they delight at the prospect of vanquishing the enemy. The expectation of a glorious victory is a powerful stimulant and arouses feelings of euphoria similar to those experienced by the supporters of a sports team that anticipates a world championship. The massive surge of popular support for an impending war may indeed push the military and political leadership to take the last fateful steps beyond the brink.[3] Children revel in war games in which they mow down armies of plastic soldiers. War movies excite passionate identification with the nation's fighting men as they cut through enemy lines, blow up bridges, and shoot down enemy airplanes.[4]

War has been recognized in almost every society except in sparsely settled groups in which warmaking would be impractical.[5] According to a popular conception, humans have engaged in war since prehistoric times. However, as an organized form of combat, war appeared in comparatively recent times, perhaps thirteen thousand years ago.[6] Wars for the purpose of conquest have been traced back between six thousand and seven thousand years ago, when the development of farming and the storage of crops became invitations to marauding bands.[7] Recent wars, however, have not been provoked by economic factors for the most part. A study by Lewis F. Richardson suggests that the majority of the wars from 1850 to 1950 were related either to religious ideology or to national pride rather than to economic or security issues.[8]

War often seems an effective way to settle disputes decisively, to determine boundaries and access to raw materials, to deter neighboring tribes or nations from aggression, to regain lost territories, or to reestablish the national honor. Wars may be pursued for such lofty ideals as abolishing slavery, overthrowing tyrants, or "making the world safe for democracy." They frequently serve the special political interests of a country as well as the personal interests of the political elite.

Above all, war seems to be the most decisive means to an end. Psychological factors, such as exacting revenge for past wrongs, enhancing the national esteem, or consolidating the power of the political elite, frequently influence the decision to go to war. Of course, the state subsequently under attack is forced to defend itself in order to survive. Sometimes a state fearing for its own security initiates a preemptive strike against a supposed enemy. Some authors believe that this motivated Germany to launch its two-front attack in World War I and Japan to bomb Pearl Harbor in World War II. In both instances, of course, the surprise attacks ultimately failed to win the war.

Although initiating wars would seem, in the words of Clausewitz, to be "merely the continuation of politics by other means," the calculations of advantages and disadvantages, profit and losses, are frequently warped.[9] Whatever the assumed benefits, the cost in lives and suffering has become so exorbitant that most wars are catastrophic even for the victors. In the twentieth century more than one hundred million lives have been sacrificed. Curiously, during this time the countries that initiated the wars have generally lost them.[10]

Some authors have proposed that the desire to make war is natural and springs from deep flaws in the human psyche, presumably passed down in our genes from ancient ancestors.[11] A popular theory in the 1960s and 1970s regarded war as an expression of Man the Predator.[12] The authors believed that in fighting wars humans are simply carrying out a genetically determined program that facilitated hunting in prehistoric environments. A more recent concept, Man the Hunted, focuses on the vulnerability of our primordial ancestors to larger predatory animals.[13] These authors cite children's fears of animals and monsters as well as the animals' appearance in dreams and myth as evidence of this primordial fear. Presumably, our ancestors developed survival strategies as a compensation for their vulnerability to lions, leopards, and other land animals, as well as strategies for hunting animals of prey.

A DIFFERENT VIEW OF WAR

Given that the desire to fight and even kill under certain conditions is widespread, does it follow that there is a specific motivation for war? As many authors have pointed out, rather than hostile aggression causing wars, war causes hostile aggression—killing, torturing, destroying homes, factories, and farms.[14] Once the leaders announce that a state of war is imminent, the population's lust for a fight is aroused. Intoxication with the idea of war spreads rapidly through the population, even though the political leaders themselves may be objective and calculating or even paralyzed with fear about the consequences.[15] At the start of the nineteenth- and twentieth-century European wars, huge crowds rioted in the streets, crying, "On to Berlin" (or Paris, or wherever the capital of the opponent was to be found).[16]

It is not necessary to consider killing and warmaking as dependent on a kind of inherited pattern. A modern view of the instigation of warfare rules out the notion of an inherited instinct for war. Natural selection provided the hardware for hostile aggression—the body and brain—and cultural selection the patterns of violence between groups.[17] There is a very practical benefit in killing or joining other groups in an anarchic world. In a lawless society, such as the Wild West of the nineteenth-century United States, posses would be formed to capture bank robbers. Cowboys would settle disputes with a shoot-out, and ranchers would hunt down and kill rustlers. These homicidal practices were a means to an end: acquisition, revenge, or punishment. The introduction of law and order put a curb on this behavior.

Until the establishment of the United Nations, there was no substantial supranational leverage against armed conflicts between nations. Conflicts were resolved by killing off a sufficient number of the enemy and depleting their resources to deter them from further military action. If only one tribe or nation takes the lead in the arms race, its neighbors must attempt to catch up in order to protect themselves. The "security dilemma" predisposes the leaders of a state to overinterpret the behavior of the rearming neighbor as malevolent and to prepare to protect their own. To do otherwise would expose them to being caught off-guard. If states respond to threats or actual aggression with nonaggression, they will not exist very long. On the other hand, overreactivity to a neighbor's behavior may precipitate an armed conflict.

The historic relationship between neighboring states influences the likelihood that one of them will initiate a preemptive strike (as Germany did in World War II). States with a history of waging war against their neighbors, however, may develop a peaceful policy. The Scandinavian countries exemplify this change in national attitudes. They abandoned war as an instrument of foreign policy well over a century ago. Given these cultural and social changes and the occasional success of mediation in preventing war, it seems unlikely that war is inevitable.

THE COGNITIVE DIMENSION

The multiple levels of causation of war include systemic factors, abstract concepts such as the "anarchic system" (the lack of regulation of relations between states), and concrete events such as the assassination of a royal family member. These causal levels have been explored by historians, political scientists, economists, and anthropologists. Exclusive analysis at the systemic level suggests that the dynamic interplay of factors such as industrial progress, nationalism, and economic competition transcends the specific motivations of the major players. This kind of analysis generally culminates in the proposition that the initiation of war is based on rational decision making.

A more comprehensive analysis deals with the interaction of the various levels.[18] The psychological level of analysis focuses on the thinking, feeling, and motivation of the individual leaders and their followers. The external factors—the arms race, previous conflicts, coalitions—have a direct impact on the psychological functions of the participants. Moreover, the causal factors can go in both directions. It may not be possible to pinpoint which comes first—the negative representation of the foe or the international conflicts. It is clear that these factors influence each other. There is evidence, for example, that not only were the political and military leaders making decisions that drew their countries closer to World War I, but they were also being pushed by "jingoistic crowd scenes" in Berlin, Vienna, Saint Petersburg, Paris, and London.[19]

It is also useful to distinguish between the predisposition to war and its precipitations. Predispositional factors such as the arms race, threatening actions by unfriendly states, the threatened dissolution of the Austro-Hungarian Empire, and the emergent power of Germany helped to shape the attitudes and beliefs of the various powers involved in the First World War. Attempts to restore the balance of power through coalitions served to destabilize relations further. With war preparations moving at a fervent pace, the scene was ready for the precipitation of war. Austria's move against Serbia, which triggered the mobilization of the Russian army, provided the final impetus for the total mobilization of the German army and the declaration of war.

In this respect, there is a similarity to the genesis of hostility between individuals and groups. Conflicts between these entities activate primal thinking and imagery, which in turn aggravate the conflict. The initiation of war flows from the interactions between the states to the images that the leaders and their followers have of their nation and of the antagonistic nation. These cognitive representations and the resulting polarized thinking activate the motivation to fight and kill. Without this motivation, the necessary mobilization and willingness to risk everything would hardly get started.

Successive insults to the nation's pride may transform a hated image of the offending state to that of an Enemy. These images may bias the meaning of the adversary's behavior and determine subsequent actions. Thus, a collective sense of vulnerability, injured pride, or grandiose dreams may lead an adversary to hostile aggression. This aggressive behavior in turn produces a more hostile image in the mind of the antagonist, who may retaliate and thus accentuate the vicious cycle leading to war.

The assassination of Archduke Ferdinand of Austria in 1914, for example, produced a sequence of psychological, political, and military operations that interacted to culminate in the First World War. Diplomatic notes, troop movements, and mobilizations exacerbated the collective national image of vulnerability and malevolent images of the adversaries and in turn facilitated the deadly progression to war.

It is important to make a distinction between the thinking, emotions, and motives of the people who fight the wars and the leaders who initiate them. As suggested by the crowds dancing in the streets after a state of war has been declared, war brings out the patriotic attitudes and motives of the ingroup members: feelings of community, generosity, and altruism. The leaders, however, are not necessarily passionate about initiating the state of war. In fact, when Europe was on the verge of the First World War, many of the national leaders in Germany, Russia, France, and Britain were extremely worried about the consequences of a general European war.[20]

The leaders generally base their decision to initiate war on what they consider to be the national interest: extending the boundaries of their state, gaining access to natural resources, or attempting to contain another aggressively expansionistic state. The national interest, of course, is colored, if not distorted, by the leaders' own wishes for self-aggrandizement, power, and prestige. On occasion, a desire for revenge also permeates the decision making.[21] The leaders of Prussia in the midnineteenth century, Serbia and the Austro-Hungarian Empire in the First World War, Hitler in the Second World War, and Saddam Hussein in his attacks on Iran and later Kuwait apparently were all strongly motivated by the urge for revenge.

In retrospect, it is possible to discern a number of incremental events or one major event that served as a catalyst to activate the image of the Enemy that needed to be attacked. There are so many interacting factors, some of which are imponderable or unknown, that it is difficult to judge the outcome of any particular provocative event. There were several confrontations in the decade prior to World War I that seemed more likely to lead to war than the actual precipitating event, the assassination of the Austrian archduke.

Some military operations, however, such as when one of the great powers judges that its vital interests are threatened, are reasonably predictable. Interventions by the United States in this century were designed to thwart the invasions of South Korea, South Vietnam, and Kuwait. Britain sent forces to the Falkland/Malvinas Islands to repel an attack from Argentina. Similarly, the Soviet Union intervened in Afghanistan to support its puppet government, and Russia took a hand in Chechnya to suppress the fight for independence.

Even though the leaders may be worried about their decision to fight, the state of war takes hold of the minds of the populace. Their images of their country and of the Enemy permeate their thinking. When patriotism, loyalty, and obedience are aroused, the participants are drawn to their appropriate places in the machinery of war. For those men in combat, the belief that they must kill intensifies their desire to kill. The power of their destructive images, beliefs, and wishes is reinforced and magnified by other group members as well as by their leaders.

THE IMAGE OF THE ENEMY

War is a psychological as well as a political state and pervades the thinking of the individual participants. The representation of the Enemy occupies a central position in their mental life. The vicious, subhuman image is reflected in such derogatory terms as the Hun or Boche (Germans) and gooks (North Koreans, Japanese, Chinese, North Vietnamese). It should be noted, of course, that the leaders employ all the propaganda resources at their disposal to create and reinforce these images.[22]

It is possible that innate factors support the perception of outsiders as enemies. The fear of strangers in early childhood may be the preliminary basis for later xenophobia. However, many children do not experience xenophobia, and there is no direct evidence that the view of strangers or outgroupers as dangerous leads to perceiving them as threats to be eliminated. Recent observations of our primate cousins, the chimpanzees, suggest that outsiders, including former members of the band, are marked for attack simply because they belong to a different band.[23]

The collective self-image of a nation's populace and the image projected onto the Enemy exemplify the kind of dualistic thinking that takes over when people's vital interests are involved. In place of the normal tendency to judge other people across a broad spectrum ranging from good to bad, aroused people make extreme categorical judgments based on the notion of "totally good us" versus "totally bad them."

- "Our cause is sacred; theirs is evil."

- "We are righteous; they are wicked."

- "We are innocent; they are guilty."

- "We are the victims; they are the victimizers."

Personifying the enemy as evil is reminiscent of people's tendency to attribute the distressing behavior of others to their "bad character" rather than to a particular situation or set of circumstances.[24] Thus, we must kill enemy soldiers because they are bad, not because they happened to be drafted into the army just as we were. The Enemy deserves to be eliminated because he is a vicious killer, not because the military situation requires that he kill or be killed. The slaughter of civilians in Vietnam, Bosnia, and Rwanda illustrates how soldiers are inclined to see evil in everyone on the other side. Our opponents must be punished because they threaten our national security, our political system, or our ideology.

A striking feature of biased thinking is the confidence not only that "we are right" but also that our goodness and righteousness will triumph over the forces of darkness. This categorical, dichotomous thinking that causes so many problems in everyday conflicts is generally adaptive when soldiers are engaged in a fight to the death against a real enemy.

The malevolent images of the Enemy are as much a creation of the imagination as are fantasies of witches, demons, and evil spirits. The individuality, the humanness of the persons on the other side is blotted out; they are visualized as representations of all that is bad in the world. The propaganda machine enhances the portrayal of the Enemy as evil in people's minds. The evil image is depicted in posters, cartoons, and magazine illustrations: a crazed killer, sadistic torturer, rapist, barbarian, gorilla, saber-toothed monster, reptile, rat, or devil.[25]

Of course, soldiers are not necessarily fanatic about destroying their counterparts on the other side. In actual combat, infantrymen often lose their taste for killing. Various studies have shown that in many engagements only a fraction of the soldiers actually fire their guns.[26] Mercenaries or professional soldiers are likely to view killing as simply part of their job and may be no more antagonistic than game hunters stalking their prey. They have no empathy for their victims, whom they view not as symbols but as targets. Similarly, the generals who pore over war maps, order troop movements, and calculate casualties are likely to view the battles in mechanical terms and to reduce the enemy troops to mere numbers rather than seeing them as symbols of evil.

THE COLLECTIVE SELF-IMAGE

The image of the enemy is linked to the communal or national self-image, a composite picture of the nation's strengths and weaknesses, goals and vulnerabilities, history and politics. In contrast to the malevolent image of an outside group or nation, citizens of the state have an image of themselves as innocent victims. Insofar as people identify with their own group or nation, their mental representation of this larger entity shapes their individual self-images. Thus, they experience their nation's defeats and triumphs as their own defeats and triumphs.[27]

The personal self-image is a blend of the characteristics—attractiveness, efficiency, intelligence—that people consider important in attaining their objectives in life. They value themselves according to how well their personal characteristics help them to reach the goals they have set for themselves and how close these achievements are to their ideals. Their evaluation of their attainment is reflected in how they feel about themselves. Depending on how they interpret their experiences and their success in reaching their goals, they may see themselves as successful or unsuccessful, popular or unpopular, triumphant or defeated.

In peacetime, people's image of their nation is generally peripheral to their personal self-esteem. Although civic-minded or politically active citizens are concerned about the discrepancy between their nation as it is and their own goals for it, most people are involved with their own personal problems and aspirations. But in a national crisis, everyone becomes intensely engaged in the problems of the country. When events start to play on the national image, when the country is threatened or involved in a confrontation, the image comes to life and begins to control what people think and feel.

In wartime the national image becomes the center of every citizen's worldview; as they rally 'round the flag, they move from an egocentric to a group-centered mode. Each person's self-image becomes attached to the image of the country. The nation's policy becomes their policy; the national vulnerability becomes their personal vulnerability; an attack on the state becomes an attack on themselves. Aroused by the specter of the malevolent enemy, they prepare to risk their lives for homeland, religion, or political movement.

The prevailing national self-image of the United States is that of a benevolent, freedom-loving, democratic nation, appreciated for its willingness to make sacrifices to protect other countries from tyranny and injustice and to provide succor for people in desperate straits. The role of the United States as the melting pot of different immigrant groups has fortified its egalitarian image. This concern for the underdog was part of the motivation for intervention in Cuba in 1898 to drive out the "tyrannical" Spanish rulers. The moral self-image expressed in the deep sympathy for the plight of the Cuban people compelled the government of the United States to attempt to mediate the conflict with the Spanish rulers and their colonial subjects. The intransigence of the Spanish government was based to a large degree on the Spaniards' pride and fear of presenting a weak image at home and to the outside world.[28]

In the meantime, various political pressure groups and the Hearst and Pulitzer newspapers created an interventionist motivation in the American people. The sinking of the battleship *Maine* during a "courtesy visit" to Havana Harbor further aroused them and the Congress, who attempted to shame President McKinley into declaring war on Spain. Theodore Roosevelt, in fact, had portrayed the president as lacking in backbone for not having intervened sooner. The popular chauvinistic image was intensified sufficiently to force the president's hand to start the Spanish-American War. Several other wars in which the United States has participated, such as the Vietnam conflict, were portrayed initially as an expression of America's lofty purpose to contain communism but evolved into a battle to defend America's honor and prestige.

In the aftermath of both world wars, the U.S. moral self-image as a generous nation was expressed in the donation of food, supplies, and financial aid to the former enemies as well as the allies. During the cold war, when the Soviet Union was perceived as the major threat to the freedom of the world, American policy provided a haven to victims of oppression. More recently, the plight of refugees in Somalia, Bosnia, and the former Zaire touched the conscience of the nation and moved it to intervene to relieve suffering. The moral self-image of the benevolent guardian of freedom and democracy throughout the world has also guided U.S. policy toward other states.

The political elite may use the national image as justification for its own goals. The Serbian leaders of Yugoslavia, for example, built up an image of persecuted Serbians to further their own goals of a homogeneous Greater Serbia. The collective image of the United States as a bulwark against foreign tyranny was juxtaposed with the image of the Communist juggernaut to justify the interventions in Korea and Vietnam. Throughout the cold war, American national policy was shaped by the view of an imperialistic and subversive Soviet Union threatening to sweep through Western Europe and the rest of the world. (During the Stalinist period this perception of the Soviet Union and its foreign policy probably had an element of truth.) Citizens of the Soviet Union had an analogous view of the United States and its international goals: imperialistic, hostile, dangerous.[29] The interplay of the threatening images on both sides tended to intensify them. The aversion to communism and the fear of world domination by the Soviet Union led to U.S. involvement in the war against the "bad" Communists in Vietnam as well as Korea. It is of interest that the image of the Soviet Union as an "Evil Empire" (in President Reagan's words) faded with the thaw in the cold war.[30]

The national self-image and the image of the adversary are manipulated by the actions of the policy makers but also serve as templates for interpreting information about the adversaries. The nation forms the most malevolent interpretations of the adversary's actions, whereas interpretations regarding the nation's own actions are positively biased. When a Korean airliner was shot down over the Soviet Union in 1982, there was widespread belief in the United States that this was a deliberate inhuman act rather than a case of mistaken identity, as the Soviets claimed.[31] Events that are consistent with the national self-image are believed and accepted as confirmation of the nation's virtue; events that might sully the image of the nation are discounted or minimized.

During a shooting war, the self-image of the United States, with its symbols of liberty, freedom, and democracy, is enhanced by the glow of righteousness and morality, a stereotype continually reinforced by chauvinist reports in the media, patriotic ballads, and stirring speeches by the leaders. The celebration of the ideological goal, the nation's basic goodness, becomes a source of inspiration to the citizenry. This ideological structure provided justification for American engagement in World War I, when the United States sent troops to France to "make the world safe for democracy." In World War II the people of the United States sought not only to defend the nation but also to protect civilization from Nazism, fascism, and Japanese imperialism. Other American wars, such as the conquest of the Philippines and the invasion of Grenada, however, have evoked a more muted image.

During the war in Vietnam, in contrast to other wars, the national self-image was divided into good and bad: some Americans steadfastly retained their faith in America's righteousness, and others perceived the continuing attacks on a smaller nation as a betrayal of basic national principles. The initial ideological impetus for the war—to stop the spread of communism throughout Asia (the "domino effect")—was eventually transformed into a desperate patriotic venture to preserve America's honor and pride. The prospect of losing the war overshadowed the original goal of saving the weak government in South Vietnam. American leaders and their followers believed that the nation's prestige, honor, and international credibility were at stake. Those who did not subscribe to this view were seen as cowards or traitors. The dissident groups, however, regarded their protests as an attempt to return the United States to its basic national values of freedom and self-determination for smaller nations.

Patriotism and *nationalism* are major ideologies that bind the population together in collective obedience to the leaders' decisions. Although similar and overlapping, they can be treated as separate entities, as pointed out by Feshbach.[32] Nationalism is centered in the glorified image of the state—its power, prestige, and possessions. By identifying themselves with this image, individuals experience a boost in their own self-esteem; they bask in the glorious past and future aspirations of their action. Of course, defeat produces a decline in self-esteem and eventually may trigger depressive feelings. The narcissistic, even grandiose, trappings of nationalism are expressed in claims to superiority over other states that may ascend to extreme racist beliefs in the status of the "master race" and the baseness of the "outsiders."

Patriotism is powered by the yearning to belong to a larger community. There is a sense of identification with and attachment to the state, and a willingness to make sacrifices in order to ensure its continued security. The patriotic image of a benevolent, empathic government contrasts with the militant, power-seeking image of the nationalists. People who are more invested in nationalism entertain hawkish, jingoistic attitudes toward other states and believe that their country should gladly initiate war to further its vital interests. They are less prone, however, than the patriots to be willing to sacrifice their lives for their country.[33]

When the survival of the state is threatened, a feeling of solidarity spreads across the population. History is replete with examples of the power of the national self-image to rally dissident factions to uphold the national security and honor. During World War I, socialist leaders in Germany and England, who had initially viewed the conflict simply as a fight among imperialist heads of state, rallied to the support of their respective countries. Similarly, in 1914 a group of leading German intellectuals issued a manifesto proclaiming Germany's total innocence in the initiation of the war. They portrayed Germany's invasions of France and Russia as defensive acts to ensure their country's survival.[34] This declaration illustrates the power of the national image to mold the thinking of even the educated elite.

Just prior to America's entry into World War II, there was substantial debate in the black community and in labor unions regarding support for the country in the event of entry into the war. When the United States did enter, they fell into line in supporting the war effort.[35] Similarly, a number of celebrated isolationists who had vitriolically opposed prewar American policies were among the first to volunteer for action.

As indicated by the political psychologist Ralph W. White, the impetus for many wars may be macho pride, fear of an antagonistic nation, or a combination of both.[36] Macho pride is based on a national self-image of superiority, toughness, and courage, as well as the implied right of the nation to impose its hegemony on other states. This belief has been the motivating force for building empires, regaining lost territory, and protecting client states or weaker nations. Examples range from the empire building of ancient Persia, Greece, and Rome to the nineteenth-century expansion and colonialism of Britain, France, Germany, and Russia. A similar macho image may be discerned in the invasion of the Philippines by the United States in 1898 and the occupation of Manchuria and Korea by Japan in the early twentieth century. Conquerors from Genghis Khan to Napoleon have been driven by grandiose fantasies to expand their domains.

Aside from deliberate wars of conquest, hostile aggression may result when nations with a combination of macho and vulnerable national images clash. A grand but hypersensitive pride combined with a shaky national image may lead to risky decisions. Beset by considerable instability at home, Napoleon III declared war against Prussia in 1879 after the publication of a note from Prussia that he and the French populace interpreted as an affront to French honor and dignity.

Ideologies—the infrastructure for the collective images—contributed to twentieth-century wars, including both world wars. Hitler molded the German national image from that of a defeated people, betrayed into surrendering in World War I and abused by the victorious powers, into that of a master race, strong enough to punish those who had conspired against it and destined for world domination.[37] The religious battles between India and Pakistan, as well as the revolutions in czarist Russia, China, and Indochina, demonstrate the enormous power of religious and political beliefs to instigate the slaughter of countless numbers of people of a different national or ethnic group.[38]

THE CLASH OF NATIONAL IMAGES: THE PRELUDE TO WAR

An analysis of the events leading to World War I reveals the importance of a clash of images: the national self-image and the image of the antagonist. The clash is analogous to the conflict between two individuals, each perceiving himself as vulnerable to the malicious intentions of the other person. It is not simply the actions of individuals or states that lead to escalating tensions, but the *meaning* ascribed to the aggressive actions. The explanation for an offensive act—whether it is construed as a bluff, a tentative feint, or a deadly serious threat—is influenced by these images and, in turn, reinforces them. Depending on the meaning attached to the act, a state may evaluate its own strengths and vulnerabilities and those of its adversary as a prelude to taking action.

The growth of economic and military power, augmented by the cult of nationalism, might tempt a state with an expansive self-image to look beyond its borders for additional territory and resources. This attitude would be perceived as a threat by its neighbors and lead to an escalation of tensions. The threatened neighbors would look for allies to compensate for their own vulnerability. Each state would become caught up in an arms race in order to avoid being unprepared. The macho nation, perceiving a shift in the balance of power, would experience a sense of vulnerability as other nations lined up against it. If that nation believed that war was inevitable, it might engage in a preemptive strike.[39]

Such a set of circumstances provided the setting for World War I. An expansionist Germany, sensitized by a long history of encirclement and invasion, perceived an increase in its own vulnerability. Its major ally, Austria-Hungary, was deteriorating and threatening alliances were forged by unfriendly France, Russia, and, later, Britain.[40] The assassination of Archduke Ferdinand, heir to the throne of Austria-Hungary, by Serb nationalists set up the catastrophic sequence of events that threatened the security of Serbia and Austria-Hungary and, consequently, Germany, France, Russia, and England.

Before World War I, the national self-image of Serbia was that of a fiercely independent kingdom vulnerable to domination, if not annihilation, by the still formidable Austro-Hungarian Empire. Serbia, although insecure in its relatively recent independence, projected itself as the macho leader of a greater South Slav kingdom formed by the neighboring provinces of the Austro-Hungarian Empire. As the tide of pan-Slavic nationalism swept over the country, splinter groups such as the Serbian Black Hand, dedicated to undermining the Austrian empire, sprang up. Ultimately, this terrorist group achieved its minimal objective: the assassination of Archduke Ferdinand.[41]

The leaders of Austria-Hungary were bent on preserving their empire under pressure from its ethnic populations that threatened to fragment its already cracked image. The various nationalities—Serbs, Czechs, Slovenes, Croats, Poles—continued to press for independence. To prevent Austria-Hungary from crumbling like the Turkish empire, the imperial government sought to suppress the dissident provinces, purportedly being agitated by Serbia toward separatism. Incorrectly attributing the assassination of the archduke to the Serbian government, Austria wanted to neutralize its subversive neighbor.

In June 1914, after some vacillation, the Austrian government decided to attack Serbia, an action apparently sanctioned by Kaiser Wilhelm of Germany.[42] Since aggression against Slavic Serbia ran the risk of inviting intervention by the Slavic Russian empire, Austria sought assurance from Germany that it would come to its aid in the event of Russian intervention. The kaiser sent a supportive note to Austria, the so-called blank check, which appeared to promise support for Austrian moves. The mobilization of the Russian army in a display of power intended to deter Austria was perceived as a major threat by Germany, whose image of vulnerability was primed by the prospect of Russian destruction of Austria, its ally. The invasion of Serbia by Austrian troops then set off the chain reaction culminating in Germany's declaration of war against Russia and its ally, France.

Germany's historical background casts some light on its disposition to initiate a "defensive" war. The pervasive image of vulnerability to outside attack had developed through its history.[43] Its lack of natural boundaries to ward off foreign invasion had made it the killing field for numerous European wars. The wholesale slaughter during the Thirty Years War had left its mark on many aspects of the German worldview, specifically a kind of claustrophobia regarding hostile and dangerous neighbors.

Under Kaiser Wilhelm, Germany sought to establish the image of a strong military power, to acquire colonies, and to surpass France and England in power, possessions, and prestige. With its aggravated fear of encirclement, it overreacted to the coalition formed in 1907 by France, Russia, and England to maintain the balance of power in Europe. Long before its declaration of war against France and Russia in August 1914, Germany anticipated that the growing strength of the Russian armed forces would pose a maximum threat by 1917. Russia, on the other hand, was intent on compensating for its diminished self-image following its humiliating defeat by Japan in 1905 and its painful acquiescence to Austria's annexation of Bosnia-Herzegovina in 1908.[44] Having previously sustained blows to its national pride, Russia was in no mood to remain on the sidelines while Serbia, its Slavic protégé, was destroyed.

The clash of the national images of Germany (vulnerable and expansionist) and Russia (humiliated but resurgent) set the stage for military confrontation. The conflict between Russia's self-image as the moral protector of its Slavic protégé and Germany's image as the protector of its Germanic protégé, Austria, played a key role in precipitating the war. France, threatened by the burgeoning German industrial and military power, posed a special danger to Germany because of its vengeful attitudes over the loss of Alsace-Lorraine in 1871 as well as its general defeat in the Franco-Prussian War. Since war seemed inevitable, Germany decided that a preemptive strike against Russia and France would strengthen its chances of defeating both countries.

While the pressure to initiate war in some instances originates with certain factions in the state, the leaders assume the responsibility for calculating the costs as well as the odds of winning and consequently whipping up or damping down the popular passions. In this respect, the leaders can make the same kinds of errors as two individuals who square off against each other in a hostile confrontation on a balloon.

The images that antagonists project onto each other often lead to hostile behavior (threat, denunciation, economic embargo), which progresses in turn to a reification of the images and further antagonistic behavior. The Japanese invasion of China in the 1930s led to increasing efforts by the United States to contain Japan. The American image of the Japanese as ruthless, grandiose, and dangerous was on a collision course with the Japanese image of the United States as controlling, intrusive, and hostile. Ultimately, the Japanese image of the United States became accentuated and fed into the policies of the war party in Japan.

During the cold war, the populace, and presumably the leaders of the Soviet Union and the United States, had mirror images of each other. As reported by Bronfenbrenner, each saw the other as a power-hungry, manipulative, and deceptive warmonger.[45] The opposite side was perceived as perpetuating a martial atmosphere, exploiting the citizenry, and controlling the media and the elections. The clash of these images pushed the adversaries to more extreme positions, which tended to confirm the images. Fortunately, there were sufficient deterrents to prevent the outbreak of a shooting war.

INSIDE THE LEADERS' MINDS

In our encounters with other people it is important to have some idea of their perspective: their thoughts, expectations, and intentions. We also need to have some notion of how our spouse or other family members, friends, and employees or colleagues perceive us: as friendly or unfriendly, weak or strong. This information may be obvious, or it may be concealed in their statements and behavior. Entering into the perspective of others is particularly important during a crisis. In our everyday relations, empathy with the other person's hurt feelings will help to mitigate the offense and restore the balance.

Whether engaged in a family conflict or an international confrontation, people apply a complex "theory of mind" to understand their antagonist's thinking—his images, misinterpretations, and plans.[46] The theory consists of an interrelated set of assumptions and rules about mind-reading. In everyday interactions, these rules may take the form of conditional statements: "If somebody stares at me, it means he's angry"; "If their voice trembles, they're afraid of me"; "If a person is silent, she probably disagrees with me."

By applying rules like these, one individual may get a sense of the perspective of another. A therapist integrates a patient's descriptions of his reactions to various events to gain insight into her basic beliefs. By understanding the perspective of the depressed patient, for example, a therapist can see the world through her eyes and then help her to evaluate the biases in her thinking.

In conflicted relations between leaders of nation-states, an analogous form of mind-reading is essential but is also more difficult, especially when there is distrust or antagonism. Leaders may relay ambiguous or deliberately distorted diplomatic messages to deceive the other side ("disinformation"). Consider how difficult it is to grasp the perspective of the opposition in a time of crisis. During the Cuban Missile Crisis of 1962, President Kennedy and his advisers, strained by the momentous decisions facing them, had to ponder the significance of contradictory messages from Premier Khrushchev and make decisions that they believed would affect the future of the world.[47]

Many factors can combine to interfere with reading the adversary's perspective and, consequently, with appropriate decision making. The inherent limitations in the capacity to process a mass of ambiguous, inadequate, and often conflicting information can make the task particularly difficult. This problem is confounded by misinformation from intelligence sources and deliberate deception by the adversary. Additionally, the intentions of the other side fluctuate in response to changing conditions and the relative influence of factions in the government such as the "hawks" and "doves." Having acquired a particular mindset, the leaders may have difficulty switching their appraisals in response to fluctuations in the adversary's intentions. The we-against-them guessing game grows more difficult when we try to appraise not only what the other side thinks of us but also what they believe we think of them.

Government leaders use their own theory of mind to obtain a feel for their adversary's or ally's perspective, but they are obviously often handicapped in ferreting out relevant and reliable information.[48] Despite their best efforts, they may arrive at an erroneous, yet firmly held, view. Such a fateful mistake can make the difference between war and peace.

History abounds with examples of such incorrect mind-reading, usually of a "wish-fulfilling" nature. Just before World War I, for example, Austrian and German leaders thought that Russian leaders would be reluctant to mobilize for war to halt Austria's attack on Serbia, and that indeed the Russians would be sympathetic over the assassination of Austria's crown prince. The Austrians incorrectly judged that Russia would be reluctant to fight, just as it had been in 1906 over the annexation of Bosnia by the Austro-Hungarian Empire. Later, German leaders concluded incorrectly that Britain would not care enough about the inviolability of Belgium to go to war if German troops invaded that small country.

Nor did the German and Austrian leaders correctly perceive the perspective of the American people when their unrestricted submarine warfare, and especially the sinking of the luxury passenger ship *Lusitania,* inflamed American opinion against them. At the time of his confrontation with Hitler in Munich prior to World War II, Prime Minister Neville Chamberlain believed he correctly read Hitler's mind and that Hitler wanted peace. And in the Korean War, General Douglas MacArthur, confident in his understanding of the "Oriental mind," believed that China would not send troops to attack the American forces as they moved toward the Chinese border.[49]

Although leaders attempt to make rational decisions regarding the advantages and disadvantages of initiating wars, misconceptions are likely. The lack of information about the military intentions and strength of the enemy may be compounded by the leaders' belief in their own propaganda. Erroneous mind-reading by the opposing sides is a crucial factor in faulty decisions. Although the leaders may have an inkling of the opposition's perspective, they are hampered in distinguishing sincerity from deviousness. The tactics of bluffing and counterbluffing, disinformation, and deception may obscure true intentions. Hitler succeeded in masking his aggressive objectives in Munich. On the other hand, in the prelude to World War I, the European powers exaggerated their adversaries' hostile intentions.[50]

When leaders are under considerable stress, they are more likely to misjudge their antagonists.[51] When deadlocked in a conflict, they are prone to. expect the worst from the enemy. Kaiser Wilhelm, already suspicious of England's intentions, misinterpreted Lord Grey's attempt to mediate the conflict between the Central Powers and Russia as a deliberate attempt to trap him.[52] In contrast to the negative bias in interpreting an adversary's perspective, leaders often have an inflated positive bias in assessing their nation's capabilities when they determine that war is inevitable. This biased thinking led the French to make disastrous errors in overestimating the probability of victory prior to initiating the Franco-Prussian War. In both world wars, Germany did not expect the entry of the United States into the war to jeopardize its chance of winning.

More recently, when Argentine troops were ordered to attack the Falkland/Malvinas Islands, their generals had an unreasonable belief that the British would not care enough to defend the islands. The generals miscalculated the likelihood of intervention by Britain, while their followers were aroused to a supreme confidence that their troops would destroy the English. When Russian troops invaded Chechnya in 1996, they overestimated their own strength and underestimated that of the resistance. It appears that when the government leaders and generals are mobilized to attack, the mind-reading of their adversaries is distorted by an optimistic bias. In their study of international crises, Snyder and Diesing found that 60 percent of the messages received from an adversary were misinterpreted or distorted in transmission.[53]

The leaders' own personal reactions to diplomatic successes and failures may play an important role in their decision to initiate war. A diplomatic triumph or defeat raises or lowers their self-esteem. Their personal elation or distress shapes the national mood and spreads through the country. The personal desire of the political elite for power and prestige often biases their determination of the best interest of their group or state. Their subjective analysis of the costs, benefits, and risks of war may override their concern for their followers' lives.

When a leader views his adversary through a "frame," he abstracts or ignores data regarding the adversary's intention, which affects his choice between war and conciliation. The frame tends to distort information about the adversary and radically limit the leader's options. The leader's images and beliefs about the Enemy are the product of the interplay of their historical relations, the balance of power, current political and economic conflicts, and their personal reactions. These factors converge to form a kind of final common pathway to the political-military decision. The outcome of these beliefs and interpretations, realistic or distorted, rational or irrational, may be—*war*.

Mobilizing Public Opinion for War

Given that the clash of national interests, goals, and self-images can set the conditions for war, how do leaders mold the final beliefs about the Enemy and initiate war? How do they engage the populace's commitment to make the necessary sacrifices? The leaders' decision to fight war may be based on the evaluation of complex political issues, the computation of relative military capabilities, and the assessment of the adversary's perspective and intentions. But they must appeal for public support on the basis of national pride and outrage over the wickedness of the Enemy. They may promote a distorted image of their nation as the victim of foreign abuse in order to garner support for their own political agenda. The citizens react in the same way as when they perceive themselves as victimized by another person: they feel impelled to punish the offender. This craving for revenge is practically a reflex reaction to being diminished in some significant way: a loss of face, security, or assets. National honor is as sacred as personal honor.

Even when leaders are invested in such political goals as establishing or maintaining national hegemony in a region, they find it expedient to project the image of a hostile foreign power and an imperiled national honor. Waging war against another state as an instrument of foreign policy has historically fulfilled such political objectives as expanding the territory of the state or responding to changes in relative power.[54] Consolidating support has also been requisite for striking preemptively at an opposing country presumed to be gearing up for attack.

Bismarck's role in the precipitation of the Franco-Prussian War of 1870 provides an example of how a leader's expansive goals are used to mobilize his people for war. His objective of consolidating the states, duchies, and principalities into the single kingdom of Germany depended in large measure on his success in provoking France to declare war against Germany. He reasoned correctly that a "defensive war" would drag in the southern German states on the side of the northern confederation headed by Prussia. Thus, military unification would lead to political unification.[55] Bismarck's personal self-image meshed with the Prussian national self-image as it waxed and waned in relation to the other European powers. The national self-image of France, threatened by rising German nationalism and Prussian militarism, provided the backdrop of the decision to initiate war.

Bismarck had recognized that France, unstable and fearful of German unification, was leaning toward war in order to curb Prussia. By portraying France as antagonistic to Germany's natural right to national self-determination, he planned to make Germany appear to be the victim and France the aggressor. Napoleon III, beset by domestic dissension, was eager to discharge the mounting current of popular distress down the lightning rod of foreign adventure.

The final deliberate provocation by Bismarck was the Ems dispatch to France from the king of Prussia. This telegram from the German resort city of Ems, having been edited by Bismarck to a more confrontational wording, was taken as an affront by the French population. Already inflamed by an alleged conspiracy by the king of Prussia to place a relative on the throne of Spain, the French were confident in the supreme power of their army to crush his army. Prussia's subsequent triumph over the French not only served Bismarck's political purpose of unifying the German nation but also fortified a powerful new national image for Germany.

Bismarck's ability to mobilize the Germans against France illustrates how a national ideology can become a tool for war. Germany's history provided fertile soil for the propaganda seeds planted by Bismarck through his public statements and his controlled press. Predominantly conservative, the people—the *volk*—felt threatened by the powerful ideology of the French Revolution. Its "subversive" message of equality and liberty was upsetting to the centuries-old Prussian aristocratic class system and to the authoritarian family. The dismemberment and consequent economic ruin of Prussia by Napoleon in 1806 contributed further to the wounded, but proud, Prussian national image. Further, the absorption of Alsace-Lorraine into France in the eighteenth century had constituted a continuous irritant to the Germans.

Bismarck traded on this to inflame the Prussian populace against France.

Also contributing to the anti-French bias were the presumptions of the French to cultural, political, and military superiority. Just as French political and geographical proximity was a threat to the security of Germany, the culture of France was an irritant to the fundamental values of the Germans. Further, the view of a "decadent" French culture was compounded by German religious differences with Catholicism, which had been displaced from the northern German states after the Thirty Years War. The repellent image of the French crystallized under Bismarck's propaganda.

To motivate a citizen army requires an inspiring ideology. The volunteers and conscripts must be taught to view their state, or at least its prestige, as diminished if they do not fight, and enhanced if they do. Bismarck was aware of the powerful effects of the revolutionary slogans of liberty and equality on the spirit of the French troops in the Napoleonic Wars. He sought to reproduce the same fervor in his own army by encouraging an analogous ideology of power, social superiority, and virtue.[56]

Gagnon has shown how the political elite can foment ethnic conflict by stirring up the latent prejudices of their followers.[57] For example, he attributes the strife in the former Yugoslavia starting in 1992 to the machinations of the powers in Serbia, whose political goal was to divert their constituents' attention from the failing economy and create a new Slavic state dominated by Serbia. The old guard sought to mobilize the Serbs to action by playing on their diabolical image of the non-Serbs (Albanians, Croats, and Muslims). A coalition of orthodox Communists, nationalists, and conservative army elements, aiming to hold the reins of power, gained the support of the Serbian population by alleging persecution of the Serbs by Albanians in the province of Kosovo and by the Croats in Croatia. The coalition then instigated mass protests by Serbs in the region and used the images they had created to incite violence by the Serbs against the non-Serbs. Concurrently the propaganda machinery successfully portrayed the non-Serbs as the villains.

Although the outside world viewed the fighting largely as the revival of ancient but persisting animosities, it appears, according to Gagnon, that the Serbian leaders deliberately primed these "memories" of past persecution.[58] Indeed, while stories of oppression by the Turkish regime in previous centuries had a historical basis, Serbs and Muslims had lived amicably as neighbors and often as allies throughout most of this era. But old stereotypes die hard, and the Serb leaders created a myth of present-day Muslims as an incarnation of their oppressive ancestors. The Serbs, ideologically bound to the dream of a greater Serbian state, feared a situation in which their former "oppressors" could once again dominate them. "Ethnic cleansing" was the solution.[59]

Similar trumped-up charges by Germany regarding alleged persecution of German residents in the Sudetenland region of Czechoslovakia served as the pretext for Germany's "annexation" of the Sudetenland in 1938, the first step in its invasion of most of Europe. Here again, the leaders deliberately constructed a persecutory image to rally support for their policies. To justify the anticipated human and material losses, the leaders inspired the image of triumphant glory and heroism.[60] The involvement in the state of war itself stirred up national pride and villainous images of an enemy that had to be beaten down.

The License to Kill

Once the political elite decides that engagement in an offensive war is desirable, they generally find it necessary to justify such engagement in the minds of those who will fight, shed their blood, and bear the economic burdens. Sometimes the proclamations of a charismatic and trusted leader are sufficient to arouse the popular will. The cloak of legitimacy may be provided by a wide variety of sacred causes: crusading to regain lost holy places or occupied provinces, rescuing a beleaguered kindred state, or establishing the self-determination of an ethnic group. People will also fight for such contradictory ideals as establishing or overthrowing a Communist state and dissolving or preserving the union. Defensive wars to protect the homeland to preserve a political or social system will similarly mobilize followers.

As the image of the Enemy hardens, the total commitment to the group or state becomes fixed. The followers are energized by two powerful emotions, love of country and hatred of the adversary. Feelings of fear and anxiety over the possibility of defeat and domination by the Enemy add to the motivation to fight. The same emotions and motivations are evident in civil wars, revolutions, and insurrections. Hatred of the imperial ruling class or of dominant political factions provided the driving force for the French and Russian Revolutions and for civil wars in the United States, Spain, and Cambodia.

The White Army and Red Army during the Russian Revolution, the south-erners and northerners during America's Civil War, and the Royalists and Republicans in France were locked into their images of the Enemy and their commitment to its dissolution.

Leaders not only stimulate the urge to kill but give it a clear direction. They manipulate the population by dramatizing the national goals and the image of the threatening Enemy, and also by playing on the people's propen-sity to obey their government's authority. In earlier times, the aura of infalli-bility conveyed by the status of the rulers gave them almost complete control of the hearts and minds of the populace.

At the same time that leaders whip up enthusiasm for humiliating the Enemy, they also suspend the taboo against violence. The moral code against murdering, plundering, and destroying property, which is characteristically more relevant to one's own kind, is further undermined during combat. Pressures such as discipline, expectations of punishment for disobedience, and demands for loyalty to the fighting unit combine to prepare the soldier for his main task—to annihilate, or at least incapacitate, the enemy. In a kill-or-be-killed mode, the soldier has no room for humane considerations that might get in the way of effective action.

The image of the Enemy extinguishes empathy and any concerns or inhi-bitions regarding the taking of human life.[61] As the malevolent image crys-tallizes, solidarity with members of one's group and devotion to the cause increase. The antagonists are no longer seen as "people like us" but as totally different, as subhuman or inhuman. The collective participation in fighting fuses the bonds between soldiers and intensifies hatred for the Enemy. As the fighters move into combat, they become more and more certain that their cause is right. Loyalty to their country and the folks back home spreads to their officers and buddies in the combat unit. Their closeness and willingness to sacrifice has been traced by some authors to the primordial bonds in Stone Age kin groups.[62]

The communal nature of killing was illustrated during the Vietnam War at My Lai when a company of American soldiers, led by Lieutenant William Calley, went on a rampage in a Vietnamese village. The driving force was the belief that, because the Enemy had killed their fellow soldiers (including a popular sergeant killed the previous day by a booby trap), these civilians—old men, women, and children—deserved to be wiped out. The craving for vengeance obliterated any sympathetic feeling for the defenseless victims. The killing continued, singly or in groups, despite the victims' obvious helplessness and cries for mercy. At his trial, Lieutenant Calley defended his action on the basis that he was "just following orders."[63] He recalled, "I pictured the people of My Lai: The bodies, and they didn't bother me. . . . I thought, *It can't be wrong, or I'd have remorse about it.*"[64] As pointed out by the political scientist Robert Jervis, if evil has been done, he cannot have done it; if he did it, it cannot be evil.[65]

While the image of the "Evil Enemy," reinforced by the self-image of righteousness, incites the soldier to commit the unspeakable atrocities associated with war, it is often difficult to harm an enemy soldier when he is perceived as human. A humane sentiment replaces hostility when the immediate threat is reduced and the humanity of soldiers on the other side is evident; for example, officers noted that joint Christmas celebrations by English and German frontline troops in 1914—singing together, exchanging presents, even playing soccer—were dangerous and forbade them from then on.

George Orwell tells an amusing but revealing story of how, as a sniper in the Republican lines during the Spanish Civil War, he was about to shoot an enemy soldier:

> A man, presumably carrying a message to an officer, jumped out of the trench and ran along the top of the parapet in full view. He was half-dressed and was holding up his trousers with both hands as he ran. I refrained from shooting at him. It is true that I am a poor shot and unlikely to hit a running man at a hundred yards. Still I did not shoot partly because of that detail about the trousers. I had come here to shoot at "Fascists"; but a man who is holding up his trousers isn't a "Fascist," he is visibly a fellow-creature, similar to yourself, and you don't feel like shooting at him.[66]

This anecdote captures a common experience in warfare. When soldiers can actually see enemy troops at close range, they are more likely to have an inner resistance to pulling the trigger or thrusting the bayonet. During both world wars and the Korean War, a large proportion of the American infantrymen did not fire their guns when engaging the enemy. The greater the physical proximity to the victim, moving from shelling and bombing to throwing grenades and fighting hand to hand, the greater the resistance to killing.[67]

When the enemy is perceived by the soldier as a real human being (especially a *similar* human being), the desire to kill is inhibited and replaced by a feeling of guilt if he decides to fire his gun. As pointed out by Grossman, much of the post-traumatic stress disorder experienced during and after the war in Vietnam was related to the guilt over killing. The capacity to experience empathy and guilt is obviously not totally extinguished.[68]

Military engagements are organized in such a way as to minimize the arousal of humane feelings toward the Enemy. The patriotic image, obedience to the commander, loyalty to other members of the unit, and rewards for killing are contrived to provide absolution for the horrors of war. The anonymity of a specific enemy soldier also tends to reduce the sense of responsibility for the death of another human being.

The statement by an English soldier in Shakespeare's play *Henry V* illustrates how soldiers can feel absolved from committing inhuman and homicidal acts: "For we know enough if we know we are the king's subjects. If his cause be wrong, our obedience to the king wipes the crime of it out of us."[69] The displacement of responsibility onto the leader helps to eliminate inhibitions against killing.[70]

Deliberate propaganda efforts are aimed at ridiculing the culture of the enemy troops, accentuating the "criminal acts" of their leaders, and assigning responsibility to the enemy soldier for criminal acts. Vengeance is legitimized in slogans and songs such as "Remember the Alamo," "Remember the *Maine*," and "Remember Pearl Harbor."

Recognition of the low firing rate in previous wars induced the American army to initiate a formal program of "deconditioning" during the Vietnam War. In addition to the standard "kill, kill, kill" shouted by drill sergeants, the program called for repeated practice in attacking the presented image of the Enemy—for example, using realistic man-shaped targets on the shooting range. The combination of physical, moral, and ideological distancing and the activation of loyalty to the unit, commander, and company attenuated the human image of the Enemy soldiers and allowed for the projection of a diabolical image onto them.

The psychological mechanisms used in disengaging the moral code in combat are similar to those seen in individual crime, intergroup violence, terrorism, and genocide. The image of the Enemies as subhuman or inhuman, the belief that they deserve to be punished, the displacement of responsibility for killing onto the leaders or the group, the perversion of the moral sense, and the belief that killing is noble contribute in varying degrees. It is conceivable that if people were to question the validity of any of these images or beliefs, they might be less willing to kill other human beings. Working from the top by the imposition of a supranational authority such as the United Nations and from below by reinforcing the moral code against killing and questioning the validity of beliefs about the enemy, the policy makers may provide an answer to the opening question in this chapter: war is *not* inevitable.

PART 3

FROM DARKNESS TO LIGHT

12

THE BRIGHTER SIDE OF
HUMAN NATURE

Attachment, Altruism, and Cooperation

Reports in the media play up the dark side of human nature (murders, muggings, rapes, riots, genocide). These stories, however, don't do justice to the brighter side of human behavior.[1] Evidence from statistical surveys, anecdotal reports, observations of children, experimental studies, and practical applications in the classroom indicate that people in general have an innate capacity for altruistic behavior that can balance or override hostile tendencies. Moreover, we all have a great capacity for rational thinking on which to draw to correct our biases and distortions.

By getting inside the head of the aggressor, we can better understand his thinking and behavior and arrive at principles for changing them. Anger and hostility thrive on rigid, egocentric beliefs and biased perspectives, but it is possible to reshape the images and beliefs that drive these feelings and consequently to weaken the disposition for violence. In an analogous way, the values and ideologies that divide people and make them mistrustful and antagonistic toward each other can be moderated.

THE POTENTIAL FOR CHANGE

The persecution, torture, and killing by the death squads in Argentina from 1975 to 1982 illustrate the extreme dichotomous beliefs and distorted moral code in an authoritarian state. As described by Erwin Staub, the sharp division between right and left in this country led each side to see the other as the personification of evil.[2] In the 1970s leftists engaged in a campaign of terror, executing high-ranking officials and bombing broadcast stations and military outposts. Starting in 1975, the military retaliated with destruction, torture, and killing. The categorical thinking was the same for both sides: our side is completely good, and the other side is totally bad.

The military leaders of the right were guided by their ideology, as well as by the need to protect their privileged status, which had become their "inalienable right." They demonized the guerrillas, whom they saw as a "new kind of enemy fighting a new type of war." Their ideology became the moral code, the transcendence of the state and religion over the individual. All possible measures to preserve the traditional morality, represented in the words *God, country,* and *home,* were justified.

The perpetrators, those who kidnapped, tortured, and murdered, were absolved from guilt according to the moral code of the new order. Further, the responsibility rested with their commander.[3] Although their initial motivation was ideological and based on obedience to orders, the perpetrators came to see themselves as the absolute rulers of their victims' lives and continued their vicious work on their own momentum. Their ideology gave them license to indulge themselves in inhumane destruction.

The eventual collapse of the Argentine military dictatorship and its ruthless policy demonstrates the effect of benevolent forces working both within a country and from the outside. In this case, the reign of terror was replaced by the reign of a nonviolent democracy. The growth of opposition to the destructive military dictatorship was sparked by the demonstrations of the mothers of executed persons (the Mothers of the Plaza del Mayo). Capitalizing on the respect accorded mothers in Argentine society, these brave women marched every week in the Plaza del Mayo and thus directed the attention of the people of Argentina and the people of other countries to the cruelty of the Argentine regime. Pressure from the American president, Jimmy Carter, also helped to soften the oppressive measures of the military regime.

In April 1982, the Argentine military, in an apparent attempt to increase its popular support, attacked the Falkland/Malvinas Islands, offshore islands then ruled by England. A counterattack by the British fleet, however, defeated the Argentine military and consequently toppled the cruel regime. A total change in the political culture occurred: a democratically elected government was installed and relative peace and harmony were established.

Like the dramatic political reversal in Argentina, the dualistic mode of thinking that assigns other nations to friendly or unfriendly categories can reverse direction. In response to a change in circumstances, the image of an adversary changes from implacable enemy to ally, from malevolent to benevolent, from dangerous to safe. America's image of Joseph Stalin and the Soviet Union changed from negative to positive when Germany attacked the Soviet Union in World War II. Indeed, our adversaries in World War II—Germany, Japan, and Italy—became staunch allies after the war was over. The Soviet Union, in contrast, was switched back to the negative track as we entered several decades of cold war; then, after its breakup, it regained a more positive image. In other parts of the world, feuding neighboring states have learned to swallow their differences and engage in constructive reciprocal relationships.

If the images and beliefs regarding other individuals, groups, or nation-states are malleable enough to be transformed by a change in circumstances, can they be influenced by preventive or interventionist strategies? The understanding of individual psychology can provide a base for formulating programs to benefit humans in general. In particular, political and social programs need to take into account the way that pernicious ideologies exploit the propensity for biased beliefs, distorted thinking, and malevolent images to bind their adherents together and make enemies of outsiders and dissidents. The effectiveness of propaganda in activating fears, paranoia, and grandiosity should be considered as well. We have tools to overcome these aberrations in individual therapy, but we must find a way to use this knowledge in a more extensive way.

BROADENING THE PERSPECTIVE

The understanding of how the benevolent side of human nature can be energized provides another base for counteracting harmful behavior. Broadening our perspective to see "alien" people as human, like ourselves, can arouse empathy for their vulnerability and suffering. For example, televised images of starving Ethiopians in the mid-1980s aroused an outpouring of sympathy for their plight that was translated into large shipments of food supplies. Obviously, education plays an effective role in engaging people in charitable acts to relieve suffering. Can the same kind of empathy be aroused to buffer antagonism toward other peoples? The presence of a bias toward adversarial groups inhibits the arousal of empathy for them.

It is important to ascertain the perspective of the oppositional group and to recognize that biases exist on both sides. If our adversary already sees us as the Enemy, he is likely to react to our actions at a primal cognitive level. If we are not aware of this, we will be more vulnerable.

Of course, international organizations like the United Nations can provide intervention programs to prevent and moderate conflicts. They will moderate more effectively, however, if they are aware of the biased thinking and images that the parties bring to the negotiating table. The mediators can adopt a broad perspective that will consider the narrow perspective of each of the opposing parties. Also, by addressing social and economic factors that lead to the development and exacerbation of biased thinking, international organizations may be able to ward off acute hostility leading to conflict. The mediators will need to be able to project into the future the possible consequences of various courses of action.

We should formulate programs based on our understanding of the brighter side of human nature: its benevolent and rational components. Thus, we may create or reinforce prosocial structures that will counteract hostility and violence. The innate qualities of empathy, cooperation, and reason, which are just as intrinsic to human nature as hostility, anger, and violence, can provide the building blocks for prosocial structures. Understanding and empathy are more readily extended to members of one's own group than to outsiders, but there is no immovable block to expanding them to all humankind.[4]

By misreading other people's impressions and intentions, a person may experience a great deal of distress and anger. Of course, he or she may at times correctly infer other people's hostility and require strategies to deal with it. The initial step, however, is to cultivate the capacity to scrutinize one's construction of other people. We need to develop the ability to envision other people's image of us with reasonable accuracy.[5]

The young child is unaware that other people view a situation differently than he does. Like the director of a play, he believes that he knows the motivations of the other players. He also assumes that, as active participants, they have the same view of the action that he does. Thus, when the child is punished, he regards himself as an innocent victim and his parents as mean and unfair. He also believes they know that they are being mean. As a child develops, of course, he realizes that different people can have differing views of a situation.[6] With socialization he becomes increasingly capable of absorbing the moral messages of his family and culture—for example, that it is wrong to hit a younger sibling for upsetting his playthings.

When people are in an aroused state—for example, the hostile mode— their thinking reverts to the level of a young child. If another person appears to run roughshod over their needs, they relive the old drama: the other person is wrong and bad and is knowingly mistreating me. As a participant in a group, the individual perceives a similar theme in response to a challenge to the group. The members of his group are innocent victims, and the challengers are wrong, bad, and immoral. And as in the medieval morality plays, the sinners must be punished.

A similar kind of thinking becomes pervasive throughout a nation when it is infected with war fever. At the start of World War I, for example, even the most highly educated individuals, the intellectual elite on both sides, were gripped by the idea that their country was completely innocent and right and the antagonists were the aggressors and all bad. Like young children, they could not or would not credit the notion that people on the opposite side felt the same way about them.[7]

An individual generally sees himself or herself as a person of goodwill. This identity is constructed around the theme "I am a good person." His conscience prods him to preserve this self-image. When he is pressed to do something that involves harming other people, he has to square this action with the dictates of conscience. Here the group ideology provides an exemption from the rule "Thou shalt not kill." The greater good calls for transforming the rule. By going off to war or participating in genocide, he is killing off evil and thus is doing good.

An important prelude to this change is a shift in the prevailing image of the opponent. It is admittedly difficult for a person, whether acting as an individual or as a member of a group, to change his narrow, highly focused outlook to a broader perspective. *Distancing* oneself from one's self-centered perspective depends on accepting the principle that although one's perspective *feels* real and legitimate, it could be biased or even totally wrong. Having acknowledged the possible fallibility of his perspective, a person can step back and raise questions about its validity:[8]

- Is it possible that I have misconstrued the apparently offensive behavior of another person (or group)?

- Are my interpretations based on real evidence or on my preconceptions?

- Are there alternative explanations?

- Am I distorting my image of the other person or group because of my own vulnerabilities or fears?

Even if the individual believes strongly in the validity of his self-centered perspective, critical questions such as these challenge his egocentric mode of thinking. Raymond, for example, was able to modify his notion that his wife's criticisms were an assault on his manhood and to assimilate the idea that she was not being malicious when she criticized him.

Successful distancing from one's egocentric construction of a conflicted situation goes hand in hand with *decentering*: reframing the meaning of a situation with the objectivity of an impartial observer.[9] Decentering can also allow for the formation of an empathetic understanding of the opponent's perspective.

Another clinical example demonstrates how understanding the perspective of an "adversary" may help resolve a conflict and the consequent harmful behavior. A father and his eighteen-year-old daughter were locked in an ongoing battle characterized by the father's strict disciplinary rules and his daughter's defiance. They were both enraged to the point that they had exchanged blows. In a joint interview, each angrily explained his or her view of the problem. Following a discussion regarding the value of role-playing the "adversary," the father assumed the role of the daughter, while the daughter took on his role. They enacted a typical conflict regarding curfew for the daughter and her use of the family automobile.

After a dramatic staging of the quarrel, the therapist asked both father and daughter how they felt. The father commented that playing the part of his daughter had brought to mind his conflicts with his own father as a teenager; he acknowledged that he could now understand how his daughter felt. He observed that while in his daughter's role, "I felt that I was being stepped on—that he [the role-played father] had no respect for my feelings, that he was only concerned with himself."

The daughter described her insight after role-playing her father: "I could see that he really did care about me. He was really worried that if I didn't come home till late that something bad might have happened. He was hard on me because he cared, not because he was trying to control me." During the role-play, she exhibited a cardinal feature of empathy, as she was actually able to experience her father's anxiety.

The experience of entering into the other individual's personal world during the reversed role-playing was a crucial factor in ending the mutual antagonism. The expanded, more objective perspective in each instance involved understanding that the two were in conflict, not because either was malicious, but because each was worried and hurt by the other's oppositional behavior. The new, more balanced understanding of the perspective of the other person undercut the anger and provided an opportunity for resolution of the conflict. The intellectual understanding and reframing also set the stage for the development of true empathy.

EMPATHY

Have you ever noticed how spectators in a football game imitate the actions of the players? As a place kicker gets set to boot the ball toward the goal post, his fans brace themselves and move their feet as though to help him get the ball aloft before it can be blocked. We observe similar automatic reactions watching people wince or writhe when they see another person in pain.

Adam Smith captured this phenomenon when he wrote as early as 1759:

> When we see a stroke aimed, and just ready to fall upon the leg or arm of another person, we naturally shrink and draw back our leg or our arm; and when it does fall, we feel it in some measure and are hurt by it as well as the sufferer. . . . Persons of delicate fibers and a weak constitution of body complain, that in looking on the sores and ulcers which are exposed by beggars in the streets, they are apt to feel an itching or uneasy sensation in the correspondent part of their own bodies.[10]

Such reactions give the appearance of being involuntary and mindless—like the knee jerk when the patellar tendon is struck with a rubber hammer. Yet the capability to experience true empathy may be a later development of these vicarious reflex reactions, which are related to the basic connectedness between people. Children and adults unconsciously mimic the facial expressions and postures of other people.[11] The spread of crying across a nursery is probably imitative and does not involve feelings for the baby who initiated the crying.[12] Nonetheless, there is strong evidence that children as young as one year can experience the sadness of another person.[13]

Richard Lazarus has elicited anxiety responses in human subjects by showing movies of industrial accidents.[14] Whether responding with body movements or emotional distress, humans seem to be hardwired to respond to other people's painful experiences. In contrast to sympathy, which may involve feeling sorry for another person without experiencing his distress, true empathy consists of sharing another person's perspective and his specific distress.

To experience true empathy it is not sufficient simply to imagine another's perspective. Psychopaths may be quite skilled at "mind-reading" as a strategy for manipulating people. True empathy requires that we care about the person in pain. An empathic perspective also involves anticipating and caring about the possibly harmful impact of one's actions on other people.

Movies such as *Schindler's List* and *Amistad* have demonstrated the innate disposition of diverse audiences to fuse their own identities with the plight of either the Jews in Nazi-controlled Europe or the Africans on slave ships. The fact that movies can elicit feelings of empathy demonstrates that we have the capacity for such social emotions even when the people and events are remote from us. In our personal experiences outside the cinema, we may have to make a concerted effort to transcend our self-centered perspective in order to become emotionally involved in the distress of people removed from us by race, religion, or geography.

In some situations, empathy needs to be inhibited. Surgeons and other medical personnel must cut, inject, and anesthetize their patients without being swamped by feelings of empathy. Doctors and nurses generally can distance themselves from the pain and distress they must impose and thus carry out their responsibilities efficiently. Unfortunately, the same kind of desensitization occurs in people who are exposed to, or participate in, torturing or killing other people. The lack of empathy for the victim during an attack holds true for security forces beating dissidents, husbands attacking wives, and mothers abusing children. This nullification of empathy can occur whether a conflict is narrowed to a family member or broadened to the community of nations.

As a result of the finding that American infantrymen showed a reluctance to fire at the enemy during the Korean War, a specialized training program was initiated for the trainees during the Vietnam War to desensitize them to killing the enemy soldiers.[15] The program achieved its purpose: when engaged in battle, the former trainees were far more consistent in firing at the North Vietnamese troops. Anecdotal accounts indicate that a subordinate who was initially repulsed by torturing a prisoner became more accustomed to and eventually even enjoyed the activity.[16]

Empathy can occur automatically when we are interacting with people whom we care about or with whom we share a common cause, but it is more difficult to experience this reaction with people who seem substantially different from us. Empathy is particularly difficult to experience when we feel disdain or anger toward another person or group. Take the example of a little girl who has slipped on the ice and injured herself. We are likely to feel genuinely sorry and come to her rescue. But what are our feelings toward an adult, obviously drunk, who slips and falls? Typically, we feel amusement or even revulsion. The little girl's accident strikes a resonant chord: we can identify with her vulnerability and her pain.

When we attribute somebody's distress to a factor beyond his control, we can empathize with that person. However, if we hold a person responsible for a harmful occurrence, especially if we attribute it to a moral or character defect, we are prone to disparage the sufferer: the drunkard *deserves* his misfortune; he brought it on himself. Similarly, members of a stigmatized group are often considered unworthy of our concern; people get what they deserve.[17] When members of one's group feel antipathy toward those in another group, during riots or war, for instance, they can inflict pain without feeling guilt or empathy. In fact, they feel that they are doing the right thing.

Nonetheless, it is possible to experience empathy even for someone whom we have devalued. When we can successfully assume the perspective of a depreciated victim, we can experience, albeit faintly, what he is feeling. If we can imagine ourselves being in the shoes of the other person and experiencing discrimination or oppression, we may be able to share the perspectives of abused outgroupers and feel their distress. The process of identifying with the victim facilitates the development of an enlightened moral perspective.

THE PERVERSION OF THE MORAL DOMAIN

Like young children, adults and groups often make self-serving interpretations and invoke an egocentric form of justice and morality. They frequently react to an actual or supposed offense with a reflex impulse to punish the transgressor, even when the latter is unaware of having transgressed. The doctrine of justice—right and wrong, good and evil—has accommodated the ideology, political aspirations, and vengeful motives of countless militant groups. The theme runs like this, "They (the government, the capitalists, the Jews) have treated us unjustly. We are therefore compelled in the name of justice to break their grip on us and punish them."[18] In publicized instances, the extremists, given the proper timing and provocation, have resorted to violence to attain their goals. The doctrine of retributive justice has historically crowded out other moral doctrines, such as "Thou shalt not kill."

Mass executions during the French and Russian Revolutions were justified by the revolutionaries on the basis that the aristocratic classes were depraved and deserved to be punished—even though the victims had no sense of having done anything wrong. They were killed because they were represented in the ideologically driven images of their captors as Enemies of the people.[19]

In their brutal attacks, security police react to the malevolent image of the dissidents planted in their minds by their political or military leaders. Although they do not fear being harmed personally, they regard the state as vulnerable and consequently feel they are doing good in assaulting, even torturing, those who would undermine the system.

Despite their ability to feel empathy for someone in pain and to be fair in sharing, children often respond with an egocentric code of justice when their own interests are involved. A six-year-old boy hits his younger sister for disturbing his carefully constructed model of a skyscraper, with the reasoning that she deserves to be punished for her offense. His notion of crime and punishment reverts to a primitive level when he believes his rights have been violated. Because he is distressed, he readily waives the rule about not hurting his sister in order to impose his own rule of justice. On another occasion, however, if she falls and hurts herself, he will feel sorry for her. When his self-esteem is reinforced by being helpful, he can bend over backward to be fair.

Typically children will complain of mistreatment if their wishes are not given priority over those of other children. If a parental verdict goes against them, they denounce the injustice of the judicial procedure and the unfairness of any punishment. Although children and adults may consciously believe that they are just, their actions are often driven by egocentric beliefs. The desire to punish, to exact justice, provides a sense of power, self-righteousness, and freedom from restraint. Consequently, individuals tend to take it on themselves to punish others for their supposed wrongdoing.

Instead of trying to clarify probable sources of conflict, their conscious rationale is that they are asserting their own right to get angry and retaliate. In actuality, their claim for justice emanates from a primal reflex pattern: their hurt feelings automatically trigger the impulse to retaliate. Their justification is the belief bridging the hurt feeling and the punitive impulse: "Since I am hurt, I am entitled to punish the offender." Further, they try to control or even eliminate others in order to alleviate their own sense of frustration or impotence.

The decision to hurt another person for alleged misdeeds more or less follows an algorithm used in the criminal justice system to determine whether a person's behavior constitutes a crime and should be punished. Although these criteria are useful in assessing legal guilt or innocence, they are often biased. When they are used in everyday interactions to evaluate another person's transgressions, they are often biased.

- He knew or should have known that what he did would hurt me.

- Therefore, his action had to be intentional.

- His major motive was to hurt me in some way.

- Since he is wrong and bad, he should be made to suffer.

- Such a punishment is just.

Society's notion of crime and punishment reflects the individual's schema of offense and retaliation. Also, the principles of the criminal justice system may influence aspects of the individual's concept of transgressions and appropriate punishment and vice versa. Just as the authorities believe that they are enforcing the law by incarcerating a thief, a man tricked in a game of cards or deceived by a spouse invokes the principle of retribution to reduce his stress at being violated.

We divide our world into different domains. Our personal, private domain includes those aspects of our life in which we have a special investment. The structure of meanings that we attach to ourselves, our close relations, and the goals and ideals of our group are incorporated into our self-concept. A complex system of rules, beliefs, and injunctions protects us and others who belong in our personal domain. This system forms the core for group morality: fairness, reciprocity, and cooperation. As we mature, our personal domain expands in time to include our allegiances: race, religion, social class, political affiliation, and country. The mantle of protection spreads to cover individuals who share membership with us in these groups. We generally acknowledge the group norms of fairness and concern and anticipate that our violations will be punished by group-induced shame, guilt, or anxiety.

Since we have such a vital investment in our personal domain, we are continuously alert to both threats and opportunities from the outside. Events that either devalue or enhance our personal domain make us feel either angry or happy; those that threaten our domain lead to anxiety. Because of the vital importance of our self-esteem and the safety and security of ourselves and significant others, we tend to overreact against a supposed Enemy.

Even when people seem to be more group-centered and are prepared to sacrifice themselves for the good of the group, they retain much of their egocentricity. When we are involved in a group process, our egoism becomes fused with the needs and evaluations of our group. Because our natural orientation tends to be egocentric, we value and have a positive bias toward members of our own circle and are prone to have a corresponding negative bias toward outsiders, especially if they are competitive with or opposed to our group. Individual egoism blows up into group egoism, or groupism. Instead of the xenophobia of "I versus the outsiders," the group member marches to the tune of "We versus the outsiders." The ideology of the group often overrides basic principles of humanism and universal morality. Indeed, as Koestler points out, groupism is potentially more devastating than individualism because it can lead to ethnic strife, persecution, and wars.[20]

The subordination of individuals to group expectations offers them so many advantages that it becomes difficult for them to take an enlightened position at odds with the ideology of the group. Cooperation, solidarity, and reciprocity with other group members are gratifying. Groups not only promote a sense of belonging but give individual members a sense of power that neutralizes the sense of inadequacy that many experience as solitary individuals. Unfortunately, the tighter the group allegiances, the greater the perceived differences between ingroupers and outgroupers. Prevailing biases stirred up by pronouncements regarding religious heresy, class warfare, or political subversion are magnified as they reverberate around the group. The demand to control, if not eliminate, the threatening antagonists becomes accentuated. Identifying an enemy greatly enhances group solidarity and provides widespread gratification. The malevolent images may lead to persecution or massacres.

The perpetrators banish the victims from the universe of moral obligation. Since these opponents are wrong, bad, evil, they are not entitled to human rights and *deserve* to be harmed or killed. "Justified" violence provides immediate satisfaction. Indeed, the very act of harming another person tends to dehumanize that person; it consolidates the image of the victim as expendable, worthless, and subhuman. In theory, the concept of universalism advocated by religious and other ideologies would provide a barrier to religious, racial, or ethnic violence. However, the philosophies of universal rights and of the sacredness of human life tend to break down as hostile pressures build up.

When the cause becomes paramount, there is frequently a total disregard for human life. Even individuals not considered dangerous enemies may be killed indiscriminately. Incidents of civilian massacres throughout the world, and bombings such as those of the World Trade Center in New York and the Federal Office Building in Oklahoma City, were perpetuated to make a political statement and undermine the government. The anonymous victims were viewed as disposable: their only significance to the terrorist is that in dying they helped the cause.

How often have wars and massacres been carried out because the perpetrators were convinced that they were obeying a higher moral code that nullified sympathy for their victims? In relations between both individuals and groups, the sense of morality is transformed into an idiosyncratic concept of justice that may exclude caring. The Serbian perpetrators of the massacre in Bosnia in 1994–96 believed they were honoring the code of justice because the Muslims allegedly supported the wholesale slaughter of Serbs by Croatians during World War II.[21] The charges were untrue, as Serb leadership knew, but their soldiers believed them. The pursuit of class warfare in the Soviet Union, China, and Cambodia was justified on the basis that the privileged classes had deprived and exploited the working class. Hence, the mass executions were based on the judicial principle of punishing people for their wrongdoing.

MORAL CONCEPTS OF JUSTICE AND CARING

The humanistic code, the notion of the universality of humankind, is an antidote to the rigid perspective characteristic of tribalism, nationalism, and self-serving morality. If the value of human life overshadows one's political or social ideology, it is more difficult to carry out harmful behavior.

People have many different moral voices. Consider a proud father who wants to praise his daughter, who received top grades in her class. Justice might decree that she should be more highly rewarded than a younger brother who did poorly. But compassion for the younger sibling would temper the father's enthusiasm for his daughter's performance if the younger sibling had to struggle even to pass and was already hypersensitive to comparisons with his older sibling. Although it may be fair to apply equal standards in evaluating performance, it is manifestly *uncaring* if this judgment unnecessarily hurts another person.

Writings on morality by developmental psychologists initially focused on the increasing sophistication of the understanding of justice as the child matured. Lawrence Kohlberg outlined a series of six stages of moral development, which culminated in the most humanistic version of justice.[22] Carol Gilligan added the concept of caring as an equally important moral voice.[23]

The usual, conventional concept of moral justice is focused on protecting the separateness of people. This *individualistic* orientation emphasizes rights and entitlements: life, liberty, and the pursuit of happiness; equal opportunity; fair treatment and justice. This orientation is centered on the assumption that people have competing claims of justice and are in conflict with each other for available resources or personal reinforcement. The *caring* orientation, in contrast, assumes a connectedness perspective. The moral precepts emanating from this orientation revolve around sensitivity to others' needs, responsibility for their welfare, and sacrifice of one's own needs for the needs of others. When confronted with a complex situation, people have to decide whether to assert their rights, to express caring, or simply to pursue their self-interest.

Research has shown that even preschool children are capable of making moral decisions. They are able to distinguish moral transgressions from other forms of misbehavior, specifically social convention violations. They know the difference between a social lapse, which causes them shame or embarrassment, and a deliberate moral transgression, which may hurt another person. The children are able to view moral deviations as wrong even when an authority is not present and there are no specific rules forbidding these transgressions. They are also able to distinguish between justice and caring. They can specify when there is a violation of rights, fairness, and justice, and when there is a default pertaining to relationship, responsibility, and feelings.[24]

Even at an early age, a child is capable of feeling glad when he is able to help a fellow student and guilty when he has hurt her. Children are able to attach the label of "wrong" to egoistic decisions such as going to the movies instead of visiting a sick friend, and "right" to prosocial actions such as finding the owner of a lost puppy. Of course, even when children know the right thing to do, their self-serving motives often win out. And they are often quite skilled at making a logical case for an exception to the moral rule.

Applying the moral code often entails costs, at least in terms of expending energy to control a harmful impulse or sacrificing a personal goal in order to help someone else. A major aspect of socialization is teaching the child the long-term value of making a deliberate effort to control impulses or cravings "that come naturally." In time the individual develops inhibitions that automatically choke off an impulse to strike another child or take a piece of candy from him. The control of hostile or egoistic impulses is grounded in appropriate beliefs learned from other people. Direct punishment or reward is generally effective in forming specific beliefs regarding acceptable and unacceptable behavior. Learning based on the observation of other significant persons, such as parents, also provides structure for enhancing social motivation and controlling harmful impulses. Finally, people learn codes of conduct by being informed of the rules. The experience of shame for publicized social lapses and guilt for harmful acts adds to the mental structure.

Learning to engage in prosocial or helping acts is facilitated by the automatic empathy one might feel. In the absence of empathy, such behavior is encouraged by the expectation of external approval or self-approval. In each case, one's self-image as a worthwhile person is enhanced.

Cooperation reduces the negative perceptions that groups may have of each other. In Sherif's study, eleven-year-old boys at summer camp were divided into two groups and engaged in intense competition with each other.[25] Each group devalued the boys in the other group and ultimately engaged in destructive acts, such as raids on the other group's bunks. The boys in the losing group were demoralized and started to fight with each other.

The experimenter then created circumstances that required cooperative action by the two groups. Their truck stalled, and the boys had to work together to push it. They also had to cooperate to fix a broken water pipe. By working together to solve problems, the boys developed positive attitudes toward those in the other group. Other studies have shown that simply joining people does not in itself reduce bias, but working toward a common goal does.

ALTRUISM

Many strands of the social fabric are woven together to form altruism: empathy, caring, identification with the underdog, and a benevolent self-image. When the pattern of another person in distress or danger becomes salient, it prods a potential helper to assist the victim. Altruism exists on a continuum ranging from providing commonplace services to making significant sacrifices or engaging in serious risks to save other people's lives. Generally, self-sacrifice and risk, earmarks of altruistic behavior, far outweigh the tangible benefits to the helper. Sometimes the only return for saving a life is the inner pleasure of having done the right thing. Key components of an altruistic act are the rescuer's respect for human life and the vicarious experience of the victim's fear or suffering. The victim and the rescuer are bound together in a display of a common humanity. A member of an inner-city gang who breaks the gang rules and helps an injured member of a rival gang risks disapproval, if not punishment, from his own gang and may not even receive thanks from the victim. His unrewarded action represents pure altruism.

Some people exhibit altruistic behavior only within the confines of their own group. A soldier volunteers for a dangerous mission behind enemy lines. A follower of a religious cult sets off a suicide bomb in a marketplace to protest the treatment of his sect. Such events are tokens of a narrow altruism. The individual limits his sacrifice to the goals of his specific group. This investment in the group's mission and well-being is generally an expression of group narcissism as well as altruism. The members of the group show commitment to the cause, loyalty to fellow members, and a willingness to risk or give up their lives, but have little concern for the lives of people on the other side.

Most militant organizations, whether political or religious, have a collective self-image of superiority, if not supremacy. They believe they have the pipeline to truth and feel disdain for the nonbelievers. Extremists, whether militant groups like the skinheads or the political elite of regimes like Hitler's Germany or Stalin's Soviet Union, often have visions of a perfect world controlled by their group or nation. Driven by their grandiose dreams, they seek to expand their power through conquest or revolution. Far from feeling compassion, they aim to eliminate their victims. The followers merge their own wishes for power and glory with those of the leader and feed on the expansion and conquest.

Popular chauvinism and imperialism may parade under the banner of a war of liberation. American troops drove the tyrannical Spanish out of the Philippines in 1898 apparently out of benign motives but proceeded to try to impose American control by force. A prolonged war resulted in the large-scale killing of rebellious Filipinos. Missionaries have sacrificed their lives throughout the world in an effort to convert the heathens to the "true religion." Such religious imperialism seems to be altruistic, but it is driven in part by the collective narcissism of the religious proselytizers.

Group narcissism is the antithesis of altruism in many ways. Militarists and revolutionary groups emphasize intangible institutionalizing values such as religion, country, and home and resort to violence to enforce their program. Pursuing the philosophy that "the end justifies the means," they declare war on adversaries within and outside of their group. Internal dissidents are persecuted as heretics or traitors, and outside opponents as enemies. The follower's identity is subordinated to the goals of the group, which are formulated by the leaders. In this context, self-sacrifice by the followers represents a naïve and circumscribed altruism.

Enlightened Altruism

Humanistic altruism, an antidote to group narcissism, is universal. Such enlightened altruism can circumvent or undermine tyrants. Returning to a previous example, the Mothers of the Plaza del Mayo marched at their own peril with white scarves bearing the names and the dates of the disappearance of their sons and daughters, who had been arrested and executed by the military dictatorship in Argentina. This act of courage helped to shake the foundation of the regime. The focus of humanistic altruism is interpersonal and global: individuals are seen as fellow human beings rather than as stereotyped members of a group. Group altruism is frequently activated to rescue large groups of people, as in the massive supply of food to starving Ethiopians in the mid-1980s. The video images of starving children with distended abdomens touched the collective conscience of the world. Similarly, people of many countries have rallied to provide aid to victims of floods and earthquakes.

In general, narcissism and altruism represent opposite modes in the dualistic personality organization. Narcissism favors the self; altruism favors others. In the narcissistic mode, the person is invested in promoting his own interests and expanding his personal domain. He is competitive with other people, asserts and defends his own rights and privileges, and fights to maintain his own individuality and identity. In the altruistic mode, the person is concerned about the welfare of others, gets gratification from subordinating his own interests to the needs of other people, and is vigilant about protecting the rights of the underdogs, the underprivileged, and the disadvantaged.

Depending on the circumstances, people can alternate between the narcissistic and altruistic modes. Situations offering an opportunity for self-enhancement or posing a challenge to the personal domain will activate the narcissistic mode. Situations in which other people are in danger or distress ("a cry for help") can activate the altruistic mode. Even though enlightened belief systems may govern the thinking and actions of people within groups or nations, problems between these entities are prone to be exacerbated by the biased thinking of the primal mode.

The narcissistic-expansive and the altruistic modes are of particular significance in relations between groups or nations. The kinds of beliefs incorporated into each of the modes give some hints about which beliefs should be undermined and which need to be reinforced. Although the *causes* of conflicts are numerous and complex, the *solutions* can be facilitated by greater attention to the psychology of the leaders and followers on both sides. As indicated earlier, representatives of supranational organizations, like the United Nations, need to take into account the biased, polarized thinking of the conflicting parties in attempting to resolve disputes. The mediators should be cognizant of the dualistic belief system, with an eye to facilitating a shift from the narcissistic-expansive orientation to the altruistic-humanistic one.

NARCISSISTIC-EXPANSIVE	ALTRUISTIC-HUMANISTIC
Our group (nation, etc.) is superior, the chosen, the elite.	All people are equal and worthwhile.
Outsiders are potential enemies.	Outsiders are potential friends.
Our rights and claims supersede others' rights.	No group has a prior claim.
Their lives are expendable.	All lives are sacred.
If I help ingroupers, it makes me a better person.	If I help outgroupers, it makes me a better person.

The acute concern for the safety or life of another human being is dramatized by acts of heroism in civilian life. Every year since 1904 the Carnegie Hero Fund Commission has rewarded individuals who have shown unusual courage by saving a stranger's life. Typical dangerous deeds include rescuing a man being attacked by pit bulls; climbing into a burning pickup truck to save the life of the trapped driver; and intervening in a rape assault at the risk of one's own life.

Researchers have been unable to pinpoint specific personality features that characterize these heroic individuals. A man who jumped onto subway tracks to rescue a child from being run over by a subway train simply stated that if he had not acted, "I would have died inside."[26] Similar statements were elicited by other individuals involved in altruistic acts of heroism. These dramatic acts depended in many cases on more than simply a spontaneous urge to save a life. Often an important factor was that the rescuer was competent or strong enough to carry out the mission. In many instances, the rescuer was a "risk-taker" who had a strong confidence in her ability to do the job and to escape with her life.

The "Just Christians" who risked their lives to save Jews during the Holocaust also exemplify pure altruism. A well-controlled study of 406 rescuers conducted by Samuel and Pearl Oliner identified a number of characteristics that differentiated rescuers from a matched control group who had not attempted to rescue Jews.[27] Interviews and questionnaires administered to both groups established that the rescuers were more empathetic and more susceptible to feeling other people's pain than the nonrescuers. A significant proportion of the rescuers recalled feeling empathy for the first person they had ever helped. They also had a greater sense of responsibility and commonality.

The authors attributed the rescuers' altruistic disposition to a number of child-rearing practices: praise for good behavior, reliance on reasoning and explanation more than on harsh discipline, presenting a caring parental role model, and inculcating a liberal attitude toward people different from them. The rescuers had incorporated the humanistic values of their parents as expressed in their responses to a question regarding "alien groups." The rescuers were more likely than the nonrescuers to endorse items indicating that they regarded Turks, Gypsies, and Jews as very much like themselves.

An interesting demographic feature was the finding that the rescuers represented a cross-section of the country: farmers and factory workers; teachers and entrepreneurs; rich and poor; single and married; Catholics and Protestants. As pointed out by Samuel Oliner, the main distinguishing characteristics of the rescuers were their connectedness to other people, their commitment and caring. Their altruistic behavior reflected their everyday patterns of perception and behavior: their belief in the sacredness of life and their realistic attitudes toward authority and mistakes and the rules for deciding right and wrong.

The spirit of volunteerism is high today, with nearly half of American adults donating blood and a majority raising funds for charitable causes, volunteering in hospitals, sponsoring youth groups, or contributing to other community causes. Even more important are the innumerable everyday kindnesses. Although these behaviors may receive a lower rating on an altruism scale than the reported heroic actions, they illustrate the human qualities of concern, generosity, and empathy. People perform many different acts of helpfulness and generosity without expecting praise or commendation. The altruistic act is its own reward. Almost anyone, for example, would help a lost child and attempt to contact his family. Most people would readily offer to guide a blind person across an intersection. A large proportion of the population would donate funds to help a sick child in need of expensive treatment.

It is possible to trace patterns of self-sacrifice throughout the animal kingdom. Although this behavior is instinctive, certain common characteristics suggest that it is probably a forerunner of human altruism. Ethologists have, for example, described innate, automatic, self-sacrificing behavior among social insects, certain species of birds, and certain vervet monkeys. Helping behaviors are commonly observed among the higher primates such as gorillas, chimpanzees, and bonobos.[28] The ethnologist Jane Goodall has described a case of a water rescue of an infant chimpanzee by an adult ape.[29] More recently, a highly publicized account described the rescue by a gorilla of a young child who had fallen into its cage.[30]

The social psychology literature strongly supports the notion of a general tendency to feel empathy and to behave in a benevolent way. There is substantial experimental evidence that the altruistic mode can be "tuned up" by appropriate interventions. Interestingly, a bystander is more likely to intervene in an emergency if he is alone. The presence of other bystanders evidently has an inhibiting effect on the motivation to be helpful. However, it is possible to train people to overcome this inhibition. Diffusion of responsibility has been the preferred explanation, but that cannot account completely for the reaction. A bystander is far more likely to behave altruistically in the presence of a friend than when he is with a stranger.[31]

When subjects were asked to imagine themselves in the situation of a person in distress or to imagine how that person experienced his situation, they showed a significant physiological response. Also, in comparison to control subjects who had not undergone the imagery exercise, the experimental group was more likely to be sympathetic and to help for purely altruistic reasons. A number of experiments have shown that when students are "primed" by a lecture on altruistic behavior, they are more likely to stop to help a sick person, even at risk of being late for class.[32]

APPLICATIONS TO SOCIETY

This kind of prosocial training has been shown to have a practical application. A program that taught younger children how to take the perspective of others increased their prosocial behaviors. Even more striking, however, was an experiment with fifteen delinquent boys. These boys received training in taking the perspective of others and in seeing themselves from the perspective of others. This intervention had a positive impact on their subsequent behavior compared to a control group not receiving this intervention.[33]

A further practical application of this model focused on empathy training with third- and fourth-grade students in Los Angeles public schools. The empathy training concentrated on identifying feelings depicted in photographs, discriminating different emotional intensities in real-life situations, and enhancing perspective-taking skills. The students showed a significant improvement in self-concept, social sensitivity, and aggressive behavior.

There is also evidence that the teaching of values can be blended into the regular curriculum of a school. The Child Development Project in San Ramon Valley, California, consists of using texts that encourage perspective-taking and prosocial behavior. Reading exercises that include narratives with a moral lesson are discussed in terms of social meanings such as friendship and feelings. Group work is designed to instill the values of cooperation. The school environment is created to encourage the children to participate productively and prosocially. Older children are expected to undertake community service projects. Initial results comparing schools with the program to those without the program showed an increase in prosocial behavior and social sensitivity.

Policy makers, educators, and parents can draw on a well of untapped psychological resources to change the beliefs embedded in individual and group egoism. Drawing on her own scientific work, as well as that of others, Dr. Leslie Brothers demonstrated how our brains have evolved a specialized capacity for "exchanging signals" with other brains.[34] She suggests that even individual neurons respond to social events. She argues that our brains working collectively create an organized social world. The capacity of the brain, twice as great as that of other primates, is available to generate rational thought and benevolent behavior. The challenge of the next millennium will be to utilize these wellsprings to provide a more benevolent climate for the human race.

COGNITIVE THERAPY FOR INDIVIDUALS AND GROUPS

Gloria asked Raymond, "When are you going to fix the faucet?" Raymond glowered at her and yelled, "Get off my back!" Gloria replied, "I would if you'd act like the man of the house and do your job!" Raymond snarled back, "I'll show you what a man is," and punched her in the mouth.

In chapter 8, I introduced Raymond, a typical wife batterer, and discussed the way he developed a hostile worldview at a very young age. This negative worldview influenced his thinking—and consequently his behavior—in adult life. Raymond's thought processes leading up to his assault on Gloria and the ways his anger may be "deactivated" serve as a prototype for understanding and dealing with the hostility complex.

Although an observer might view Gloria's complaint and Raymond's assault on her as a simple cause-and-effect sequence, his harmful behavior can best be understood in terms of his underlying beliefs. These beliefs are organized as a kind of algorithm that determines his response to a perceived threat or challenge. This algorithm takes the form of a series of decision rules for identifying an offense, evaluating its nature, and responding to it. The interpretations of the relevant features of the offense—who is responsible for its occurrence, whether he or she intended to do harm, and the pros and cons of retaliation—are integrated into a composite assessment of the situation. This global evaluation then determined the nature of Raymond's response, in this instance, a physical assault.

The understanding of the psychology of hostile interactions can be applied to problematic situations to reduce the likelihood of harmful behav-

Figure 13.1

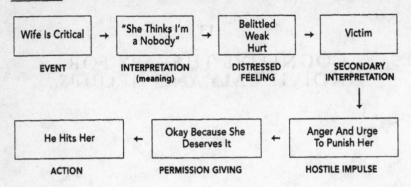

ior. In reviewing the hostility complex, the therapist can focus on the crucial psychological and behavioral components and design specific interventions for any or all of them. A recent study by Eric Dahlen and Jerry Deffenbacher demonstrated the efficacy of such an approach. Compared to a control group, cognitive therapy significantly lowered ididuals' anger and enhanced more positive adaptive forms of self-expression. Figure 13.1 illustrates these components. The specific strategies that could be utilized will be discussed in the context of their components.[1]

IDENTIFYING THE MEANING OF THE EVENT

It is obvious that becoming angry is not an invariable response to being scolded. Whether a hostile reaction occurs depends on how all the factors relevant to the event are processed: the nature of the current relationship, memories of previous conflicts, and the specific vulnerabilities and behavior patterns of the partners. Overreactions such as Raymond's can occur when an external event (such as Gloria's criticism) strikes at the specific vulnerability of the partner. Uncovering the meaning of the stimulus can clarify the partner's inappropriate or excessive reactions.

The cognitive therapist has a variety of strategies available for eliciting the meaning of an event. From my earliest work with cognitive therapy, I concentrated on teaching the patient to recognize the automatic thoughts that preceded a specific feeling or impulse. These thoughts occurred automatically, without prior reflection or cognition, and were fleeting in nature. Initially, when experiencing sadness, anger, anxiety, or elation in response to a stimulus situation, the patient would be instructed to try to capture the thought that initially preceded the feeling or impulse. The content of the thoughts revealed the meaning, often idiosyncratic, of the activating event and made sense of the emotional response.[2]

The content of the automatic thoughts in depression generally revolve around themes of failure, pessimism, and self-criticism; in anxiety the dominant theme is danger, and in anger, being wronged. Eventually I observed that the automatic thoughts could occur at several points in the hostility complex: preceding a feeling of emotional or somatic distress, preceding and following the experience of anger, and finally, following an impulse to attack the other person. The initial disturbing thought had a content such as, "She's putting me down," or, "He doesn't care about me." The distressed feeling might be emotional, such as a flash of anxiety or sadness, or a physical sensation such as a feeling of tightness in the chest, a lump in the throat, or discomfort in the abdomen.

The patients were not generally aware of having either the distressing automatic thought or the distressed feeling until they were asked to focus on each of the thoughts and feelings they experienced prior to feeling angry and wanting to retaliate. If patients showed any signs of annoyance, such as a grimace or a testy tone of voice during a therapy session, I would inquire, "What is going through your mind *right now?*" They might say, "I thought, 'You don't understand me.'" When I would press them to recall a feeling they had following such a thought, they usually were able to pinpoint feelings of hurt prior to feeling angry. The observation of the hurt feeling turned out to be a fruitful guide to the specific intervention.

The initial meanings ascribed to events that trigger the hostility complex could be grouped according to their content. One set of meanings revolved around the theme of interpersonal loss: withdrawal of caring, love, support, or help. The other theme was concerned with devaluation: being ignored, disparaged, ridiculed. In normal situations, these meanings would represent a reasonable interpretation of another person's behavior. However, people sometimes attach exaggerated meanings to situations that affect either relationships they value or their self-esteem, and consequently overreact. The evoking of anger depended on the inference that the other person had *wronged* the patient in some way and was therefore blameworthy.

Consequently the patient experienced an impulse to punish the offender and made a rapid evaluation of the ways and means for retaliating.

The primal thinking activated in response to threat or harm also is responsible for assigning absolute meanings to the problematic events. Thus, Raymond was prone to thoughts such as, "Gloria *always* criticizes me," or, "Gloria *never* shows me respect" (overgeneralization). He also exaggerated the degree and significance of Gloria's criticisms (magnification), which he interpreted as a harmful assault on him. Finally, he experienced a malevolent image of her as a hostile adversary.

Raymond's exaggerated reaction should not be considered simply as a unilateral behavioral pattern. Such reactions are only one face of the ambivalence present in many, if not most, close relationships. Most of the time, Raymond had a positive conception of Gloria. His negative image of her with its hostile consequences came to the fore only when an interaction touched on his specific vulnerability.

APPLICATION OF COGNITIVE STRATEGIES

The targeted interventions begin by addressing the meaning of the provocative event. By showing the patient how to reframe his conclusion about the supposedly noxious behavior of another person, the therapist can help him to reduce the intensity of his excessive and inappropriate anger and his impulse to retaliate.

The following techniques were utilized in helping Raymond to consider the meanings he attached to Gloria's criticisms.

- *Applying Rules of Evidence:* Patients are encouraged to look at *all* the evidence, pros and cons, relevant to the interpretation. I asked Raymond whether Gloria had shown caring or respect on other occasions. When he took time to consider the question, he recalled that Gloria had been respectful much of the time. She valued his opinion on a broad range of subjects and was supportive in social situations. Focusing on the contradictory evidence undermined his generalizations about Gloria and softened his negative image of her.
- *Considering Alternative Explanations:* There are many possible reasons for a person's apparently abrasive behavior. People with specific vul-

nerabilities like Raymond's are prone to arrive at biased explanations. I was able to raise a number of questions that prompted Raymond to reconsider his conclusions: "Are there other reasons Gloria may be criticizing you? For example, could the explanation that your procrastination frustrates her fit better?" On reflection, Raymond could recognize that Gloria was more impatient than critical. He could relate to this because he had similar experiences at work: he felt let down when his assistants did not get their work done on time.

- *Problem Solving:* A person may make a number of different interpretations of another person's statements. He can take the message at its face value, or he can ignore its content and react primarily to the subjective "surplus meanings" he attaches to it. In the setting of a distressed relationship, an individual is likely to center his interpretation on the nonverbal aspects, such as the partner's tone of voice and facial expression, and ignore the manifest message. The history of noxious previous interactions and a negative image of the partner will help to frame his interpretation.

  While traveling on a super highway, Gloria yelled at Raymond, "You're driving too fast." He responded angrily, "If you don't like the way I drive, you can get out." She responded, "You're making me crazy," and they continued the rest of the trip in silence. What happened here? The answer lies in the meanings Raymond attached to the emotional communication: "She doesn't have confidence in me," "She's trying to control me," "She enjoys criticizing me." Any of the interpretations could be correct, but they were irrelevant to the message she was trying to communicate: she was worried. He did not address the central questions: Was he driving too fast (for safety's sake)? Was he making Gloria uncomfortable?

  By attending to their own subjective meaning, people aggravate rather than solve problems. In therapy, I pointed out to Raymond that if he directed his attention to Gloria's worries or complaints and tried to solve the real problem, he would be less likely to feel hurt. He protested that doing so would be "giving in." I suggested that it would *feel* like giving in only if he viewed his relationship as a kind of contest or power struggle.

  I proposed that he try to substitute the concept of "cooperation" for the notion of "control": get away from the focus on who's right and who's wrong, who wins and who loses. This change in orientation was helpful. He became less concerned that Gloria was "getting her own way" and more involved in the challenge of solving the problem. He also was able to relate to the notion that compromising did not mean losing or subordinating himself and that he would be rewarded by a better relationship.

People are generally good at solving problems. Their procedure becomes stuck when hurt feelings, anger, suspicion, and mistrust intervene. The parties, whether spouses or national leaders, need practice in putting aside the subjective meanings they attach to a communication and focusing on the objective content. By responding to the manifest problem, they have taken the first step to solving it.

- *Examining and Modifying Beliefs:* The meanings attached to a specific situation are shaped by specific beliefs. These beliefs, incorporated in cognitive structures (schemas), are triggered by the stimulus situation. They generally have a conditional form of "if . . . then." The match of the conditions specified by the "if" to those observed in the stimulus produces the meaning (the "then").

Beliefs leading to distress are important components of the hostility algorithm but are also found in a depressive state. The kind of explanations that a person assigns to the offender's behavior determines whether a person proceeds to feel angry or depressed. If he attributes the cause of the distress to another person or persons, he is likely to become angry; if he attributes it to his own shortcomings, he is prone to feel depressed.

Examples of distress-producing conditional beliefs are:

- If a person criticizes me, it means she doesn't respect me.
- If I don't receive respect, then I'm open to further assaults.
- If my partner doesn't comply with my wishes, it shows that she doesn't care for me.
- If a person procrastinates attending to things, it means I can't depend on him.

When the belief touches on a significant issue, the interpretation tends to be overgeneralized or extreme. Thus, when Raymond applied his belief regarding disrespect to a specific critical comment, he spread the conclusion to cover his global conception of Gloria's view of him ("She thinks I'm a piece of dirt"). Similarly, Gloria generalized from Raymond's procrastinating: "He never assumes responsibility for the house." With repetition, these kinds of generalizations led to more categorical beliefs, such as, "He's irresponsible," and, "He's basically lazy." Over time these verbal labels harden into a negative image of the other person. When this occurs, the negative image becomes the basic structure for interpreting each other's behavior. Each subsequent experience with that person is filtered through this negative image.

- *Modifying the Rules and Imperatives:* The *rules,* another class of beliefs, have a more imperative quality than the conditional beliefs since they are directly involved in promoting or preventing behavior. This system of injunctions and prohibitions transforms the more passive wishes, such as, "I wish my wife was more respectful," into an absolute rule: "My wife *should* be more respectful." In normal circumstances, the "shoulds" are valuable aids in motivating ourselves or other people to do things that we consider important. Similarly, the "should-nots" are helpful in blocking wishes that could lead to undesirable behavior. People with a shaky self-esteem, however, like Raymond, apply the shoulds and should-nots excessively as a way of protecting themselves from being hurt. When other people do not comply with these injunctions or prohibitions, they usually trigger frustration and criticism.

The injunctions are counterposed by the prohibitions.

PEOPLE SHOULD BE . . .	PEOPLE SHOULD NOT BE . . .
Caring	Uncaring
Kind	Unkind
Sensitive	Insensitive
Accepting	Rejecting
Respectful	Disrespectful
Permissive	Strict

Shoulds may take a highly specific form, such as, "My spouse should be more affectionate." When he came home from work, Raymond would have the thought, "Gloria should be loving (since I worked so hard)." When she acted "bossy" rather than loving, Raymond felt let down and then became angry. She had violated an important rule: "*Show me* appreciation and affection." People are generally not aware of how frequently they have such automatic "should" thoughts and how egocentric they are. When I asked some of my chronically angry patients to check off their automatic thoughts on a wrist counter that I had provided, they were amazed to find that in the course of a day they had checked the counter between one hundred and two hundred times. They noted that in numerous situations they had thoughts like, "He should not bother me," "She should smile when I talk to her," and, "He should not interrupt me when I'm talking to him." Simply becoming aware of those thoughts was therapeutic: the imperatives lost much of their impact on the patient's mood.

Evaluating and correcting automatic interpretations and imperatives not only helps a person to reframe hostile inferences but can also modify the underlying belief structure. Ascertaining that a negative interpretation is incorrect provides valuable feedback into the system of beliefs. The repetitive disconfirmation of the interpretation, "My spouse has no respect for me," can modify the image of oneself that one attributes to the mate.

A durable impact on the underlying belief system may also be achieved by subjecting the belief itself to the same type of logical and empirical evaluation that we apply to inferences. For example, the therapist might propose an empirical test of a husband's belief that his wife does not respect him by suggesting that he attend to his wife's request promptly and observe whether she responds with a sign of appreciation. A positive finding would then undermine the strength of his negative belief.

Another strategy for dealing with dysfunctional beliefs is to subject them to pragmatic criteria. For example, the therapist can raise the question: What are the advantages of equating criticism with disrespect? What are the disadvantages? I found that as the patients became more aware of the injunctions, they increasingly realized how unreasonable they were. I attempted to provide further perspective by asking: "Is it reasonable to expect people to behave all the time as you would like them to? How would you feel if they made the same demands on you?" Raymond came to realize that he would always open himself to being hurt by expecting Gloria to "speak nicely" to him all the time.

Raymond had developed an image of himself as a "wimp" when he was young. His older brothers and the other neighborhood boys picked on him. To compensate for this he put up a strong front, but even as an adult he was hypersensitive. When someone was critical, the latent images of tormenting siblings and peers became activated. Thus, in his hypersensitive reactions to Gloria, she appeared as his persecutor and he as the victim.

The images that became magnified during the hostile encounters may also be subjected to analysis and ameliorated. I asked Raymond to visualize himself after being criticized by Gloria. He saw himself shrink in size, with a scared look on his face. His image of Gloria grew in size and became increasingly more menacing. By replacing this with a more benign image of Gloria, he was able to draw back from their confrontations and reflect that she was not his Enemy but was simply upset by his procrastination.

DEALING WITH DISTRESS

Unpleasant feelings immediately following the negative interpretation of an experience are generally glossed over by the patient, who is more aware of subsequent angry feelings. Probing for the earlier hurt, anxious, or other distressing feelings provides a valuable opportunity to understand the hostility complex and is an important instrument for therapeutic interventions. Questioning the patient systematically can tease out these subjective feelings. The meanings ascribed by the patients to the provocative event can provide a rough guide to the kind of distress they experienced.

If the patient, for example, recognizes that the provocative situation represented a threat to him, he might then become aware of anxious feelings. If he felt let down, he might pinpoint a sad or hurt feeling. Alternatively, his subjective feeling might be an unpleasant somatic sensation. The interpretation of being held down or blocked is often associated with pressure in the chest, a smothered feeling. An unexpected disappointment could be associated with a lump in the throat or a sinking feeling in the abdomen.

Focusing on the hurt feelings tends to short-circuit the obsessive concern about being wronged and undercuts the need to retaliate. Working backward from his sad, anxious, or smothered feelings enables the patient to recognize more fully his own sense of vulnerability and the belief system that predisposes him to feel threatened or disparaged by other people.

A transcript of Raymond's interaction with Gloria illustrates this point.

Therapist: What did you feel when Gloria called you down for not fixing the faucet?

Raymond: I felt like hitting her.

Therapist: What did you feel before then?

Raymond: Mad as hell.

Therapist: Let's go back a little ways. Picture Gloria scolding you. What thoughts do you have?

Raymond: She thinks I'm dirt, like she can say whatever she wants.

Therapist: What are you feeling as she says that?

Raymond: I feel hurt, I guess.

Therapist: Where do you feel it?

Raymond: In my gut.

Therapist: The next time you get into a fight with Gloria, do you think you can stop and see what you are feeling—before you get angry—just as you did now?

Raymond: I can try I guess.

By taking a pause like this in his confrontations with Gloria, Raymond could interrupt the progression of the hostile sequence and at the same time become aware of his own sensitivity. This provided the basis to explore his beliefs about being devalued. We then went on to discuss the notion that he was upset not only by his perception that she was putting him down but also by the feeling that *she didn't care for him*. From this we were able to frame an additional belief: "If Gloria really cared for me, she wouldn't criticize me." The notion of her not caring elicited a transient feeling of depression. It became clear subsequently that Raymond also feared that Gloria might leave him. This discovery opened up a whole new avenue for discussion and further evaluation: his concern about rejection. Raymond was then able to see the paradox in his reactions: by physically abusing Gloria, he was likely to bring about what concerned him the most—being rejected by her.

In most situations, fixing responsibility for distress on another person is sufficient for an anger-prone individual to turn that person's behavior into an offense and to experience anger. How often have we seen a driver, caught in a traffic tie-up, roll down his window and yell at the driver in front of him—even though the other driver is just as jammed in as he is! Being trapped this way evokes feelings of helplessness; blaming another person helps to reduce one's sense of powerlessness.

The question of whether another person's problematic actions were accidental or intentional is another important feature of the hostile reaction. Even accidental behavior can elicit intense anger in the offended person. In particular, if the misdeed was due to carelessness, negligence, or ignorance, the victim may be aroused to punish the offender. When there is serious harm, the arousal is even greater. On the other hand, if an offense appears relatively harmless and excusable, then there may not be any hostility directed at the offender.

Anger-prone people are more likely to read hostile intent into another person's aversive behavior than are non-anger-prone people, who are willing to explain an ambiguous noxious incident as accidental. Similarly, mates in a distressed marriage are more prone to ascribe unpleasant behavior to the bad nature of the spouse, whereas nondistressed couples will ascribe the same incident to the nature of the situation. An individual's bias in explaining other people's behavior may reside in the way she generally processes information, or it may occur only in conflicted relationships. In any event, when people are aroused to blame the other person, they are also more likely to make "characterological" diagnoses of that person: malicious, manipulative, deceptive.

RIGHTS AND ENTITLEMENTS

How often have we heard somebody complain indignantly, "You have no right to treat me that way!"? We all develop a system of expectations as a form of protection against the incursions or derelictions of other people. When we determine that one of our rights has been violated, the degree of anger that we experience is generally out of proportion to the actual loss: the idea of somebody disregarding our wishes, for example.

Anger-prone people in particular place great emphasis on the protection of their rights. On occasion, I have noted somebody's rage when another person stepped in front of him in a line at the theater or supermarket. Similarly, having to wait for service enrages some people even though the "lost" time is insignificant. The "protection" of their rights is obviously an integral part of the way people deal with problematic situations. Many people live by an idiosyncratic bill of rights as though they are obvious and generally accepted. They are apparently oblivious to the fact that this only causes conflicts with other people who have a different view of their entitlements.

Among the rights asserted by anger-prone people are:

- I have a right to do what I want to do.

- I have the right to express my anger if I feel annoyed.

- I have the right to criticize other people if I think they are wrong.

- I expect people to do what I consider reasonable.

- People do not have the right to tell me what to do.

For people who have such a fixation on their rights, I raise the following questions:

- Does the other person know that you have this right?

- What is actually subtracted from you if the other person does not comply with your expectations?

- Did you experience a real loss, or are you angry as a matter of principle?

- What do you gain by retaliating? What do you lose by retaliating?

I have found that helping patients to recognize and articulate their claims, demands, and expectations is a valuable first step in their attaining self-objectivity. Simply becoming aware of the egocentricity of the presumed entitlements enables them to gain greater objectivity toward them. Further, they begin to realize how the self-oriented rights inevitably lead to conflicts with other people. Although everyone needs to establish personal boundaries with other people, the rigid assertion of their rights actually leads to more frequent personal distress and nonproductive anger.

The composite of blaming, attributing negative intentions to others, and making negative generalizations about the "offender's" character is a prescription for the generation of intense anger and the urge to punish the offender. By becoming aware of each of these components, the person is more likely to realize that offenses may be accidental and not due to the bad character of the offender and consequently not blameworthy.

DEALING WITH ANGER

The subjective feeling of anger may vary from mild irritation to rage. Many authors believe that anger is not only experienced subjectively as an emotion but is also expressed somatically in facial expressions, tightening of muscles, and racing pulse. Alternatively, the experience of anger and the somatic expression may be viewed as separate but integrated components of the fight reaction. The subjective feeling of anger and the bodily reactions occur when the individual is mobilized to retaliate against an offender.

In a more restricted sense, anger has information value; like pain, it acts as a signal or stimulus to alert the person to a threat. The experience of anger impels a person to identify the source of the interpersonal problem so that he can do something about it. Generally, anger exerts a pressure to take corrective action and persists until the irritant is neutralized or removed. The function of anger can be likened to that of a smoke detector: it captures the attention and directs it to the noxious agent. Because the threshold for activation of the mechanism is generally low, there are bound to be false alarms. From an evolutionary standpoint, it is better to overreact than underreact to a threat. After only one failure to respond to a life-threatening situation, the individual and his further contribution to the gene pool are eliminated.

Although it is useful to think of anger as a warning that something is wrong, the potential for harm in everyday life comes not from the provocative agent but from the anger itself. In addition to its negative medical effects, the powerful distraction produced by anger draws attention away from the specific nature of the problem. It also creates a pressure to act against the provocative agent rather than to engage in constructive problem solving. Under these circumstances, the anger itself becomes the problem and requires the application of pragmatic principles.

In dealing with interpersonal conflicts that can rapidly escalate to physical abuse, the therapist can reformulate the person's anger as a warning *not* to take action. In my therapy with Raymond, I described the intensity of anger in terms of "zones." Mild degrees of anger were like a yellow light that he could interpret as a signal to back off. More intense anger was represented as the red zone: a sign to withdraw from the scene. My initial work with Raymond consisted of educating him regarding the necessity of preparing himself to react adaptively when he felt that he was entering into one of the danger zones. It was obviously necessary to provide a terminus to his destructive behavior before we could get to the more cognitively oriented strategies. In the next interview, he reported that an argument with Gloria escalated until he realized he was in the red zone. He then left the room, went into the next room, paced back and forth, and then went upstairs. He found he was still too angry, so he went for a long walk for almost an hour until he had cooled off enough to return home. The duration of his anger indicated how highly mobilized he was for aggressive action.

Another symptomatic aid for dealing with anger is to divert discussion from a hot topic to a neutral one and postpone further discussion of the highly charged topic. The utilization of relaxation techniques is another potentially useful method for reducing anger. The therapist can train the patient using Jacobson's Progressive Relaxation Method or induce the patient to practice relaxation at home using one of the many available tapes designed for this purpose.[3]

When the patient has acquired some leverage against the progression into destructive behavior, he can utilize his angry feelings as a marker to identify and deactivate the automatic hostile thoughts. Raymond reported a series of such automatic thoughts that he experienced when he was angry. "She enjoys dumping on me. . . . She treats me like dirt. . . . She's always leaning on me." I then induced Raymond to construct reasonable responses to each of these thoughts: (1) "I don't know if she enjoys it. She says she hates being upset." (2) "Her criticisms aren't really that bad. I just don't like being criticized." (3) "She doesn't *always* lean on me."

Raymond found that by taking time out to write down his automatic thoughts when he was angry with Gloria and responding to them either at that time or later when he was cooler, he was able to increase his objectivity toward his overreaction to Gloria. Concomitantly, the intensity of his anger diminished. At this stage of therapy, it was important for Raymond to recognize the hurt he had felt prior to becoming angry. He was able to capture the hurtful thoughts, such as, "She thinks I'm a creep. . . . She has no respect for me." It certainly was possible that Gloria had a negative view of him, particularly in view of his physical assault on her. Hence, it was important to check on the accuracy of his interpretations.

Our adaptation to other people does require that we be able to "read their minds," as Raymond attempted to do with Gloria. From an early age, children develop a "theory of mind" that helps them to make a reasonable guess regarding the beliefs and intentions of other people.[4] In highly conflictual situations, however, the theory of mind becomes negatively skewed and the guesses become biased. In reviewing his automatic thoughts, Raymond concluded that there was no evidence that Gloria generally considered him a creep or lacked respect for him. We then pursued the notion that his hurt feelings were also due to his interpretation that she didn't care for him, that she was rejecting him. Again applying the rules of evidence, he decided these interpretations did not hold water. As a result of this intervention, much of the sting was removed from Gloria's criticism.

Of course, a person's negative mind-reading may be accurate. A wife may indeed regard her husband as defective and lack respect for him. The therapeutic approach then has to be directed at the meanings he attaches to these characterizations. For example, even if his wife regards him as a creep, does it follow that he *is* a creep? If she lacks respect for him, does it necessarily mean that he doesn't deserve respect? I explained this idea to another patient: "Suppose your wife does think that. You know she can be wrong, just as you've been wrong about thoughts you've had about her. . . . Her regarding you as a loser doesn't make you a loser. It's in her head, not in reality. But you aren't going to be able to knock it out of her head by attacking her. You have to decide for yourself what you are. If you decide you're not a loser, that's what really matters. If you get that into your head, we can get on with checking out what you do that gave your wife that image of you in the first place."

I have found that directing the therapeutic focus to the patient's hurt feelings and the meanings behind them helps the patient avoid focusing on the spouse as the Enemy. The more he can understand his basic beliefs, the less he is driven to attack his wife. He starts to incorporate the idea that the hurt he feels comes from the meanings he attaches to his wife's behavior rather than from her behavior itself.

Couples' therapy is often helpful in forestalling excessive reactions in marital conflicts. Training in communication skills is particularly important for partners like Raymond and Gloria. In particular, Gloria realized that it was counterproductive to scold him for his lapses. She practiced speaking to him in a neutral tone of voice, and he practiced responding nondefensively. They both agreed that she could post a list of things for him to do on their bulletin board, and he would indicate a date for attending to each task.

The escalation from being criticized by Gloria to punching her in the mouth did not occur in a vacuum, of course. The memory of past interactions with Gloria controlled Raymond's current thinking as much as they would if they were occurring in the present. Typically, in previous episodes that ended up in heated verbal fights, Gloria would express dissatisfaction with Raymond's behavior. His response would be to criticize her for "picking on him." She would defend herself, and the argument would become more heated until one of them stomped off in a rage or Raymond struck her.

Gloria, of course, came into the marital relationship with her own set of expectations, beliefs, and vulnerabilities. She wanted more than anything else to have a harmonious relationship, but she felt let down by Raymond on numerous occasions and, of course, could not tolerate his rage reactions. In my work with them in couples' therapy, I attempted to give each of them an appreciation of the other's difficulties and sensitivities. A typical event at a restaurant delineated Gloria's sensitivity and Raymond's obtrusiveness—the mixture of the two leading to an unpleasant outcome.

> *Gloria*: Could you ask the waitress to change tables? It's awfully noisy here.
>
> *Raymond*: It won't make any difference. It's noisy all over.
>
> *Gloria*: Well, you certainly can ask.
>
> *Raymond*: It won't do any good.
>
> *Gloria (angrily)*: You never do anything I ask you.

Raymond decided that Gloria was "giving him a hard time" and felt devalued by the interchange. First, she challenged his judgment. Then she accused him of never attending to her wishes. In therapy, he became aware of Gloria's perspective. The reason she was upset was the meaning of his refusal: "He doesn't care." She attached the same meaning to his not attending to household chores. His "pragmatic" answer, that he would get around to things eventually, did nothing to modify her basic belief: "If he really cared, he would do what I asked." She would feel hurt and then angry at him for his lack of caring. Her upset was compounded by his angry reaction, culminating ultimately in an exchange of verbal abuse. She fought against feeling intimidated by him and would answer him back, becoming more critical with each interchange. The memory of these previous critical episodes took charge of Raymond's thinking and led to the final denouement of his hitting her. These episodes were responsible for their decision to seek therapy.

Dealing with the Violent Impulse

The impulse to harm or kill another person may be expressed so suddenly following a provocation that it might appear to be a kind of reflex. On the other hand, the interval between the arousal of a destructive impulse and its execution may be prolonged for a considerable period of time. The nature of the therapeutic strategies aimed at preventing violent episodes may be the same for immediate and delayed acts of violence.

Raymond's sudden assault on Gloria following her criticism represented the activation of the total hostility constellation. By taking that constellation apart, it is possible to see the potentially explosive combinations of beliefs, vulnerabilities, imperatives, and images that disposed him to punish her for hurting him. The memories of past altercations incorporated into the constellation also contributed to his decision, "I've got to shut her up" (by hitting her in the mouth). The strength of his impulse was proportional to the degree of activation of this constellation, which had become more strongly charged with each hostile interaction.

The characteristics of the constellation controlling Raymond's behavior may be compared to that of Billy (described in chapter 8), who was insulted in a bar and obsessed about getting back at his tormentor for an hour while he looked for a loaded gun so that he could return to shoot him.[5] Both

Raymond and Billy were "reactive offenders." The psychological constellations triggered by the provocative situations were similar.

1. Vulnerable, unstable self-image

2. Drop in self-esteem following apparent devaluation

3. Pain and helplessness following drop in self-esteem

4. Image of the insulting person as an Enemy

5. Disposition to punish or eliminate the Enemy to neutralize pain and helplessness.

Given Raymond's and Billy's personal susceptibilities, their violent behaviors were acts of desperation. Only a violent act would be sufficient to neutralize their deep sense of humiliation. Hitting and killing are strong forms of empowerment and powerful antidotes to a depreciated self-image. Whether the violent act was immediate or delayed following the provocation depended on practical considerations. Raymond did not have to wait to get a weapon—he had his fists. Also, the private setting was more conducive to violence, free of interference from bystanders.

In order for Raymond to gain control over his impulsive explosions, it was necessary to reduce the intensity of the underlying psychological constellation. Acquiring insight into the labile nature of his self-esteem and how it affected his feelings was an important step. When he examined the fluctuations in his mood over days or weeks, he could see how it was responsive to how he regarded himself—his self-esteem. When receiving support and appreciation from other people, he noted that he felt better about himself. When exposed to criticism or indifference, he felt worse about himself.

He also became cognizant of the fact that when his mood was low, he felt weak and helpless. At times like this, he would feel a strong urge to do something aggressive, like driving a nail into the wall, socking a punching bag, or getting into a fight. Aggressive action was a "fix" for his low mood. In this respect, offenders are like substance abusers: they resort to a maladaptive behavior to reduce dysphoria.[6]

Raymond realized that attacking Gloria for criticizing him was motivated not only by a wish to punish her for hurting him, but by a need to improve his mood by elevating his self-esteem. We then discussed the value of violence as a mood-normalizer. Although he realized its negative consequences, he needed further insight into why he it relieved his immediate distress. His initial reaction was that Gloria made him feel like "less of a man" when she criticized him. I then proposed the following problem: Is he more of a man by hitting a weaker person? Or is he more of a man by being cool: taking insults without flinching and maintaining control of himself and the problematic situation?

Raymond was intrigued by this question and was able to create an image of himself as imperturbable, responding reasonably to Gloria's criticisms. He could then change his underlying formula from, "A man doesn't take any crap from his wife," to, "A man can take the crap without allowing it to get to him." Therapy thus was aimed not only at undermining his dysfunctional beliefs but at substituting more adaptive beliefs. Of course, for the new attitudes to become consolidated, it was necessary to apply them to actual situations. Initially I set up practice sessions—domestic scenarios in which Gloria and Raymond would re-create typical conflictual situations. Without necessarily using better communication skills, Raymond found that he could listen to Gloria's complaints with a minimal drop in his mood or desire to shut her up. He later told me he was able to evoke the image of her as an upset, but not threatening, woman.

We were then able to review the methods of control he could use if his impulses became too intense. The first safety measure was to take time out until he settled down, to leave the room. The second was to rehearse the image of Gloria as vulnerable and upset, rather than as hostile and menacing. The third was to remind himself that the way to feel more manly was to be cool and masterful.

PERMISSION GIVING

Even when a person is highly aroused to engage in antisocial behavior, he usually has to contend with an inner deterrent to such behavior. From time to time, most people experience a desire to hit somebody or even kill that person but are usually restrained by an automatic inhibition. Under certain circumstances, however, they are able to disengage the moral code, to use Bandura's term.[7] When there is social pressure or simply approval for antisocial behavior, many people can overcome moral restraints. This phenomenon is readily apparent in combat infantry, street gangs, and lynch mobs. Indeed, once a person has engaged in a destructive act, it is generally easier to repeat it the next time. Security police who engage in torturing victims initially out of submission to orders eventually find they can perform the same or more gruesome acts of their own free will.

When a person is motivated to engage in an antisocial act as an individual, rather than as a member of a group, he has to devise special justification for that harmful behavior. He has to override the moral code, reduce concern about social disapproval, repress empathy for the victim, and ignore the negative consequences for himself if he is apprehended.

In previous heated encounters with Gloria, Raymond had a number of thoughts that condoned his hurting her. Among these were:

- "She deserves it": This fits in with what has been termed "the just world philosophy"—people get what they deserve, good or bad.[8]

- "She's asking for it": This translates into, "She would not act that way unless, for some perverse masochistic reason, she wants to be hit."

- "It's the only way to get her to shut her mouth." This is Raymond's "pragmatic" solution to the problem.

- "There may be bad consequences, but I'll be able to handle them."

During his rage, Raymond would forget about previous resolutions to control himself. These rationalizations, having occurred previously to Raymond, were compressed into a concrete justification: "It's okay to hit her." Thus, when he experienced the urge to strike her, the permission-giving justification was already in place.

14

PERSPECTIVES AND PROSPECTS

Applying Cognitive Approaches to the Problems of Society

In the preceding chapters, I have tried to map out a conceptual model of the development of hostility, how it plays out in various domains, and what resources may be used to reduce it. If this theory is substantiated, it could provide a basis for further clarification and a framework for applying a wide variety of strategies for intervention and prevention.

It would be valuable at this point to review the cognitive theory of hostility and the evidence supporting it and to consider further investigations that might be undertaken to test it and apply it to the pressing problems listed in chapter 1.

In exploring the various approaches to such complex phenomena as hostility, persecution, and war, it is essential to use a broad frame of reference. General Systems Theory provides a framework for analyzing a phenomenon at various conceptual levels.[1] The precise level of analysis usually varies according to the discipline and the special interest of the investigator. The conceptual system, or level, can be as minuscule and concrete as a segment of DNA or as broad and abstract as the balance of power between nation-states. Despite the conceptual differences between the levels of analysis, the systems are related to one another and directly or indirectly affect each other. Among the levels to be considered in attempting to understand a phenomenon like hostility are the biological, psychological, interpersonal, cultural (or sociological), economic, and international. The investigative methods that are used for each of these systems vary enormously. This approach avoids the reductionism reflected in assertions such as, "It's all in the genes," or, "It's due to society (or economic conditions, and so on)."

Take, for example, Billy, the man described in chapter 8 who got in a barroom brawl with another man and went home to get a gun to kill him. To understand this kind of reactive violence at the *biological level,* investigators might, for example, analyze the neurotransmitters in the brains of aggressive rats, assay the neurochemicals in the spinal fluid or the hormone levels of aggressive apes or humans, or obtain images of the regions or structures that are hyperactive during violence.

Neuropathologists examine the brains of violent offenders to identify specific cerebral lesions that might underlie violence. Pharmacologists investigate the effects of alcohol on the inhibitory areas of the brain. Geneticists look for family patterns and specific chromosomal deviations to determine the hereditary factors. At the *psychological level,* the investigator would focus on Billy's idiosyncratic information processing, his basic beliefs about the cause of personal offenses, his biased interpretation of the victim's behavior, and so on.

Social psychologists focus on the *interpersonal level:* What was the nature of the interactions, the verbal exchanges between Billy and his drinking companion, that eventually escalated into violence? At the *societal level,* sociologists look at the values and norms of the particular culture or subculture, the social institutions that endorse or proscribe violence. They examine the economic and social stressors that affect the social role of offenders and explore the social rituals that reinforce drinking.

Cross-cultural comparisons by anthropologists point up the commonalities and differences between literate and preliterate societies to clarify the molding of attitudes toward violence. Finally, *evolutionary* psychologists and biologists search for analogues in the animal kingdom for clues regarding the adaptive value of revenge seeking in our ancestral environment. To understand other phenomena, such as genocide or war, additional conceptual systems are studied.

A systems analysis is often crucial in order to understand group or individual behavior. A striking example is the change in the relation between declining economic conditions and hate crimes in the United States. From 1882 to 1930, the incidence of hate crimes increased whenever there was a worsening of the economy. Scholars have generally attributed this relationship to a mechanism whereby economic hardship increases personal frustration and aggression, which is then directed toward a scapegoat, such as a member of a vulnerable minority. This theoretical explanation fits the data regarding the relation of decreased economic performance and lynchings of blacks in the pre-Depression South. However, economic fluctuations since then have not been associated with lynchings. In order to account for this change, it is important to examine another "lower" level in systems analysis.

Analysis of the behavior of prejudiced political elites and organizations showed that they played an important role in attributing blame and fomenting public resentment toward blacks and other minority groups during periods of economic contraction. They would actively try to instigate anti-black violence in depressed areas by exaggerating the economic competition from blacks. Propaganda about black laborers is credited with fostering race riots in urban areas.[2] Changes in societal attitudes regarding anti-black prejudice eventually stripped the racist groups of much of their power.

The systems approach also suggests an alternative hypothesis to the conventional frustration-aggression hypothesis. When people are upset by economic (or other stressful) conditions, they are prone to respond to easy explanations of causality. Viewing economic distress in terms of complex, abstract, and often unknowable variables required considerable effort. If they already have a biased image of a vulnerable minority, they are readily moved to view that group as at least partly responsible for their distress.

An illustration of systems theory is its application to the initiation of war by an aggressive state. Although wars at this scale have been largely superseded by ethnopolitical wars, the typical European wars of the twentieth century can serve as examples of a systems analysis. At the "top" level is the unstable international system. This system has an impact on the next level, namely, the psychology of the ruling power that perceives the instability either as a threat to its nation's security or as an opportunity to take advantage of a weaker state. In any event, states determine that their best interest resides in building up their military establishment and forming coalitions with other kindred states. Leaders with expansionist dreams may decide to take advantage of the state of affairs by attacking a neighboring country.

In order to accomplish this, they need to rally the support of the political elite, the armed forces, and the citizenry. Typically, the leaders use the news media to whip up popular support. In some cases, like Hitler and Milosevic, they may also conjure up stories of persecution of their own people who are minorities in neighboring countries. Hitler used this strategy for mobilizing support to invade Czechoslovakia in order to protect the German minority in the Sudetenland. Milosovic borrowed this tactic when he asserted that the Serbian minority in Kosovo was being persecuted by ethnic Albanians.

At the next level down, the populace is moved by their leaders to image the neighboring ethnic population as their Enemy. The individuals each identify with the goals of the commander.

For an even more comprehensive understanding of the causality of hostility and violence, a scholar would examine the *interactions* between the various systems. For example, it would be interesting to investigate not only how cerebral deficits and changes in brain chemistry produce changes in information processing but also how the latter affects the former: whether, for example, excessive conflicts due to misattributions can aggravate the organic deficits.

Organic brain deficits might affect the psychological system—information processing—in a number of ways. A person who sustained a specific brain lesion as a child tends to use the least effortful ways to explain problematic situations. Thus, he is more likely than a normal person to assimilate only the salient aspects of a situation, at the expense of the entire context. He is also more likely to stick to egocentric, dualistic (either/or) reasoning. Since it is easier to attribute interpersonal problems to another person's hostile intentions than to survey the situation for more complex, neutral causes, he is prone to make personal attributions that lead to anger or violence. In more extreme cases, the psychological reaction to his original organic deficit may be paranoia.[3]

An important role is also played by the interaction between increased biases in information processing and the escalation of an interpersonal conflict, say, a marital dispute or a barroom argument. The psychological and the social systems become progressively more disjointed. At this point in the analysis, one could raise the question: How do community values affect the individual's attitudes toward interpersonal violence and the permissive belief systems that condone it? The impact of the subculture on attitudes toward violence was illustrated in chapter 8 in the section on the southern "Code of Honor" and the inner-city "Code of the Streets." Briefly, beliefs regarding the importance of the threat of violence in order to maintain one's status in the group tend to perpetuate physical assaults and murder. The interactive model I have proposed can also be applied to other major problems such as prejudice, persecution, and ethnopolitical violence.

COGNITIVE CONTINUITIES

As shown in the following list, individuals who either actively initiate or endorse violence—whether against another individual or against ethnic or national groups—present a similar profile of attitudes. An abusive husband like Raymond (described in chapter 13) perceives himself as the innocent victim of his wife's unwarranted criticism. Due to a cognitive blockade at the peak of his rage, he is unable to understand that she might have some rationale for her criticisms other than to "torment" him. He further believes that striking her is not only permissible but necessary in order to protect his self-esteem.

The reactive offender, typified by Billy (described earlier in this chapter and in chapter 8), believes that a person who insults him during a verbal altercation deserves to be punished and that physical violence, even homicide, is justified. Blinded by his egocentric mindset, he doesn't recognize that the other person may have been traumatized by his insults. Also, like Raymond, he sees his antagonist as the Enemy.

As noted in the list, there is a similarity in the way a hostile aggressor perceives himself and his victim in a broad range of settings. Contrary to the interpretation of an objective observer, he tends to view himself as the innocent victim and the true victim as the victimizer, the Enemy. His information processing is stuck at the primal egocentric level, and consequently he sorts his perceptions into dichotomous categories (friend or foe, good or evil). Further, his destructive impulses are not checked by the usual social restraint because of his permissive attitudes about violence (for himself).

	Spouse/ Child Abuser	Reactive Offender	Persecutors	Populace of Aggressive Nation
Image of Self	Blameless Victim	Victim	Victims	Victims
Image of Victim	Victimizer	Victimizer	Victimizers	Victimizers
Orientation	Egocentric	Egocentric	Group Egoism	Group Egoism
Attitude to Violence by Self	Permissive	Permissive	Permissive	Permissive
Thinking	Dualistic	Dualistic	Dualistic	Dualistic

There is a continuity in the cognitive characteristics of violence across the various domains: family abuse, street crime, persecution, genocide, war. Whether participating in individual or mass violence, a person tends to have the same dichotomous view of himself or herself and the other. Spouses engaged in physical abuse tend to see themselves as the victims and the other spouse as wrong and bad; they may even use the word *enemy* to describe the other. Similarly, violent offenders see themselves as innocent victims of the other's hostile behavior. The persecutors engaged in pogroms, lynchings, and ethnic massacres believe that they are protecting themselves by attacking the Enemy. Individuals participating in persecution are gripped by ethnocentric or nationalistic bias, which allows any degree of physical or psychological abuse. Similarly, the population in most wars believe they or their brethren have been endangered or mistreated by a neighboring state.

The Enemy of the group is "my Enemy" and needs to be subjected to discrimination, humiliation, and segregation and ultimately purged. The citizenry of states initiating wars see their nation as totally justified. Having complete confidence in the reliability of their leaders, they are moved by the disinformation and exaggerations they are fed. Groupism, like egoism, projects an image of their leaders and nation as benevolent and threatened and therefore justifies "counterattack." Numerous examples, ranging from the world wars to the mass slaughters in Bosnia and Rwanda support this thesis. Germany's entry into World War I, for example, was enormously popular among the German people, who accepted the idea that their army was simply defending its nation from an invasion. This notion was accepted by all classes. In *Rites of Spring*, Ecksteins describes its acceptance by even the intellectual elite: "In the course of the war, thirty-five of forty-three holders of chairs in German Universities averred that Germany had become involved in the war only because she had been attacked."[4] Greek and Turkish Cypriots have used the same vicious language to describe each other. Judging from newspaper accounts, the same is true of Serbs and Croatians, Israelis and Palestinians, Hindus and Moslems.

SEQUENCE OF HOSTILITY

To this point, I have reviewed the ways of exploring a particular phenomenon at various levels and have discussed the interactions of the various systems. I have also illustrated the cognitive commonalities in various domains (family conflict, political violence, and so on). It is also valuable to consider the sequential development of hostility and the relevant research at each level. The development of hostility may be viewed as proceeding in a sequence of steps from predisposition to precipitation to reaction. These steps are dependent on the activation of specific psychological structures and processes.

Predisposition

We can readily observe that some people react with hostility (initially at least) to certain situations, especially those that require defense or counterattack. This predisposition is embedded in specific beliefs such as, "If a person raises his fist at me, it means he may be ready to attack me." Some of these predisposing beliefs are so clearly related to real threats that they will produce a predictable reaction to the provocative situation. Other beliefs are more idiosyncratic; for example, "If my wife does not respond to me, it means she is rejecting me," "If somebody contradicts me, it means he dislikes me," or, "If people keep me waiting, it means they don't have respect for me." If rigid and applied indiscriminately, such beliefs constitute a specific vulnerability that can lead to excessive or inappropriate anger.

When an event penetrates this vulnerability, it activates the belief, which then shapes the interpretation (or misinterpretation): "[Since she kept me waiting], she doesn't care." This meaning, or interpretation, which appears in the form of automatic thoughts, produces distress. If the injured party blames another party for having caused his distress, he feels angry and becomes aroused to punish her. On the other hand, if he devalues himself, he is more likely to feel sad.

Of greater clinical importance is the development of a more enduring hostile mindset. This psychological state is illustrated by a person's continually hostile interactions within the family or with outsiders.[5] Let's take a commonplace example of a husband and wife whose initially positive images of each other have gradually turned into negative images. These negative images arose initially from marital conflicts, then gradually became reinforced by continuing negative interpretations of each other's behavior. Assume that the negative image portrays the other mate as the Enemy. When partners "make up" after a fight, the negative image loses its potency (for the time being), and they may perceive each other more or less realistically.

Now assume that a specific incident or series of incidents strikes directly at a partner's vulnerable self-image: say an argument degenerates into physical abuse. Then the image of the Enemy becomes charged and shapes the interpretations of each other's subsequent behavior. The negative images therefore become more forceful and are triggered by increasingly less serious interactions. Eventually the image of the Enemy becomes fixed and takes control of the mates' reactions to each other. Ultimately, the fixed images lead to divorce, serious physical abuse, or, in rare cases, homicide.

Experiments have shown that mates in a distressed marriage are predisposed to attribute the *cause* of a problem to their spouses; in contrast, they explain an identical problem in a different marriage as due to the situation.[6] Variants of the "causal beliefs" are the "characterological beliefs." A problem arises. The distressed spouse not only blames it on the partner but also attributes it to the partner's bad character: manipulative, devious, evil. The feelings range from disgust to anger—to hate.

Further work in this area would include developing inventories of the beliefs that predispose individuals to hostile reactions. Of course, some beliefs may be consciously denied by a respondent, or he may not be aware of them. However, by using priming procedures an investigator may be able to infer its presence. A number of different experimental procedures may be used to prime a particular belief. One method is induced imagery. The investigator could suggest an inflammatory scenario, for example: "Imagine that a friend arrives late and doesn't even apologize." If the subject becomes angry, one could infer the predisposing belief: "If somebody keeps me waiting unnecessarily, he has wronged me." Further corroboration would come from automatic thoughts such as, "He doesn't have any respect for me," or, "She doesn't really care (about our friendship)." Once the subject has been coached in the use of imagery, then a more idiosyncratic scenario can be induced. Such scenarios may be used to confirm the findings on the belief inventory or to identify covert beliefs and images.

Similar scenarios may also be used on videotape or film. Kenneth Dodge created an ambiguous videotape of a child bumping into another child.[7] Predelinquent children were more likely than others to call the act intentional than accidental. It is of interest that the responses to this test could predict hostile behavior several years later. Scenarios like these evoke hostility if the subject has a specific predispositional belief that such acts are intentional. Following the film or videotape, questions could focus on the meaning of the incident and the evoked emotions: the intention of the "offender"; whether the act was intended, whether it was malicious or playful; whether the offender should be punished; how *much* the offender should be punished; the *intention* of the intended punishment (to educate, retaliate, and so on); whether the person observing the film felt angry or some variant of this emotion (irritated, annoyed, enraged).

Research that will test this formulation could be carried out in the context of marital therapy and could have therapeutic benefits. A questionnaire consisting of a series of negative and positive adjectives could be administered to each spouse and the responses analyzed to determine their images of each other. The spouse would also be asked to indicate how frequently the adjective fits (from "sometimes" to "always") and how fixed it is ("can it be changed?"). An open-ended questionnaire consisting simply of spouses' spontaneous descriptions of one another could also be administered. From these data, the investigator could compile a profile of the partners' images of each other.

Having determined the spouses' images of each other, the researcher could proceed to investigate the other components in the hostility sequence. The next stage consists of beliefs regarding punishment for offenses. A questionnaire could list beliefs regarding punishment for offenses, such as:

- If somebody (or "my spouse") bothers me, that person should be punished.

- I should not let him get away with treating me that way.

- He hurt me, so I should hurt him.

- I have to teach her a lesson.

- If she provokes me, she deserves to be hit.

The next stage in the sequences addresses the question: What cognitive process translates the concept "I *should* retaliate" into "I *will* retaliate," and then into violent action?

A questionnaire designed to access the permissive beliefs could be used to determine the factors that lift the moral restraint against physically harming another person. The permissive belief questionnaire would have items such as the following:

- If I am really angry at somebody (my spouse), it is okay to let him (her) have it.

- There is no other choice. I *must* teach her (him) a lesson.

- It will be better for our relationship in the long term if I attack in some way.

- I can't stand all the tension. Hitting her (him) will release it.

It would be useful to prime these beliefs through mental imagery, videotape, or film. The questionnaire could be administered both prior to and following the stimulation by self-induced imagery or films showing marital strife. This would demonstrate the changes in the beliefs toward the acceptability of violence. Such studies should be carried out in the controlled setting of a laboratory. They also should be directed to the goal of increasing the subjects' understanding of themselves, leading to increasing control of their impulsive behavior.

Questionnaires could be devised to assess more general permissive beliefs related, for example, to group violence. Sample items might include:

- "Our leader said we should do it—so it's okay" (diffusion of responsibility).

- "All the others in the group are doing it" (group empathy).

- "Violence in a good cause is no crime" (ideological reasoning).

- "If we don't get them now, they will get us first" (fear of incipient attack).

Beliefs like these can be primed by carefully selecting scenarios.

A careful scrutiny of other hostile reactions indicates the important role of the fearful beliefs. What seems to be a normal reaction to a provocation may actually be based on a hidden fear or self-doubt. The mother who abuses her rebellious child may have the predisposing belief, "If my child misbehaves, I cannot function." Spouse abusers like Raymond have a secret fear that their wife will demolish them if allowed to criticize them. The convict who attacks the prison guard has a fear of being rendered totally helpless. Finally, participants in persecution believe, usually as a result of propaganda or myth, that the persecuted group is dangerous.

Precipitation

Up to this point, I have discussed how hostile beliefs can be activated, how the rapid judgment of a brief stimulus can lead to a hostile response, and how the specific meaning of a stimulus can be assessed through automatic thoughts or other investigative strategies. These studies can help to pin down the psychological processes in a typical episode of anger and hostility.

The predisposing beliefs generally do not exert a significant influence over thinking and feeling until they are activated. An external event that fits a particular belief is likely to activate it. When Louise (described in chapter 3) observed that her assistant had made a mistake, she became enraged and thought, "He is careless and should be chastised." Her belief was, "Mistakes are a sign of carelessness." However, a less obvious belief centered on the fear that "if my subordinates default, they threaten my authority."

At times a relatively trivial event may be sufficient to evoke intense anger if the predisposing belief is very robust. A politician who had developed a very high level of self-esteem was generally able to shrug off political reverses and attacks. However, when a former but insignificant supporter defected to his opponent, he became enraged and had a rush of thoughts about physically attacking the former supporter. The belief that was triggered was, "If anyone is disloyal, it's a stab in the back." Because of this belief, even a minor defection was a major threat.

It may be necessary to circumvent some pitfalls in order to ascertain the kinds of beliefs that lead to excessive anger with or without violence. One problem is that people may be unwilling to acknowledge having certain beliefs that would be considered immature, prejudiced, or otherwise socially undesirable. Further, they may not even be aware of having them because when they are activated, they are expressed in the form of what seems to be a valid interpretation of a situation. Therefore, they are not seen as beliefs but as realistic. The awareness of an interpretation may itself be preempted by the rapid experience of anger and an impulse to act. People frequently automatically downgrade a person of a different race, ethnic group, or political group—without knowing that their thinking is controlled by a groupish "we-versus-them" attitude.

At times it is difficult to define the actual precipitating event precisely. The spark that is generally acknowledged to have ignited the First World War was the assassination of the Archduke Ferdinand by Serbian terrorists. However, a broader view of that conflict would include subsequent as well as previous events. The assassination precipitated an attack on Belgrade by Austria, which felt threatened by a Slavic challenge to its empire. The attack on Serbia posed a threat to Russia's pan-Slavic aspirations and instigated full mobilization of its armed forces. Also, the Russian leaders were very worried about Germany's allegiance to Austria-Hungary. Russia's mobilization presented a threat to Germany and precipitated the full mobilization of Germany to make a preemptive strike against Russia before it became too strong.

The series of provocations leading to war, though ostensibly motivated by hostility, can be better understood in terms of the perception of danger to the state. In the events leading up to World War I, each state perceived itself as directly or indirectly threatened. Their reactions centered on beliefs regarding their vulnerability rather than primarily on goals of aggrandizement. The beliefs of the populace, however, revolved around themes of patriotism, self-righteousness, and national pride. Unlike the leaders, who were worried about a chain reaction, the citizenry were euphoric over the opportunity to avenge the insults to their national pride and to enjoy the power and glory of victory.

Automatic Reactions

The reactions to an upsetting event tell us a great deal about information processing. According to the cognitive theory, the activated beliefs produce the automatic cognitive response to the precipitating event. The information processing consists of several stages. The initial processing occurs too rapidly to be within the sphere of awareness. A "microscopic" approach to the priming of hostility stems from the work of Bargh and his associates. They found that a person evaluates any stimulus very rapidly, usually within one-third of a second. In one of their experiments, a stimulus perceived by respondents as "good" facilitated pulling a lever toward themselves, and a "bad" evaluation facilitated pushing the lever away from themselves.[8] A plausible inference from this study is that the subject processes threat very rapidly and mobilizes to act against it before becoming consciously aware of the threat.

Future experiments could test whether other kinds of threat stimuli (for example, the picture of an angry face) facilitate the push response. Experiments like these can help trace the progression of a hostile response from the very beginning. After the initial "unconscious" responses, the subject is usually in a "good" or "bad" mindset. At this point, an ambiguous scenario similar to that used in the Dodge studies could be presented and the subject queried regarding his thoughts. If a hostile mindset is induced, I would predict that he would provide automatic thoughts regarding the malicious intent of the offender.

This experiment would then proceed from the earliest to the next stage of information processing, which involves conscious evaluation. It would test the hypothesis that one's conscious interpretations can be biased on the basis of a preexisting mindset. Positive findings would support the notion of the cognitive continuum in information processing.

The early conscious or preconscious phase is manifested in the automatic thoughts. Ordinarily, people rapidly evaluate the automatic thought and either reject or put aside the interpretation or elaborate on it. Raymond, for example, reacted to Gloria's "reminders" initially with the thought, "She's always putting me down." This was followed by a number of elaborations: "She doesn't have any respect for me. She'll cut me up if I let her. I can't let her get away with it!" A number of studies support these clinical observations.

In any study like this, it is important that the aggressive beliefs of the offender be "primed." As mentioned, an appropriate videotape or film, the induced imagination of a provocative event, or even a narrative told in the subject's first-person voice, may be used. This induction procedure ideally should activate the hostile beliefs.

Experimental confirmation of the priming of automatic thoughts and cognitive biases has been demonstrated by Christopher Eckhardt and his collaborators.[9] Essentially the team had subjects repeat their thoughts out loud during anger-inducing audiotapes that depicted an imagined interaction with their wives. The main thrust of Eckhardt's work was to see whether cognitive biases could differentiate *violent males* in distressed marriages from *nonviolent males* in distressed marriages, and males in satisfactory marriages. Specifically, subjects' articulated thoughts to anger-arousing and non-anger-arousing scenarios were coded by trained raters. The results showed that the maritally violent males were more likely than the nonviolent males to show various types of cognitive distortions that are ascribed to primal thinking (chapter 4): magnification, dichotomous thinking, and arbitrary inference.

Eckhardt's procedure could be used to test a variety of hypotheses relevant to anger. By changing the nature of the scenarios, it would be possible to determine the exact thought content of individuals exposed to various anger-inducing situations. For example, a researcher could present scenarios relevant to interpersonal conflict in the workplace, with family members, with friends and acquaintances, with strangers, and then record and classify the automatic thoughts.

Wickless and Kirsch conducted a study of the cognitive theory of emotion. Seventy-two hundred graduates recorded their thoughts and emotions whenever they felt angry, anxious, or sad over a three-day period. Structured interview data were collected after each day, and subjects were asked to write down their thoughts and feelings. The interview data were then scored for thought content. Analyses strongly supported the hypothesis that anger is associated with thoughts of transgression, anxiety with thoughts of threat, and sadness with thoughts of loss. The theme of being "wronged" was most frequently associated with anger.[10]

When simple questionnaires are not likely to reveal basic biased beliefs, there are a number of "implicit" or indirect methods available. For example, in a typical experiment, photographs or words descriptive of a particular stigmatized group (for example, African Americans) are displayed at a speed just below the threshold for conscious recognition. Subsequently, the time required by an individual to recognize and enunciate the negatively or positively biased words is measured. Subjects not belonging to the stigmatized group (Caucasians) recognize the negatively valenced words (for example, *unfriendly* or *aggressive*) faster than they recognize positively valenced words (such as *friendly* and *helpful*). Thus, the reaction time for positive words is faster after exposure to members of the same race.[11]

INTERVENTION AND PREVENTION

When seriously considering what type of intervention or prevention program to undertake to remedy hostile behavior, it is important to adapt the strategies to the characteristics of the specific problem, for example, child abuse, criminal assault, or ethnic conflict.

Therapy of Angry Patients

As a clinician, I have found it fruitful to investigate the information-processing (or cognitive) and interpersonal systems. I have been able to collaborate with patients in correcting their various distortions and modifying the belief systems that predisposed them to anger and violence. Similarly, in working with members of a family or couples, I have been able to induce them to focus on changing their individual egocentric frame of reference, to become more sensitive to their partner's perspective, feelings, and goals, and to develop greater empathy for each other.

I have already described the effective cognitive therapy for anger by Deffenbacher (chapter 2). Beck and Fernandez conducted a meta-analysis of 50 studies incorporating 1,640 subjects to determine the efficacy of cognitive behavior therapy of anger. They found that the average cognitive therapy recipient was better off than 76 percent of the untreated subjects in terms of anger reduction. The cases surveyed covered a wide variety of anger and hostility-related domains, including prison inmates, abusive parents, abusive spouses, juvenile delinquents, adolescents in residential treatment, aggressive children, and college students with reported anger problems.[12]

Anger reduction may be life-saving in many instances. Schwartz and Oakley, for example, found that a cognitive behavior program was as effective as blood pressure medication in reducing elevated blood pressure, a frequent precursor of strokes and heart attacks.[13]

Child Abuse

It is obviously important to design effective programs to reduce child abuse, not only to help the child but also to prevent the transmission of child abuse from one generation to the next. Abusing parents were generally abused themselves as children.

An outcome trial was designed by Whiteman, Fanshel, and Grundy to test the efficacy of a cognitive behavioral intervention in reducing the anger of an abusive parent. The cognitive behavior package consisting of cognitive restructuring, problem solving, and relaxation showed a significant improvement in anger as compared with a control group. The cognitive intervention was directed toward reframing the meaning of a child's "provocation."[14] When the parent believed the provocation was intentional—as most of these parents did most of the time—the degree of anger was substantially greater than if the provocation was viewed as unintentional.

The parent also learned to look for explanations for the child's behavior other than that he or she was "just a rotten kid." An interesting part of this study utilized role-play within the session to simulate a provocative act by the child. Presumably, improvement was measured by the degree of anger reduction noted in the provocative challenge.

Spouse Abuse

Abusive husbands show a unique sensitivity to any kind of threat to their self-esteem. One study showed that batterers were far more likely to perceive minor, ambiguous acts by the wife as attacks on their pride than were non-battering husbands. "Romantic jealousy" was another source of hypersensitivity. Violence was often triggered between spouses or dating partners by a very trivial act on the part of the woman: if she simply smiled in a friendly way at another man, her partner might conclude that she was attracted to him—which he then interpreted as an attack on his own sexual adequacy.[15] As he became mobilized, the urge to strike her became paramount. Having pinpointed precisely the beliefs underlying the urge to strike the wife, the therapist tailored his intervention to remedy the specific vulnerability (feelings of rejection, fears of abandonment, unstable self-esteem) and to teach the husband more adaptive coping mechanisms.

Initially, a therapist would concentrate on controlling the expression of hostility and protecting the wife from further attacks. Then the focus would shift to the predispositional beliefs, helping the patient to deal better with the precipitating elements and helping the couple to solve their problems in a rational way (including separation, if indicated).

Convicted Offenders

Promising results have been obtained in the preventive treatment of incarcerated offenders.[16] The investigators compared the recidivism rates for a group of 55 male offenders participating in a "cognitive self-change program" with a similar group of 141 offenders from the same facility. Although still high, the recidivism rate was significantly lower in the treatment group (50 percent) than in the untreated group (70.8 percent).

Treatment groups consisted of five to ten offenders who met three to five times a week. The therapeutic model focused on "criminogenic thinking errors." The convicts showed the typical thinking problems described in chapter 8. Typical cognitions of a robber were, "I deserve to make a couple of bucks after all the cops put me through last time." The program also drew on techniques described in Ross and Fabiano's 1985 manual *Time to Think*.[17]

Prevention of Delinquency

The work of Dodge and his associates in defining the cognitive predisposition to delinquency in preschool children has culminated in a large-scale prevention study using cognitive strategies within their broad programs.

Their Fast Track Project is a randomized clinical trial to test the efficacy of a comprehensive intervention in preventing serious conduct disorder in high-risk children. Following the screening of 10,000 kindergarten boys and girls from high-poverty schools in four regions of the country, 892 were identified as highly aggressive and were randomly assigned to receive intervention or to serve as nontreated controls. Intervention is planned to last from first through tenth grades and to include social cognitive skills training for children, parent training in behavior management, academic tutoring in reading skills for children, peer pairing for children to improve peer relations, classroom curricula in social cognitive development, mentoring by adult volunteers, and teacher consultation.

After the first four years of intervention, the treated group is performing more favorably than the control group in the proximal targeted areas of intervention (including social cognitive skills and academic performance) and in their control of their aggressive behavior. Furthermore, and significantly, sophisticated analyses indicate that positive outcomes in aggressive behavior are mediated by changes in *social cognitive skills*. Long-term outcomes in adolescent conduct disorder await further intervention and evaluation.[18]

Prevention of Ethnopolitical Violence

With the end of the cold war, the world entered a new phase. The pattern of warfare has shifted from interstate to intrastate wars—intercommunal wars fought by rival ethnic groups. Since the end of the cold war, there have been approximately thirty such conflicts each year. Researchers studying ethnopolitical warfare have a growing awareness of the need to use a multidisciplinary mode that incorporates the cognitive, social, and emotional elements involved in ethnopolitical conflict. With the increase of bitter ethnopolitical conflicts, there is a pressing need for well-trained professionals who are equipped to work in war zones and understand the local culture and situation. A crucial challenge is to prepare a new generation of psychologists to participate in this important applied work.

With these facts in mind, the Institute of Ethnopolitical Warfare has been established at the University of Pennsylvania. The purpose of the initiative is to stimulate scholarship and practice that will increase understanding of the processes involved in ethnopolitical warfare and improve methods of prediction, intervention, and prevention. The goal of the institute is to prepare psychologists and other social scientists to address psychosocial problems and to meet human needs in countries torn by armed conflict and political violence.

The cognitive model can serve a useful purpose in understanding the dysfunctional thinking of ethnic or national leaders. Certainly, mediators need to be aware of the primal thinking (personalization, magnification, all-or-nothing thinking, and so on) of the opposite sides. When an individual on either side feels threatened, for example, he is more likely to revert to these cognitive distortions. It is, of course, often difficult to separate dramatic rhetoric intended to bully or deceive the other side from true distortion. Clues to distortion would be an increase in defensiveness or withdrawal into silence. Negotiators can be expected to picture their side as the victim and the other as the victimizer. Considerable skill is required to redirect the focus toward the question of what benefits they would gain from an agreement.

THE FUTURE

I believe that considerable progress has been made in understanding the various factors leading to excessive anger and violence. As we move away from the notion of the "inner furies" and the "demonic male" to a cognitive model, we can find many points for beneficial intervention. It is not possible to exorcise these mythical inner demons, but we can address and modify dysfunctional attitudes and erroneous thinking. Automatic interpretations and inferences are potentially destructive if they are not corrected. But we can draw on the immense powers of the mind to modify them before they damage the individual himself or others.

There are obviously many systems that need to be analyzed thoroughly before sophisticated prevention and intervention programs can be implemented. We already know a great deal about the psychology of temperamental employers and abusive spouses and parents. Our knowledge of the aberrations of power-driven leaders and their naïve followers needs to be expanded. We know a great deal about the nature of prejudice, but we have not yet been able to convert this information into effective programs to prevent large-scale massacres. The most successful form of intervention to date has been in the superimposition of controls from above by supranational organizations like NATO and the United Nations.

There are many positive features of human nature that can be used in future programs. As Sober and Wilson have shown, natural selection has endowed us with many benevolent capacities: empathy, generosity, altruism.[19] We can draw on these to energize beliefs such as, "The ends do *not* justify the means." However, corrective programs need to be directed at the kinds of beliefs that justify violence: egocentrism and group egoism; punishment and retribution; diffusion of responsibility; permissive attitudes toward violence.

Although many authors—such as Jervis, Snyder and Diesing, Vertzberger, Tetlock, and White—have pinpointed the errors in the thinking of national leaders, this knowledge has not yet been put to practical use.[20] Similarly, only rudimentary steps have been taken to modify the thinking of members of warring ethnic groups.

In the final analysis, we have to depend on our rich resources of rationality to recognize and modify our irrationality. The "voice of reason" is not necessarily quiet if we use appropriate methods to amplify it. We can recognize that our own interests are best served by applying reason. In this way, we can help to provide a better life for ourselves, others, and the future children of the world.

NOTES

INTRODUCTION

1. The tendency to judge other people's behavior as though it were solely determined by their internal attributes rather than by situational factors has been labeled the Fundamental Attribution Error. F. Heider, *The Psychology of Interpersonal Relations* (New York: Wiley, 1958). See also chapter 5, note 7.

2. K. Horney, *Neurosis and Human Growth: The Struggle Toward Self-realization* (New York: Norton, 1950).

3. A. Ellis, *Reason and Emotion in Psychotherapy* (1962; reprint, New York: Carol Publishing Group, 1994).

4. A. T. Beck, *Love Is Never Enough* (New York: HarperCollins, 1988); A. T. Beck and G. Emery, with R. L. Greenberg, *Anxiety Disorders and Phobias: A Cognitive Perspective* (New York: Basic Books, 1985); A. T. Beck, A. Freeman, and Associates, *Cognitive Therapy of Personality Disorders* (New York: Guilford, 1990); A. T. Beck, F. W. Wright, C. F. Newman, and B. Liese, *Cognitive Therapy of Substance Abuse* (New York: Guilford, 1993).

5. D. A. Clark and A. T. Beck, with B. Alford, *The Scientific Foundations of Cognitive Theory of Depression* (New York: John Wiley & Sons, 1999); K. Dobson, "A Meta-analysis of the Efficacy of Cognitive Therapy for Depression," *Journal of Consulting and Clinical Psychology* 57, no. 3 (1989): 414–19.

CHAPTER I

1. State-sponsored mass murder such as the massacres of the Tutsis in Rwanda, the Muslims in Bosnia, and the Jews in Hitler's Europe have been studied in terms of the interplay of ethnopolitical, cultural, and socioeconomic factors. Nonetheless, a complete understanding cannot be attained without focusing sharply on the leaders who initiate the genocidal programs and on those who implement them. All of the contributing group processes and the individual factors such as the power drives of the leaders and their manipulations of the imaginations and emotions of their adherents converge on the final common pathway: the decision of the perpetrators to carry out their assigned roles.

    The governing elite exploit all channels of communication to demonize the victims. In Rwanda, for example, official propaganda created the image of the Tutsis as "vipers and drinkers of untrue blood" who were plotting to massacre innocent Hutus. D. N. Smith, "The Psychocultural Roots of Genocide: Legitimacy and Crisis in Rwanda," *American Psychologist* 53, no. 7 (1998): 743–53.

2. S. Baron-Cohen, *Mindblindness: An Essay on Autism and Theory of Mind* (Cambridge, Mass.: MIT Press, 1995).

3. I have capitalized the initial letter of the words *Enemy* and *Evil* because they have a special significance in the context in which I use them—separate from their metaphysical or theological meanings. The moral concept of Evil as an adjective (rather than a noun as in the theological doctrines) is used by individuals or groups to describe other individuals. They use the concept of the Enemy with all its pejorative connotations in a similar fashion. Both words are abstractions that transcend the actual characteristics of the Other (another abstraction designating a homogeneous entity—the outsiders, the aliens) and impose the most deadly, absolute, categorical devaluation of the Other.

    Despite the transcendental nature of these words, they are reified by people: they become a "verifiable" fact, a reality. The behavior of the objects of hatred is automatically interpreted to conform to the image—thus confirming the validity of the image. The subjective response is revulsion or hate—or fear. The hater feels compelled to castigate or eliminate the hated persons who have been incarcerated in these categories. Both the hater and the hated become the prisoners of this primitive mode of thinking.

The lynch mob or rampaging soldiers do not recognize that in attacking the Evil Enemy they are actually attacking other human beings like themselves. As with Evil and Enemy, the Other is compressed into a monolithic category but is perceived as a real entity rather than an abstract concept.

4. A. T. Beck, *Love Is Never Enough* (New York: HarperCollins, 1988).

5. The term "framing" is also appropriate in the vernacular sense. On the basis of distortions and erroneous conclusions, an aggrieved person can build up a case against a supposed adversary based on minimal or no evidence of wrongdoing.

6. Beck, *Love Is Never Enough.*

7. The persecuted minority, however, generally have a reasonably accurate perception of the hostile, biased perspective of their persecutors.

8. K. A. Dodge, "Social Cognitive Mechanisms in the Development of Conduct Disorder and Depression," *Annual Review of Psychology* 44 (1993): 559–84.

9. K. Horney, *Neurosis and Human Growth: The Struggle Toward Self-realization* (New York: Norton, 1950).

10. D. G. Kingdon and D. Turkington, *Cognitive-Behavioral Therapy of Schizophrenia* (New York: Guilford, 1994). There have been a number of published cases of delusional persons killing the object of their paranoia and then themselves. For example, on May 5, 1998, Cedrich Tornay, a noncommissioned officer of the Swiss guard at the Vatican, killed the chief of the pope's guards and his Venezuelan wife before turning his gun on himself. *New York Times,* May 5, 1998, p. 1.

11. I assume that a therapist would take the proper precautions, for example, notifying the authorities if the threat is real.

12. Proper precautions need to be given priority over psychotherapy. In most instances, however, both are feasible.

13. D. P. Barash, *Beloved Enemies: Our Need for Opponents* (Amherst, N.Y.: Prometheus, 1994). I disagree with Barash on this point. There is no

evidence that people have a "need" for opponents or enemies. Misperceiving others as enemies is a cognitive problem.

14. L. Silber and A. Little, *Yugoslavia: Death of a Nation* (New York: Penguin, 1996).

15. E. K. Fromm, *The Anatomy of Human Destructiveness* (New York: Holt, Rinehart, and Winston, 1973).

16. C. Browning, *The Path to Genocide* (Cambridge: Cambridge University Press, 1992); D. Goldhagen, *Hitler's Willing Executioners: Ordinary Germans and the Holocaust* (New York: Alfred A. Knopf, 1996).

17. A. Huxley, *Ends and Means.* (New York: Harper & Brothers, 1937).

18. See chapter 1, note 43.

19. B. Tuchman, *Guns of August* (New York: Macmillan, 1962).

20. C. R. Mann, *When Women Kill* (Albany: State University Press of New York, 1982).

21. Paul Hollander discusses four models of the specialists in coercion and political violence, including those involved in both the Nazi and Soviet apparatuses. The first group is represented by the ideologically driven, supposedly incorruptible, puritanical executioners exemplified by Heinrich Himmler in the Nazi case. The second group embodies the "banality of evil," of which Adolf Eichmann, described by Hannah Arendt in *Eichmann in Israel*, is the prototype. This group is composed of supposedly very ordinary human beings who simply follow orders without being driven by strong convictions. Often money and privilege are factors. The third category comprises well-educated careerists who find satisfactory employment and mobility opportunities in the organization. The fourth group is composed of individuals who gravitate toward organizations of violence and coercion. Many of the notable torturers belong to this group. Their personalities are congenial to sadistic and repressive actions. P. Hollander, "Revisiting the Banality of Evil: Political Violence in Communist Systems," *Partisan Review* 64, no. 1 (1997): 56.

22. R. F. Baumeister, A. M. Stillwell, and T. K. Heatherton, "Guilt: An Interpersonal Approach," *Psychological Bulletin* 115 (1994): 243–67.

23. D. Grossman, *On Killing: The Psychological Cost of Learning to Kill in War and Society* (Boston: Little, Brown, 1995).

24. Browning, *The Path to Genocide*.

25. R. Robins and J. Post, *Political Paranoia: The Psychopolitics of Hatred* (New Haven, Conn.: Yale University Press, 1997).

26. D. Maybury-Lewis, and U. Alamagor, Eds., *The Attraction of Opposites: Thought and Society in the Dualistic Mode* (Ann Arbor: University of Michigan Press, 1989).

27. R. Baumeister, *Evil: Inside Human Cruelty and Violence* (New York: W. H. Freeman, 1997).

28. Religious people in the northern United States, for example, are less prone to violence than are nonreligious people. R. Nisbett and D. Cohen, *Culture of Honor: The Psychology of Violence in the South* (Boulder, Colo.: Westview, 1996).

29. J. A. Bargh, S. Chaiken, P. Raymond, and C. Hymes, "The Automatic Evaluation Effect: Unconditional Automatic Attitude Activation with a Pronunciation Task," *Journal of Experimental Social Psychology* 32, no. 1 (1996): 104–28.

30. Albert Bandura provides a comprehensive social learning theory of hostility that takes into account biological factors and learning through direct experience or observation. He proposes that aggression is instigated by the influence of models (by attack or frustration), incentives such as desire for money or admiration, instructions (for example, orders from a superior), and delusions. He also notes that aggression can be regulated by external rewards and punishments; vicarious reinforcements, such as observing other people's rewards and punishments; and self-regulatory mechanisms, such as pride and guilt. A. Bandura, *Aggression: A Social Learning Analysis*. (Englewood Cliffs, N.J.: Prentice Hall, 1983). Robert Baron and Deborah Richardson expand on this model and apply it to the systematic study of aggression. R. A. Baron & D. R. Richardson, *Human Aggression*. 2nd ed. (New York: Plenum Press, 1994).

31. J. H. Barkow, L. Cosmides, and J. Tooby, *The Adapted Mind: Evolutionary Psychology and the Generation of Culture* (New York: Oxford University Press, 1992); P. Kropotkin, *Evolution and Environment* (Montreal: Blackrose Books, 1995).

32. P. Gilbert, *Human Nature and Suffering* (Hillsdale, N.J.: Erlbaum Associates, 1989).

33. J. H. Barkow, L. Cosimides, and J. Tooby, *The Adapted Mind: Evolutionary Psychology and the Generation of Culture* (Oxford: Oxford University Press, 1992).

CHAPTER 2

1. There is increasing evidence that people respond very rapidly to new stimuli or to changes in their immediate environment. I refer to these changes as "events." These new bits of information are subjected at first to an immediate crude evaluation that attaches a positive or negative valence to the stimulus. Translated into language, the evaluation conveys the notion of "good for me" or "bad for me," or "right" or "wrong." This initial process occurs very rapidly—within one-third of a second—and is unconscious.

As additional information from the environment (or from bodily sensations) is taken in, more complete processing takes place. This allows for an evaluation of the total context of the event and a determination of whether the initial evaluation was relevant to the individual's interests and significant. If this second appraisal indicates that the event is significant and relevant, then a more elaborate response, in the form of an image or in words, occurs. The imagery or verbal experience conveys the more complete meaning of the event. Although they occur very rapidly, the imagery and verbal ideation are accessible to introspection. Since these mental responses occur like a reflex, without volition, I refer to them as "automatic thoughts," or "preconscious cognitions." Although the automatic thoughts and images are at the periphery of awareness, people can become more aware of them with training and can describe them in detail. In therapy these cognitions provide the basic material we use to understand the personal meaning of an event. Patients can be trained not only to identify these cognitions but also to reality-test them. If the cognitions do not appear to be valid, reasonable interpretations, the patient can learn to reframe them.

The stages of information processing are implemented by specific structures labeled "schemas." Crude schemas simply assign a positive or negative valence to a stimulus at the initial stage of information processing. More elaborate schemas have a content labeled "beliefs," which facilitate the progression of the initial evaluation into a more comprehensive assignment of meaning to the event. These "meaning assignment schemas" are not necessarily rigid, and the beliefs are subject to disconfirming experiences or corrections on the basis of empirical testing and rational analysis. Correction of inaccurate automatic thoughts, for example, can "filter down" to the schemas and modify the incorporated beliefs.

At a different level, identification of the beliefs can facilitate their conscious, "rational" modification. Cognitions are *not* the only products of information processing—behavior and affect are also activated. There is evidence that the initial activation of the behavior apparatus occurs automatically and immediately after the initial automatic evaluation. In experiments, stimuli evaluated as good facilitated the automatic pull of a lever by the subject; bad stimuli facilitated the automatic *push* of the lever. This automatic behavioral tendency may be manifested in the musculature: mobilization to attack or flee; freezing; an inclination to attack passive demobilization; sadness or depression. J. A. Bargh, S. Chaiken, P. Raymond, and C. Hymes, "The Automatic Evaluation Effect: Unconditional Automatic Attitude Activation with a Pronunciation Task," *Journal of Experimental Social Psychology* 32, no. 1 (1996): 104–28.

2. R. Hofstadter, *The Paranoid Style in American Politics and Other Essays* (New York: Vintage, 1967).

3. A. T. Beck, *Love Is Never Enough* (New York: HarperCollins, 1988).

4. J. L. Deffenbacher, E. R. Oetting, M. E. Huff, G. R. Cornell, and C. J. Dallager, "Evaluation of Two Cognitive-Behavioral Approaches to General Anger Reduction," *Cognitive Therapy and Research* 20, no. 6 (1996): 551–73; J. L. Deffenbacher, E. R. Dahlen, R. S. Lynch, C. D. Morris, and W. N. Gowensmith, "Application of Beck's Cognitive Therapy to Anger Reduction," paper presented at the 106th annual convention of the American Psychological Association, San Francisco (November 1998).

5. A. Koestler, *The Ghost in the Machine* (1967; reprint, London: Pan Books, 1970).

6. R. Robins and J. Post, *Political Paranoia: The Psychopolitics of Hatred* (New Haven, Conn.: Yale University Press, 1997).

7. K. Lorenz, *On Aggression* (New York: Routledge, 1966).

8. C. Helm and M. Morelli, "Stanley Milgram and the Obedience Experiment: Authority, Legitimacy, and Human Action." *Political Theory* 7 (1979): 321–46.

9. W. B. Cannon, *Wisdom of the Body* (New York: Norton, 1963).

10. L. Berkowitz, "Frustration-Aggression Hypothesis: Examination and Reformulation." *Psychological Bulletin* 106 (1989): 59–73.

11. A. Bandura, *The Social Foundations of Thought and Action: A Social Cognitive Theory* (New York: Paramus Prentice-Hall, 1985).

12. E. Anderson, "The Code of the Streets," *Atlantic* (May 1994): 81–92; R. Nisbett and D. Cohen, *Culture of Honor: The Psychology of Violence in the South* (Boulder, Colo.: Westview Press, 1996).

13. J. H. Barkow, L. Cosmides, and J. Tooby, *The Adapted Mind: Evolutionary Psychology and the Generation of Culture* (New York: Oxford University Press, 1992); D. P. Barash, *Beloved Enemies: Our Need for Opponents* (Amherst, N.Y.: Prometheus, 1994).

14. F. DeWaal, *Good Natured* (Cambridge, Mass.: Harvard University Press, 1996); F. B. M. DeWaal, L. M. Luttrell, "Mechanisms of Social Reciprocity in Three Primate Species: Symmetrical Relationship Characteristics or Cognition," *Ethological Sociobiology* 9, nos. 2–4 (1988): 101–18.

CHAPTER 3

1. A. T. Beck, *Love Is Never Enough* (New York: Harper & Row, 1988).

2. N. Roese, *What Might Have Been: The Social Psychology of Counterfactual Thinking* (Hillsdale, N.J.: Erlbaum Associates, 1995).

3. For a review of the experimental literature on threats to the self-esteem, see R. Baumeister, *Evil: Inside Human Cruelty and Violence* (New York: W. H. Freeman, 1997).

CHAPTER 4

1. A. T. Beck and G. Emery, with R. L. Greenberg, *Anxiety Disorders and Phobias: A Cognitive Perspective* (New York: Basic Books, 1985).

2. C. Sagan and A. Druyan, *Shadows of Forgotten Ancestors: A Search for Who We Are* (New York: Random House, 1992).

3. A. Ellis, *Reason and Emotion in Psychotherapy* (New York: Carol Publishing, 1994).

4. M. Daly and M. Wilson, *Homicide* (New York: Reed Elsevier, 1988).

5. M. R. Leary, *Understanding Social Anxiety: Social, Personality, and Clinical Perspectives* (Beverly Hills, Calif.: Sage Publications, 1983); J. Birtchnell, *How Humans Relate: A New Interpersonal Theory* (Westport, Conn.: Praeger, 1993).

6. Sagan and Druyan, *Shadows of Forgotten Ancestors.*

7. I. L. Janis, *Victims of Groupthink: A Psychological Study of Foreign Policy Decisions and Fiascoes* (Boston: Houghton Mifflin, 1982).

8. K. Williams, "Social Ostracism," in *Aversive Interpersonal Behaviors,* Plenum Series in Social/Clinical Psychology, ed. R. Kowalski (New York: Plenum Press, 1997), pp. 133–70.

CHAPTER 5

1. False positive sightings have led to tragedies. On September 1, 1983, a Soviet fighter was ordered "to stop the flight" of a Korean Air Lines Boeing 747 over the island of Sakhalin after it had failed to obey the fighter's demands that it land at a Soviet airfield. The fighter pilots involved did not know they were dealing with a civilian airliner. A. D. Horne, "U.S. Says Soviets Shot Down Airliner," *Washington Post,* September 2, 1983, A1. Also, a U.S. Navy Aegis-class cruiser—the USS *Vincennes*—fired the missile that destroyed an Iranian civilian airplane over the Persian Gulf on July 3, 1988, killing all 290 aboard. The civilian aircraft had been misidentified as a warplane. G. Wilson, "Navy Missile Downs Iranian Jetliner over Gulf," *Washington Post,* July 4, 1988, A1.

2. J. Piaget, *The Moral Judgement of the Child,* translated by Marjorie Gabain (1932; reprint, Glencoe, Ill.: Free Press, 1960).

3. S. Freud, *The Basic Writings of Sigmund Freud,* translated and edited by A. A. Brill (New York: Modern Library, 1938).

4. M. E. Oakley and D. Shapiro, "Methodological Issues in the Evaluation of Drug-Behavioral Interactions in the Treatment of Hypertension," *Psychosomatic Medicine* 51 (1989): 269–76.

5. I. E. Sigel, E. T. Stinson, M.-I. Kim, *Socialization of Cognition: The Distancing Model* (Hillsdale, N.J.: Erlbaum Associates, 1993).

6. A. Ellis, *Anger: How to Live With and Without It* (New York: Carol Publishing, 1985).

7. The Fundamental Attribution Error—the tendency to judge other people's behavior as though it were solely determined by their internal attributes rather than by situational factors—is particularly evident in individuals who are prone to experience inappropriate or excessive outbursts of anger or violence. The tendency is further accentuated in conflicts between individuals or groups. It is particularly apparent in distressed marriages when difficulties are attributed exclusively to faults of the other mate. It also has been credited to the misreading of the intentions of the other side, as in the events leading up to World War I; B. Tuchman, *Guns of August* (New York: Macmillan, 1962).

   It has been suggested that the Fundamental Attribution Error—blaming unpleasant events on other people—is the simplest, most satisfying, and least effortful explanatory strategy. Although there is some controversy as to how "fundamental" the error is, there is substantial evidence of its ubiquity, in the positive direction as well as the negative one. Happy people, for example, attribute their success to their personal qualities and their failure to external causes. The reverse is true of depressed individuals. This kind of biased thinking has also been labeled "the correspondence bias." D. T. Gilbert and P. S. Malone, "The Correspondence Bias," *Psychological Bulletin* 117, no. 1 (1995): 21–38; J. P. Forgas, "On Being Happy and Mistaken: Mood Effects on the Fundamental Attribution Error," *Journal of Personal and Social Psychology* 75 (1998): 318–31.

8. G. M. Buchanan and M. E. P. Seligman, eds., *Explanatory Style* (Hillsdale, N.J.: Erlbaum Associates, 1995).

CHAPTER 6

1. For another example of instant evaluation, see J. A. Bargh, S. Chaiken, P. Raymond, and C. Hymes, "The Automatic Attitude Evaluation Effect: Unconditional Activation with a Pronunciation Task," *Journal of Experimental Social Psychology* 32, no. 1 (1996): 104–28.

2. A. T. Beck, and G. Emery, with R. L. Greenberg, *Anxiety Disorders and Phobias: A Cognitive Perspective* (New York: Basic Books, 1985).

3. N. Roese, *What Might Have Been: The Social Psychology of Counterfactual Thinking* (Hillsdale, N.J.: Erlbaum Associates, 1995).

4. Ibid.

5. This is another example of people being drawn to "personal" explanations more readily than to "situational" explanations.

6. K. Horney, *Neurosis and Human Growth: The Struggle Toward Self-realization* (New York: Norton, 1950); A. Ellis, *Reason and Emotion in Psychotherapy* (New York: Carol Publishing, 1994).

7. J. H. Barkow, L. Cosmides, and J. Tooby, *The Adapted Mind: Evolutionary Psychology and the Generation of Culture* (Oxford: Oxford University Press, 1992).

CHAPTER 7

1. A. T. Beck, *Love Is Never Enough* (New York: HarperCollins, 1988).

2. P. Noller, *Nonverbal Communication and Marital Interaction* (New York: Pergamon Press, 1984); N. Jacobson and J. Gottman, *When Men Batter Women: New Insights into Ending Abusive Relationships* (New York: Simon & Schuster, 1998).

CHAPTER 8

1. A distinction should be made between the "reactive offender," the focus of this chapter, who has a very labile self-esteem, and the "primary psychopath," who coasts on a tide of grandiosity and perceives others as weak and malleable.

2. Neil Jacobson and John Gottman, after painstaking research with more than two hundred couples, categorized male batterers as falling

into one of two categories, which they named "Pit Bulls" and "Cobras." Although they seem to present some important clinical insights in their volume, the authors fall into the pitfall of stereotyping their male clients. It would seem to be difficult for a patient, knowing that he has been categorized as a Pit Bull or a Cobra, to be able to relate to the therapist, and it might be equally difficult for the therapist who entertains such an image of the patient to work with him. Another pitfall here is not only stereotyping but demonizing individuals who have engaged in antisocial behavior. N. Jacobson and J. Gottman, *When Men Batter Women: New Insights into Ending Abusive Relationships* (New York: Simon & Schuster, 1998).

3. K. D. O'Leary, "Physical Aggression in Intimate Relationships Can Be Treated Within a Marital Context Under Certain Circumstances," *Journal of Interpersonal Violence* 11, no. 3 (September 1996): 450–52.

4. The psychology of abuse is the same irrespective of whether or not the domestic partners are married.

5. C. I. Eckhardt and D. J. Cohen, "Attention to Anger-Relevant and Irrelevant Stimuli Following Naturalistic Insult," *Personality and Individual Differences* 23, no. 4 (1997): 619–29.

6. A. Holtzworth-Munroe and G. Hutchinson, "Attributing Negative Intent to Wife Behavior: The Attributions of Maritally Violent Versus Nonviolent Men," *Journal of Abnormal Psychology* 102 (1993): 206–11.

7. C. I. Eckhardt, K. A. Barbour, and G. C. Davison, "Articulated Thoughts of Maritally Violent and Nonviolent Men During Anger Arousal," *Journal of Consulting and Clinical Psychology* 66, no. 2 (April 1998): 259–69; R. Serin and M. Kuriychuk, "Social and Cognitive Processing Deficits in Violent Offenders: Implications for Treatment," *International Journal of Law and Psychiatry* 17 (1994): 431–41.

8. M. Daly and B. Smuts, "Male Aggression Against Women: An Evolutionary Perspective," *Human Nature* 3, no. 1 (1992): 1–44; B. Smuts, "Male Aggression Against Women: An Evolutionary Perspective," in *Sex, Power, Conflict: Evolutionary and Feminist Perspectives,* ed. D. M. Buss and N. M. Malamuth (New York: Oxford University Press, 1996), pp. 231–68.

9. Although alcohol does not cause violence, it is obvious that it is a

contributing cause. Many studies of murder, rape, and assault have demonstrated that the majority of violent crimes occur when the assailants or offenders have been drinking. The propensity of alcohol to intensify aggressive tendencies has also been supported by laboratory experiments. Evidently alcohol makes individuals more egocentric and therefore more likely to interpret events as personal and to revert to the type of primal thinking that calls for revenge for insults. R. Baumeister, *Evil: Inside Human Cruelty and Violence* (New York: Freeman, 1997), p. 140.

10. Eckhardt, Barbour, and Davison, "Articulated Thoughts. . . . "

11. A. Holtzworth-Munroe, G. L. Stuart, G. Hutchinson, "Violent Versus Nonviolent Husbands: Differences in Attachment Patterns, Dependency, and Jealousy," *Journal of Family Psychology* 11, no. 3 (1997): 314–31.

12. M. I. Wilson and M. Daly, "Male Sexual Proprietariness and Violence Against Wives," *Current Directions in Psychological Science* 5, no. 1 (1996): 2–7.

13. K. A. Dodge, "Social Cognitive Mechanisms in the Development of Conduct Disorder and Depression," *Annual Review of Psychology* 44 (1993): 559–84.

14. Ibid.

15. Ibid.

16. J. P. Newman, W. A. Schmitt, and W. D. Voss, "The Impact of Motivationally Neutral Cues on Psychopathic Individuals: Assessing the Generality of the Response Modulation Hypothesis," *Journal of Abnormal Psychology* 106, no. 4 (1997): 563–75.

17. H. Cleckley, *The Mask of Sanity: An Attempt to Clarify Some Issues About the So-called Psychopathic Personality* (St. Louis: Mosby, 1950).

18. R. D. Hare, L. M. McPherson, and A. E. Forth, "Male Psychopaths and Their Criminal Careers," *Journal of Consulting and Clinical Psychology* 56 (1988): 710–14.

19. D. T. Lykken, "Psychopathy, Sociopathy, and Crime," *Society* 34, no. 1 (1996): 29–38; B. B. Wolman, *The Sociopathic Personality* (New York: Brunner/Mazel, 1987).

20. M. H. Stone, ed., "Antisocial Personality and Psychopathy," in *Abnormalities of Personality: Within and Beyond the Realm of Treatment*

(New York: Norton, 1993), pp. 277–313; Hare, McPherson, and Forth, "Male Psychopaths and Their Criminal Careers."

21. M. R. Burt, "Cultural Myths and Supports for Rape," *Journal of Personality and Social Psychology* 38 (1980): 217–30.

22. D. L. L. Polaschek, T. Ward, and S. M. Hudson, "Rape and Rapists: Theory and Treatment," *Clinical Psychology Review* 17, no. 2 (1997): 117–44.

23. N. M. Malamuth and L. M. Brown, "Sexually Aggressive Men's Perceptions of Women's Communications—Testing Three Explanations," *Journal of Personality and Social Psychology* 67, no. 4 (1994): 699–712.

24. Burt, "Cultural Myths and Supports for Rape."

25. J. B. Pryor and L. M. Stoller, "Sexual Cognition Processes in Men High in the Likelihood to Sexually Harass," *Personality and Social Psychology Bulletin* 20, no. 2 (1994): 163–69.

26. T. Ward, S. M. Hudson, L. Johnston, and W. L. Marshall, "Cognitive Distortions in Sex Offenders: An Integrative Review," *Clinical Psychology Review* 17, no. 5 (1997): 479–507.

27. Baumeister, *Evil.*

28. Ward et al., "Cognitive Distortions in Sex Offenders."

CHAPTER 9

1. M. Ridley, *The Origins of Virtue: Human Instincts and the Evolution of Cooperation* (New York: Viking, 1997), pp. 166–67.

2. Ibid.

3. S. Asch, *Social Psychology* (Englewood Cliffs, N.J.: Prentice-Hall, 1952).

4. C. Haney, C. Banks, and P. Zimbardo, "Interpersonal Dynamics in a Simulated Prison," *International Journal of Criminology and Penology* 1, no. 1 (1973): 69–97.

5. M. L. Simner, "Newborn's Response to the Cry of Another Infant," *Developmental Psychology* 5, no. 1 (1971): 136–50.

6. E. Hatfield, J. T. Cacioppo, and R. L. Rapson, *Emotional Contagion* (New York: Cambridge University Press, 1994).

7. Ibid.

8. J. Bavelas, A. Black, N. Chovil, C. Lemery, and J. Mullet, "Form and Function in Motor Mimicry: Topographic Evidence That the Primary Function Is Communication," *Human Communication Research* 14 (1988): 275–99.

9. J. Victor, *Satanic Panic: The Creation of a Contemporary Legend* (Chicago: Open Court, 1993), pp. 91–105.

10. Ibid., pp. 113–14; R. Ofshe, and E. Watters, *Making Monsters: False Memories, Psychotherapy, and Sexual Hysteria* (Berkeley: University of California Press, 1996).

11. C. W. Upham, *Salem Witchcraft*, vol. 2 (New York: Frederick Ungar, 1959).

12. M. Harris, *Cows, Pigs, Wars, and Witches: The Riddles of Culture* (New York: Random House, 1975), p. 207.

13. For example, the Albigensians were persecuted at the end of the thirteenth century for expressing a heretical theology.

14. N. Cohn, *Europe's Inner Demons: An Enquiry Inspired by the Great Witch-Hunt* (New York: Basic Books, 1975); E. Staub, *The Roots of Evil: The Origins of Genocide and Other Group Violence* (New York: Cambridge University Press, 1989); Harris, *Cows, Pigs, Wars, and Witches.*

15. Harris, *Cows, Pigs, Wars, and Witches.*

16. W. Lippmann, *Public Opinion* (New York: Harcourt, Brace, & Co., 1922).

17. G. Allport, *The Nature of Prejudice* (Cambridge, Mass.: Addison-Wesley, 1954), p. 20.

18. The term "prejudice" has been loosely used to refer to biased inter-group perceptions, judgments, or attitudes. But it does not always mean the same thing to various scholars who either associate preju-dice with cognitive distortions or link it to injustice. The former may be labeled "cognitive prejudice" and the latter "moral preju-dice." Cognitive prejudice includes stereotypical judgments of a group, erroneous generalizations, formation of social attitudes despite contradictory objective evidence, and the Fundamental Attribution Error. Moral prejudice consists of the designation of a different set of rights, principles of justice, and judgment of basic value depending on one's social status, race, ethnicity, or other group

membership. Of course, moral prejudice generally is based on the same biased thinking as cognitive prejudice, requiring separate explanations. Revenge, which can lead to feuding, may continue long after the original cause of conflict has vanished. K. Sun, "Two Types of Prejudice and Their Causes," *American Psychologist* 48, no. 11 (1993): 1152–53.

19. P. G. Devine, D. L. Hamilton, and T. M. Ostrom, *Social Cognition: Impact on Social Psychology* (San Diego: Academic Press, 1994).

20. H. Tafjel, *Human Groups and Social Categories: Studies in Social Psychology* (Cambridge: Cambridge University Press, 1981).

21. M. Sherif, O. J. Harvey, B. J. White, W. R. Hood, and C. W. Sherif, *The Robbers Cave Experiment: Intergroup Conflict and Cooperation* (Middletown, Conn.: Wesleyan University Press, 1988).

22. W. Doise, *Groups and Individuals: Explanations in Social Psychology,* translated by Douglas Graham (Cambridge: Cambridge University Press, 1978).

23. Tafjel, *Human Groups and Social Categories.*

24. J. P. Forgas and K. Fiedler, "Us and Them: Mood Effects on Intergroup Discrimination," *Journal of Personality and Social Psychology* 70 (1996): 28–40.

25. Devine et al., *Social Cognition.*

26. R. Fazio, J. Jackson, B. Dunton, and C. Williams, "Variability in Automatic Activation as an Unobtrusive Measure of Racial Attitudes: A Bona Fide Pipeline?" *Journal of Personality and Social Psychology* 69, no. 6 (1995): 1013–27.

27. The tendency to blame difficulties on other people's defective character rather than on the situation that caused the problem has been labeled "the correspondence bias." D. T. Gilbert and P. S. Malone, "The Correspondence Bias," *Psychological Bulletin* 117, no. 1 (1995): 21–38.

28. A very clear-cut example of cognitive distortion is cited by Elliott Aronson. A 1951 football game between Princeton and Dartmouth has been described as the roughest and dirtiest in the history of either school. One of the players on the Princeton team, an All-American, was gang-tackled, piled on, and mauled every time he carried the ball and was finally forced to leave the game with a bro-

ken nose. Soon after his injury, the Princeton team became more aggressive toward the opponent, and subsequently a Dartmouth player was carried off the field with a broken leg. In addition, several fistfights broke out on the field, causing many injuries. E. Aronson, *The Social Animal,* 7th ed. (New York: W. H. Freeman, 1995).

In a psychological study by Albert Hastorf of Dartmouth and Hadley Cantril of Princeton, films of the game were shown to a number of students on each campus. The students were asked to be completely objective while watching the film and to take notes of each infraction of the rules, including how it started and who was responsible. The investigators found a huge difference in the way the game was viewed by the students at each university. The students tended to see their own fellow students as victims of illegal infractions rather than as the perpetrators of such acts of aggression. It was also found that Princeton students saw twice as many violations on the part of the Dartmouth players as the Dartmouth students saw. The authors conclude that the manner in which individuals view and interpret information depends on how deeply committed they are to a particular belief or course of action. A. Hastorf and H. Cantril, "They Saw a Game: A Case Study," *Journal of Abnormal and Social Psychology,* 49 (1954): 129–34.

29. Struch and Schwartz showed how the instigation of a conflict of interest can produce hostility where there was no hostility previously. When individuals in a group perceive a conflict from another group, they then go through the cognitive cycle of viewing the outsiders in negative ways—demeaning them, dehumanizing them, and erecting psychological boundaries. The authors applied this theory to an incident in Jerusalem in which ultra-orthodox Jews were moved into an area where non-orthodox Jews were living. A good deal of anger was aroused in the insiders, and it took a number of forms of aggressive actions, such as organizing a boycott against stores owned by the ultra-orthodox group members and irritating them by playing their radios very loudly. N. Struch and S. H. Schwartz, "Intergroup Aggression: Its Predictors and Distinctness from Ingroup Bias," *Journal of Personality and Social Psychology,* 56 (1989): 364–73.

30. P. G. Devine, "Prejudice and Outgroup Perception," in *Advanced Social Psychology,* ed. A. Tesser (New York: McGraw-Hill, 1994), pp. 467–524.

31. D. Maybury-Lewis and U. Alamagor, Eds., *The Attraction of Opposites: Thought and Society in the Dualistic Mode* (Ann Arbor: University of Michigan Press, 1989).

32. C. Lévi-Strauss, "Do Dual Organizations Exist?" In *Structural Anthropology* (New York: Basic Books, 1963).

33. M. Rokeach, *The Open and Closed Mind: Investigations into the Nature of Belief Systems and Personality Systems* (New York: Basic Books, 1960).

34. I. L. Janis, *Victims of Groupthink: A Psychological Study of Foreign Policy Decisions and Fiascoes* (Boston: Houghton Mifflin, 1982); C. McCauley, "The Nature of Social-Influence in Groupthink: Compliance and Internalization," *Journal of Personality and Social Psychology* 57, no. 2 (1989): 250–60.

35. C. Strozier, *Apocalypse: On the Psychology of Fundamentalism in America* (Boston: Beacon Press, 1994).

36. R. Grossarth-Maticek, H. J. Eysenck, and H. Vetter, "The Causes and Cures of Prejudice: An Empirical Study of the Frustration-Aggression Hypothesis," *Personality and Individual Differences* 10, no. 5 (1989): 547–58.

37. Egocentric thinking among foreign leaders has been well illustrated by the political scientist Robert Jervis. He points out that important national leaders tend to believe, without substantial basis, that a foreign leader acts in response to their own prior decisions or in order to elicit a response to them. For example, many Americans attributed the breakup of the Soviet Union to the negative impact on the Soviet economy of President Ronald Reagan's exorbitant military spending program rather than to the economic and structural problems of the Soviet Union that had been going on for many years. Jervis underscores one of the unfortunate parts of such egocentric thought: the leader's belief that he has been the exclusive or major cause of the behavior of other nations leads to an exaggerated confidence in deterrence. Believing that he can prevent future offense with threats of punishment, the leader cannot see that, as in the case of the Soviet Union, the determinants are often internal rather than

external. R. Jervis, *Perception and Misperception in International Politics* (Princeton, N.J.: Princeton University Press, 1976).

38. McCauley, "The Nature of Social-Influence in Groupthink."

39. Sometimes, of course, the government is indeed tyrannical, for example, Argentina in the period 1976–83.

40. R. S. Robins and J. Post, *Political Paranoia: The Psychopolitics of Hatred* (New Haven, Conn.: Yale University Press, 1987).

41. Ibid.

42. W. Reich, ed., *The Origins of Terrorism: Psychologies, Ideologies, Theologies, States of Mind* (Cambridge: Press Syndicate of the University of Cambridge, 1990).

43. R. Hofstadter, *The Paranoid Style in American Politics and Other Essays* (New York: Vintage Books, 1967); Robins and Post, *Political Paranoia*.

44. Robins and Post, *Political Paranoia*.

45. K. S. Stern, *A Force upon the Plain: The American Militia Movement and the Politics of Hate* (New York: Simon & Shuster, 1996).

46. R. Nisbett and D. Cohen, *Culture of Honor: The Psychology of Violence in the South* (Boulder, Colo.: Westview Press, 1996).

47. In 1996 the U.S. murder rate was 7.4 per 100,000 people. The next closest was Finland at 3.2, France at 1.1, Japan at 0.6, and Britain at 0.5. Twelve former slaveholding states of the old Confederacy ranked in the top twenty states for murder, with Louisiana first, showing a rate of 17.5. The ten states with the lowest homicide rates were in New England and the Northwest. F. Butterfield, "Southern Curse: Why America's Murder Rate Is So High," *New York Times,* July 26, 1998, pp. D1, D16.

48. E. Anderson, "The Code of the Streets," *Atlantic* (May 1994): 81–92.

49. K. A. Dodge, "Social Cognitive Mechanisms in the Development of Conduct Disorder and Depression," *Annual Review of Psychology* 44 (1993): 559–84.

CHAPTER 10

1. P. Du Preez, *Genocide: The Psychology of Mass Murder* (London and New York: Boyars/Bowerdean, 1994).

2. A somewhat similar scenario is evident in many wars, as indicated in Chapter 11.

3. War against certain factions to achieve political goals, such as clearing out the indigenous inhabitants of a land to make room for colonists (Germany versus the Hereros in Uganda in 1904, for instance), is *expedient genocide.*

4. Du Preez, *Genocide.*

5. R. Hofstadter, *The Paranoid Style in American Politics and Other Essays* (New York: Vintage Books, 1967).

6. A. Bandura, B. Underwood, and M. E. Fromson, "Disinhibition of Aggression Through Diffusion of Responsibility and Dehumanization of Victims," *Journal of Research in Personality* 9 (1975): 253–69.

7. H. Arendt, *Eichmann in Jerusalem: A Report on the Banality of Evil* (New York: Viking Press, 1963); F. Alford, "The Political Psychology of Evil," *Political Psychology* 18, no. 1 (1997): 1–17.

8. D. Goldhagen, *Hitler's Willing Executioners: Ordinary Germans and the Holocaust* (New York: Alfred A. Knopf, 1996); J. Weiss, *Ideology of Death: Why the Holocaust Happened in Germany* (Chicago: I. R. Dee, 1996); G. Fleming, *Hitler and the Final Solution* (1984; reprint, Berkeley: University of California Press, 1994).

9. N. Cohn, *Warrant for Genocide: The Myth of the Jewish World-Conspiracy and the Protocols of the Elders of Zion* (Chicago: Scholars Press, 1980).

10. Goldhagen, *Hitler's Willing Executioners;* Weiss, *Ideology of Death.*

11. After World War I, the threatened takeover of the Berlin government by a Communist group headed by a Jewish revolutionary, Rosa Luxemburg, and the assumption of control of the Bavarian government by a revolutionary Communist group with a significant number of Jews exacerbated the fears. The specter of Jewish domination became even more real when Jews assumed a major role in the left-wing government of numerous cities. The final denouement was the revolution in Hungary led by a Jew, Bela Kun. Weiss, *Ideology of Death.*

12. R. J. Lifton, *The Nazi Doctors: Medical Killing and the Psychology of Genocide* (New York: Basic Books, 1986), p. 16.

13. J. Glass, "Against the Indifference Hypothesis: The Holocaust and

the Enthusiasts for Murder," *Political Psychology* 18, no. 1 (1997): 142.

14. Quoted in D. LaCapra, *Representing the Holocaust: History, Theory, Trauma* (Ithaca, N.Y.: Cornell University Press, 1994), p. 109.

15. D. Chirot, *Modern Tyrants: The Power and Prevalence of Evil in Our Age* (New York: Free Press, 1994).

16. S. Keen, *Faces of the Enemy: Reflections of the Hostile Imagination* (San Francisco: Harper & Row, 1986).

17. Lifton, *The Nazi Doctors.*

18. Ibid., p. 16.

19. Arendt, *Eichmann in Jerusalem.*

20. This argument, an expression of the social psychological theory of "situationism," was apparently supported by the "obedience experiments" devised by the Yale psychologist Stanley Milgram. In these studies, most volunteers obeyed instructions to apply increasingly painful (actually sham) shocks to subjects of the experiment. Despite crucial differences between the experimental situation and the actual context of genocide, scholars have extrapolated a clear message from these studies: anyone, even good Americans, could be induced by authority figures to perform inhuman acts.

　　More recent evaluations of the methodology of these experiments casts doubt on their validity. It has been shown, for example, that experimental subjects are not deceived by the sham nature of the deception experiments, even though they indicate later that they were. They are more frequently "wiser" to the goals of the experiment than the experimenter ascertains. K. M. Taylor and J. A. Shepperd, "Probing Suspicion Among Participants in Deception Research," *American Psychologist* 51, no. 8 (1996): 886–87.

　　In rebuttal to the situationist argument, Daniel Goldhagen offers a cognitive explanation. Yes, ordinary Germans were indeed perpetrators, but not necessarily out of obedience to orders. The orders fit into their genocidal ideology. They believed that they were doing the right thing by stamping out the evil Jew. The machinery of the Holocaust was fueled by the eliminationist ideology that had been in place for decades and was shaped into the homicidal imperative. What the perpetrators perceived was filtered by and integrated into their ideology. What they did was a logical consequence of what they saw. If they saw evil, they had to eliminate it. Goldhagen, *Hitler's Willing Executioners.*

21. Goldhagen, *Hitler's Willing Executioners;* C. Browning, *The Path to Genocide* (Cambridge: Cambridge University Press, 1992), p. 142.

22. The middle managers, the bureaucrats, and the low-level functionaries did not all hate the Jews, nor would they have pressed for genocide on their own. They appear to have been involved in low-level procedural thinking in carrying out their tasks without reflecting on the consequences to an entire population of people. As pointed out by Browning, they waited for signs from the decision makers to dictate their actions: "It was their receptivity to such signals and the speed with which they aligned themselves to the new policy, that allowed the Final Solution to emerge with so little internal friction and so little formal coordination." Browning, *The Path to Genocide,* p. 143.

23. I. Kershaw, *The "Hitler Myth": Image and Reality in the Third Reich* (Oxford: Clarendon Press, 1987).

24. J. P. Stern, *Hitler: The Führer and the People* (London: Fontana, 1975), p. 36.

25. Goldhagen, *Hitler's Willing Executioners.*

26. Kershaw, *The "Hitler Myth".*

27. Ibid.

28. Explaining why Hitler appeared to be possessed by such a virulent hatred of the Jews is a formidable challenge and perhaps impossible. Dollersheim, an Austrian village, the site of Hitler's "ancestral home" and potentially the source of relevant documents, was destroyed during World War II. The diversified speculations regarding the origins of his personality have been neatly summarized by Ron Rosenbaum in *Explaining Hitler* (New York: Random House, 1998). Irrespective of its origins, the evidence of Hitler's hatred for the Jews and his responsibility for the ultimate decision to murder them seems inescapable.

29. The final statement in Hitler's political testament dictated on April 29, 1945, the day before he shot himself, read: "Above all, I obligate the leaders of the nation and their following to a strict observance of

the racial laws, and to a merciless resistance to the poisoners of all people, international Jewry." Quoted in Fleming, *Hitler and the Final Solution*, p. 188.

30. Ibid.

31. E. A. Zillmer, M. Harrower, B. A. Ritzler, and R. P. Archer, "The Quest for the Nazi Personality," *Psychological Record* 46, no. 2 (1996): 399–402.

32. Browning, *The Path to Genocide;* Goldhagen, *Hitler's Willing Executioners.*

33. Ibid.

34. Du Preez, *Genocide.*

35. Quoted in P. Hollander, "Revisiting the Banality of Evil: Political Violence in Communist Systems," *Partisan Review* 64, no. 1 (1997): 56.

36. Ibid., p. 62.

37. J. W. Young, *Totalitarian Language: Orwell's Newspeak and Its Nazi and Communist Antecedents* (Charlottesville: University Press of Virginia, 1991).

38. G. Orwell, *Nineteen Eighty-four* (New York: Harcourt, Brace & World, 1949).

39. H. Arendt, *The Origins of Totalitarianism* (New York: Harcourt Brace Jovanovich, 1973).

40. Chirot, *Modern Tyrants.*

41. J. H. Barkow, L. Cosmides, J. Tooby, *The Adapted Mind: Evolutionary Psychology and the Generation of Culture* (Oxford: Oxford University Press, 1995).

42. T. C. Brock and A. H. Buss, "Effects of Justification for Aggression and Communication with the Victim of Post-aggression Dissonance," *Journal of Abnormal and Social Psychology* 68, no. 4 (1964): 403–12. After committing a harmful act, offenders are able to recall the supposed benefits of their behavior more readily than its harmful effects.

CHAPTER 11

1. R. A. Hinde and H. E. Watson, *War, a Cruel Necessity?: The Bases of Institutionalized Violence* (New York: St. Martin's Press, 1995).

2. In contrast to previous wars, the military leaders in the Vietnam War were not elected to high political offices—probably a testament to the relative unpopularity of that war.

3. M. Eksteins, *Rites of Spring: The Great War and the Birth of the Modern Age* (Boston: Houghton Mifflin, 1989).

4. This glamorous picture, of course, fades away when the realities of war are experienced: destruction, disease, and dirt at the front lines, rebellion of the infantry (in World War I), and antidraft riots (in the Civil War).

5. It has been pointed out that the Buid of the Philippines developed a value system that emphasizes the value of generosity and avoiding conflict, despite the proximity of other tribes that practice warfare. Similarly, although the Semai of Malyasia suffer from jealousy, theft, and marital infidelity, these conflicts never result in violence. N. Saunders, "Children of Mars" [review of *The Anthropology of War*], *New Scientist* 18 (1991): 51.

6. S. Kull, *Minds at War: Nuclear Reality and the Inner Conflict of Defense Policymakers* (New York: Basic Books, 1988), p. 307.

7. Some authors see more primitive warfare as the result of cultural selection, emphasizing the gain obtained by the group: land, water, and food. Presumably, a particular society weighed the costs against the benefits of such an activity before committing to war. The leaders, however, could manipulate the whole process of aggression for power and prestige. Saunders, "Children of Mars," p. 51.

8. S. Kull, M. Small, and J. D. Singer, *Resort to Arms: International and Civil Wars, 1816–1980* (Beverly Hills, Calif.: Sage Publications, 1982); L. F. Richardson, *Arms and Insecurity: A Mathematical Study of the Causes and Origins of War* (Pittsburgh: Boxwood Press, 1960).

9. P. Paret, *Clausewitz and the State: The Man, His Theories, and His Times* (Princeton, N.J.: Princeton University Press, 1985), p. 398.

10. R. K. White, "Why Aggressors Lose," *Political Psychology* 11 (1990): 227–42.

11. S. Freud, "Mourning and Melancholia," *Essential Papers on Object Loss, Essential Papers in Psychoanalysis,* ed. Rita V. Frankiel (New York: New York University Press, 1994), pp. 38–51; K. Lorenz, *On Aggression* (New York: Routledge, 1966); D. Morris, *The Naked Ape:*

A Zoologist's Study of the Human Animal (New York: McGraw-Hill, 1967).

12. R. Ardrey, *The Territorial Imperative: A Personal Inquiry into the Animal Origins of Property and Nations* (New York: Atheneum, 1966); Lorenz, *On Aggression.*

13. B. Ehrenreich, *Blood Rites: Origins and History of the Passions of War* (New York: Henry Holt & Co., 1997).

14. Hinde and Watson, *War, a Cruel Necessity?*

15. O. R. Holsti, *Crisis, Escalation, War* (Montreal: McGill-Queen's University Press, 1972).

16. B. Tuchman, *Guns of August* (New York: Macmillan, 1962).

17. J. Haas, ed., *The Anthropology of War* (Cambridge: Cambridge University Press, 1990).

18. Political scientists have proposed that the causation of war involves an analysis at a minimum of three levels: the international system, the national state subsystem, and the individual. In the view of Kenneth Waltz, the total international system, involving the relations between states (for example, balance of power), is far more important in the genesis of war than the psychology of the individual involved (the proximate level). The internal workings of the national state with its own particular self-interest and economic and political tensions occupies an intermediate level. As pointed out by David Singer, the importance of the psychology of the individual has been underestimated by Waltz and many other political scientists. Perhaps a different approach to individual psychology could be demonstrated to play a more significant role. In theory at least, the perceptions of the individual players in the drama of international relations can affect their decisions—and their decisions can affect their psychology. K. Waltz, *Man, the State, and War: A Theoretical Analysis* (1959; reprint, New York: Columbia University Press, 1969); J. D. Singer, "International Conflict: Three Levels of Analysis," *World Politics* 12, no. 3 (1960): 453–61.

Prior to the Spanish-American War, the war fever of the populace and Congress (instigated in part by the press) put enormous pressure on President McKinley to declare war on Spain. In World War I, the wild enthusiasm for war helped to tip the hand of the national leaders. Satisfactory explanations for war—or any military strikes, for that matter—need to take into account individual phenomena, such as national self-images, images of the Enemy, and so on. In addition, interpersonal factors such as group empathy and "we-versus-them" reactions obviously play a role. M. Eksteins, *Rites of Spring: The Great War and the Birth of the Modern Age* (Boston: Houghton Mifflin, 1989).

19. Eksteins, *Rites of Spring.*

20. Tuchman, *Guns of August;* Holsti, *Crisis, Escalation, War.*

21. It is important to make a distinction between the thinking, emotions, and motives of the people who fight the wars and the leaders who initiate them.

   Robert Jervis has applied the principles of cognitive psychology and information processing to diplomatic decision making. He proposes a number of hypotheses:

   1. The leaders' belief system has a greater impact on the interpretation of data when there is ambiguity.

   2. Pronounced confidence in the beliefs will also have a greater impact on decisions.

   3. Information that means one thing to the other side means something else to the receiver of the information.

   4. Messages from the other side are shaped to conform to expectations. The closer a "message" is to one's belief system, the more readily it will be accepted. More important, it may be distorted to produce spurious conformity even if it does not conform.

   5. The decision makers use historical analogies or past experiences as though they were a completely reliable way of determining the meaning of a present event.

   R. Jervis, *Perception and Misperception in International Politics* (Princeton, N.J.: Princeton University Press, 1976), p. 300.

22. S. Keen, *Faces of the Enemy: Reflections of the Hostile Imagination* (San Francisco: Harper & Row, 1986), is a thorough review, including propaganda posters of twentieth-century war. See also R. Rieber, *The*

Psychology of War and Peace: The Image of the Enemy (New York: Plenum Press, 1991).

23. R. Wrangham and D. Peterson, *Demonic Males: Apes and the Origins of Human Violence* (Boston: Houghton Mifflin, 1996); J. Goodall, *Through a Window: Thirty Years with the Chimpanzees of Gombe* (Boston: Houghton Mifflin, 1992).

24. D. T. Gilbert and P. S. Malone, "The Correspondence Bias," *Psychological Bulletin* 117, no. 1 (1995): 21–38.

25. Keen, *Faces of the Enemy*.

26. D. Grossman, *On Killing: The Psychological Cost of Learning to Kill in War and Society* (Boston: Little, Brown, 1995).

27. R. N. Lebow, *Between Peace and War* (Baltimore: Johns Hopkins University Press, 1981); M. S. Hirshberg, "The Self-perpetuating National Self-image: Cognitive Biases in Perceptions of International Interventions," *Political Psychology* 14 (1993): 77–98; N. Kaplowitz, "National Self-images, Perception of Enemies, and Conflict Strategies," *Political Psychology* 11 (1990): 39–82.

28. Lebow, *Between Peace and War*.

29. Ibid.

30. R. K. White, *Fearful Warriors: A Psychological Profile of U.S.-Soviet Relations* (New York: Free Press, 1984).

31. Ibid.

32. S. Feshbach, "Individual Aggression, National Attachment, and the Search for Peace," *Aggressive Behavior* 13 (1986): 315–25.

33. Ibid.

34. O. Nathan and H. Norden, *Einstein on Peace* (New York: Schocken Books, 1968).

35. P. C. Stern, "Nationalism as Reconstructed Altruism," *Political Psychology* 17, no. 3 (1996): 569–72.

36. White, *Fearful Warriors*.

37. J. P. Stern, *Hitler: The Führer and the People* (London: Fontana, 1975).

38. E. Staub, *The Roots of Evil: The Origins of Genocide and Other Group Violence* (New York: Cambridge University Press, 1989).

39. R. Smoke, *War: Controlling Escalation* (Cambridge, Mass.: Harvard University Press, 1977).

40. Lebow, *Between Peace and War*.

41. V. Dedijer, *The Road to Sarajevo* (New York: Simon & Schuster, 1966).

42. White, *Fearful Warriors*.

43. G. Craig, "Making Way for Hitler" [review of *How War Came: The Immediate Origins of the Second World War, 1929–1939*], *New York Review of Books*, October 12, 1989, pp. 11–12.

44. Dedijer, *The Road to Sarajevo*.

45. U. Bronfenbrenner, "The Mirror Image in Soviet-American Relations," *Journal of Social Sciences* 17 (1961): 45–56.

46. S. Baron-Cohen, *Mindblindness: An Essay on Autism and Theory of Mind* (Cambridge, Mass.: MIT Press, 1997).

47. A. Fursenko and T. Naftali, *One Hell of a Gamble: The Secret History of the Cuban Missile Crisis* (New York: Norton, 1997).

48. White, *Fearful Warriors*.

49. Ibid.

50. Ibid.; Lebow, *Between Peace and War*.

51. Holsti, *Crisis, Escalation, War*.

52. J. G. Stoessinger, *Why Nations Go to War* (New York: St. Martin's Press, 1993).

53. G. H. Snyder and P. Diesing, eds., *Conflict Among Nations: Bargaining, Decision Making, and System Structure in International Crisis* (Princeton, N.J.: Princeton University Press, 1977).

54. L. F. Richardson, *Arms and Insecurity: A Mathematical Study of the Causes and Origins of War* (Pittsburgh: Boxwood Press, 1960); P. Paret, *Clausewitz and the State: The Man, His Theories, and His Times* (Princeton, N.J.: Princeton University Press, 1985), p. 398; Lebow, *Between Peace and War*.

55. A. J. P. Taylor, *Bismarck: The Man and the Statesman* (New York: Alfred A. Knopf, 1955). Taylor differs from other historians in that he disputes that Bismarck actually intended to provoke war with France.

56. O. Pflanze, *Bismarck and the Development of Germany*, 2nd ed., vol. 1 (Princeton, N.J.: Princeton University Press, 1963).

57. V. P. Gagnon, "Ethnic Nationalism and International Conflict: The Case of Serbia," *International Security* 19, no. 3 (1995): 130–66.

58. Ibid.

59. L. Silber and A. Little, *Yugoslavia: Death of a Nation* (New York: Penguin Books, 1996).

60. Keen, *Faces of the Enemy.*

61. N. Eisenberg and P. A. Miller, "The Relation of Empathy to Prosocial and Related Behaviors," *Psychological Bulletin* 103 (1988): 324–44; N. Eisenberg and S. Mussen, *The Roots of Prosocial Behavior in Children* (New York: Cambridge University Press, 1989).

62. P. C. Stern, "Why Do People Sacrifice for Their Nations?" *Political Psychology* 16, no. 2 (1995): 217–35.

63. W. L. Calley, *Lieutenant Calley: His Own Story,* as told to John Sack (New York: Viking Press, 1970).

64. Ibid.

65. R. Jervis, *Perception and Misperception in International Politics* (Princeton, N.J.: Princeton University Press, 1976), p. 300.

66. S. Orwell and I. Angus, eds., *An Age Like This: The Collected Essays, Journalism, and Letters of George Orwell, Vol. I* (New York: Harcourt, Brace & World, 1968).

67. Grossman, *On Killing.*

68. Ibid.

69. *Henry V,* act 4, scene 1, line 140 (Oxford edition).

70. A. Bandura, B. Underwood, and M. E. Fromson, "Disinhibition of Aggression Through Diffusion of Responsibility and Dehumanization of Victims," *Journal of Research in Personality* 9 (1975): 253–69.

CHAPTER 12

1. A. Kohn, *The Brighter Side of Human Nature: Altruism and Empathy in Everyday Life* (New York: Basic Books, 1990).

2. E. Staub, *The Roots of Evil: The Origins of Genocide and Other Group Violence* (New York: Cambridge University Press, 1989).

3. A. Bandura, B. Underwood, and M. E. Fromson, "Disinhibition of Aggression Through Diffusion of Responsibility and Dehumanization of Victims," *Journal of Research in Personality* 9 (1975): 253–69.

4. There is considerable evidence that altruism, generosity, and kindness are biologically based aspects of human nature. The philosopher Elliott Sober and the biologist David Sloane Wilson present a

detailed view of altruism throughout the animal kingdom. Their examples range from self-sacrificing parasites to social insects, to instances of the human capacity for self-sacrifice and altruism. In order to demonstrate that altruism could arise through natural selection, the authors revive the concept of "group selection," in which the unit of evolutionary development is the group rather than the individual. The concept of group selection, which had been discarded for many years, now seems to find an important role as an explanatory construct for these "prosocial behaviors." E. Sober and D. S. Wilson, *Unto Others: The Evolution and Psychology of Unselfish Behavior* (Cambridge, Mass.: Harvard University Press, 1998).

5. W. Ickes, ed., *Empathic Accuracy* (New York: Guilford Press, 1997).

6. R. Selman, *The Growth of Interpersonal Understanding: Developmental and Clinical Analyses* (New York: Academic Press, 1980).

7. R. N. Stromberg, *Redemption by War: Intellectuals and 1914* (Lawrence: Regents Press of Kansas, 1982).

8. In some emergency situations, however, the automatic reflex interpretation may be life-preserving but is revised if it is found on reflection to be exaggerated or incorrect.

9. I. E. Sigel, E. T. Stinson, and M. Kim, *Socialization of Cognition: The Distancing Model* (Hillsdale, N.J.: Erlbaum Associates, 1993).

10. A. Smith, *The Theory of Moral Sentiments* (1759; reprint, Oxford: Clarendon Press, 1976).

11. M. L. Hoffman, "Empathy and Justice Motivation," *Empathy and Emotion* 14 (1990): 151–72.

12. E. Hatfield, J. Cacioppo, and R. Rapson, *Emotional Contagion* (Cambridge: Cambridge University Press, 1994), pp. 82–86.

13. C. Zahn-Waxler, E. M. Cummings, and R. Iannotti, eds., *Altruism and Agression: Biological and Social Origins* (New York: Cambridge University Press, 1986).

14. R. S. Lazarus, *Emotion and Adaptation* (New York: Oxford University Press, 1991).

15. D. Grossman, *On Killing: The Psychological Cost of Learning to Kill in War and Society* (Boston: Little, Brown, 1995).

16. See also Chapter 11, page 254. I. L. Janis, *Victims of Groupthink: A*

Psychological Study of Foreign Policy Decisions and Fiascoes (Boston: Houghton Mifflin, 1982).

17. M. Lerner, *The Belief in a Just World: A Fundamental Delusion* (New York: Plenum Press, 1980).

18. Despite the wholesale perversion of this doctrine, we need to bear in mind that genuine grievances do exist; groups of people are oppressed, exploited, and abused, and violence may seem to be the only solution.

19. It seems likely that the conceptual category "Enemy," although temporarily empty, remains in a latent state, ready to be filled again by a different, or perhaps the same, antagonist.

20. A. Koestler, *The Ghost in the Machine* (1967; reprint, London: Pan Books, 1970).

21. In 1986 Slobodan Milosevic created an imaginary "physical, political, legal, and cultural genocide" of the Serbian population by ethnic Albanians in Kosovo. Although only a few of the small minority of Serb residents in Kosovo have actually died in violent incidents with the ethnic Albanians there, the majority of Serbs in Serbia believed it. The manifesto by Milosovic led to the Serbian massacres in Bosnia in 1992 and to the destruction of Kosovo villages in 1998. R. Cohen, "Blood Stains in the Balkans; No, It's Not Just Fate," *New York Times,* October 4, 1998, p. D1.

22. L. Kohlberg, *The Psychology of Moral Development: The Nature and Validity of Moral Stages* (San Francisco: Harper & Row, 1984).

23. C. Gilligan, *In a Different Voice: Psychological Theory and Women's Development* (Cambridge, Mass.: Harvard University Press, 1982).

24. K. W. Cassidy, J. Y. Chu, and K. K. Dahlsgaard, "Preschoolers' Ability to Adopt Justice and Care Orientations to Moral Dilemmas," *Early Education and Development* 8 (1997): 419–34.

25. M. Sherif, O. J. Harvey, B. J. White, W. R. Hood, and C. W. Sherif, *The Robbers Cave Experiment: Intergroup Conflict and Cooperation* (Middletown, Conn.: Wesleyan University Press, 1988).

26. Kohn, *The Brighter Side of Human Nature.*

27. S. Oliner and P. Oliner, *The Altruistic Personality: Rescuers of Jews in Nazi Europe* (New York: Free Press; London: Collier Macmillan, 1988).

28. F. DeWaal, *Good Natured* (Cambridge, Mass.: Harvard University Press, 1996).

29. J. Goodall, *Through a Window: Thirty Years with the Chimpanzees of Gombe* (Boston: Houghton Mifflin, 1992).

30. K. Hamilton, "The Winners; Newsmakers of 1996: Hero of the Year," *Newsweek*, Winter 1997 special edition, p. 40.

31. M. Hunt, *The Compassionate Beast: The Scientific Inquiry into Human Altruism* (New York: Anchor Books/Doubleday, 1991).

32. Kohn, *The Brighter Side of Human Nature.*

33. N. Feshbach, "Empathy Training: A Field Study in Affective Education," in *Aggression and Behavior Change: Biological and Social Processes,* ed. Seymour Feshbach and Adam Fraczek (New York: Praeger, 1979), pp. 234–250; N. Feshbach, S. Feshbach, M. Fauvre, and M. Ballard-Campbell, *Learning to Care: Classroom Activities for Social and Affective Development* (Glenview, Ill.: Scott, Foresman, 1983).

34. L. Brothers, *Friday's Footprint: How Society Shapes the Human Mind* (Oxford: Oxford University Press, 1997).

CHAPTER 13

1. One can raise a significant question regarding the validity of the various cognitive techniques described in this chapter. Over the past two decades, 50 studies incorporating 1,640 angry subjects treated with cognitive therapy were analyzed by R. Beck and E. Fernandez. It was found that cognitive behavior therapy had a mean effect size of .70, which indicated that the average individual treated with cognitive behavior therapy had a better outcome than 76 percent of untreated subjects in terms of anger reduction. The techniques used in these studies overlap some of the methods described in chapter 8. They focus predominantly on Novaco's adaptation of Meichenbaum's stress inoculation training, which had been initially developed for the treatment of anxiety. R. Beck and E. Fernandez, "Cognitive-Behavioral Therapy in the Treatment of Anger: A Meta-analysis," *Cognitive Therapy and Research* 22, no. 1 (1998): 63–74; R. W. Novaco, *Anger Control: The Development and Evaluation of an*

Experimental Treatment (Lexington, Mass.: D.C. Heath, 1975); D. H. Meichenbaum, *Stress Inoculation Training* (New York: Pergamon Press, 1975). Also see the recent study by Eric Dahlen and Jerry Deffenbacher that demonstrated the efficacy of cognitive therapy in reducing anger. E. R. Dahlen and J. L. Deffenbacher, "A Partial Component Analysis of Beck's Cognitive Therapy for the Treatment of General Anger," *Cognitive Therapy and Research* (in press).

2. A. T. Beck, "Thinking and Depression: Idiosyncratic Content and Cognitive Distortions," *Archives of General Psychiatry* 9 (1963): 324–33.

3. E. Jacobson, *Progressive Relaxation: A Physiological and Clinical Investigation of Muscular States and Their Significance in Psychology and Medical Practice* (1938; reprint, Chicago: University of Chicago Press, 1968).

4. S. Baron-Cohen, *Mindblindness: An Essay on Autism and Theory of Mind* (Cambridge, Mass.: MIT Press, 1995).

5. Zillman emphasizes the role of excitation in the development of hostility. He considers cognition and excitation to be independent but capable of influencing one another. When an individual perceives a situation as a threat and *ruminates* about the threat or insult and the potential retaliation, then he remains in a high level of excitement. On the other hand, when the individual is able to reframe the situation or recognize mitigating circumstances, then the urge to punish the offender is reduced.

   As pointed out by Zillman, simply teaching people to think carefully about potential danger from apparently threatening situations or people is generally not sufficient to reduce hostility. In a therapeutic situation it is therefore important to re-create provocative situations and to teach the patient to focus on the "hot reactions" and reconstrue the situation during the period of frustration. This method has been demonstrated to be particularly effective in reducing anger and hostility in abusive parents. D. Zillman, "Cognition-Excitation Interdependencies in Aggressive Behavior," *Aggressive Behavior* 14 (1988): 51–64.

6. J. Bush, "Teaching Self-Risk Management to Violent Offenders," in *What Works: Reducing Reoffending—Guidelines from Research and*

Practice, ed. J. McGuire et al., Wiley Series in Offender Rehabilitation (Chichester, Eng.: Wiley, 1995), pp. 139–54.

7. A. Bandura, *Aggression: A Social Learning Analysis* (Englewood Cliffs, N.J.: Prentice-Hall, 1973).

8. M. Lerner, *The Belief in a Just World: A Fundamental Delusion* (New York: Plenum Press, 1980).

CHAPTER 14

1. J. D. Singer. "The Level-of-Analysis Problem in International Relations," *The International System: Theoretical Essays,.* eds. Klaus Knorr and Sidney Verba (Princeton, N.J.: Princeton University Press, 1961), pp. 77–92.

2. E. P. Green, J. Glaser, and A. Rich, "From Lynching to Gay Bashing: The Elusive Connection Between Economic Conditions and Hate Crimes," *Journal of Personality and Social Psychology* 75 (1998): 109–20.

3. P. Kinderman and R. Bentall, "The Clinical Implications of a Psychological Model of Paranoia, in *Behaviour and Cognitive Therapy Today: Essays in Honour of Hans J. Eysenck,* ed. E. Sanavio (Oxford: Elsevier Press, 1998).

4. M. Eksteins, *Rites of Spring: The Great War and the Birth of the Modern Age* (Boston: Houghton Mifflin, 1989), p. 159.

5. One reason to expect aggressive impulses to persist over time is that cognitive mediation (that is, interpretation, rehearsal, and rumination about the instigation) justifies and maintains the impulse; L. Berkowitz, "Frustration-Aggression Hypothesis: Examination and Reformulation," *Psychological Bulletin* 106 (1989): 59–73. Individual differences in the tendency to ruminate have been shown to moderate aggressiveness. On the other hand, Zillmann provided evidence that cognition can diminish aggression through reasessment of the situation. D. Zillman, "Cognition-Excitation Interdependencies in Aggressive Behavior," *Aggressive Behavior* 14 (1988): 51–64.

6. A. T. Beck, *Love Is Never Enough* (New York: HarperCollins, 1988).

7. K. A. Dodge, "Social Cognitive Mechanisms in the Development of Conduct Disorder and Depression," *Annual Review of Psychology* 44 (1993): 559–84.

8. J. A. Bargh, S. Chaiken, P. Raymond, and C. Hymes, "The Automatic Attitude Evaluation Effect: Unconditional Activation with a Pronunciation Task," *Journal of Experimental Social Psychology* 32 (1996): 104–28.

9. C. I. Eckhardt, K. A. Barbour, and G. C. Davison, "Articulated Thoughts of Maritally Violent and Nonviolent Men During Anger Arousal," *Journal of Consulting and Clinical Psychology* 66, no. 2 (1998): 259–69; C. I. Eckhardt and M. Dye, "The Cognitive Characteristics of Maritally Violent Men: Theory and Evidence," *Cognitive Therapy and Research* (in press).

10. C. Wickless and I. Kirsch, "Cognitive Correlates of Anger, Anxiety, and Sadness," *Cognitive Therapy and Research* 12, no. 4 (1988): 367–77.

11. M. Chen and J. A. Bargh, "Nonconscious Behavioral Confirmation Processes: The Self-fulfilling Consequences of Automatic Stereotype Activation," *Journal of Experimental Social Psychology* 33, no. 5 (1997): 541–60; Eckhardt and Dye, "The Cognitive Characteristics of Maritally Violent Men"; P. G. Devine, "Prejudice and Outgroup Perception," in *Advanced Social Psychology,* ed. A. Tesser (New York: McGraw-Hill, 1994), pp. 69–81.

12. R. Beck and E. Fernandez, "Cognitive-Behavioral Therapy in the Treatment of Anger: A Meta-analysis," *Cognitive Therapy and Research* 22, no. 1 (1998): 63–74.

13. D. Shapiro, K. K. Hui, M. E. Oakley, J. Pasic, and L. D. Jamner, "Reduction in Drug Requirements for Hypertension by Means of a Cognitive-Behavioral Intervention," *American Journal of Hypertension* 10, no. 1 (1997): 9–17.

14. M. Whiteman, D. Fanshel, and J. Grundy, "Cognitive-Behavioral Intervention Aimed at Anger of Parents at Risk of Child Abuse," *Social Work* 32, no. 6 (1987): 469–74.

15. Eckhardt, Barbour, and Davison, "Articulated Thoughts."

16. K. R. Henning and B. C. Frueh, "Cognitive-Behavioral Treatment of Incarcerated Offenders: An Evaluation of the Vermont Department of Corrections' Cognitive Self-change Program," *Criminal Justice Behavior* 23, no. 4 (1996): 523–41.

17. E. A. Fabiano and R. R. Ross, *Time to Think* (Ottawa: T3 Associate, Training & Consulting, 1985).

18. K. A. Dodge, letter to author, October 1998.

19. E. Sober and D. S. Wilson, *Unto Others: The Evolution and Psychology of Unselfish Behavior* (Cambridge, Mass.: Harvard University Press, 1998).

20. R. Jervis, *Perception and Misperception in International Politics* (Princeton, N.J.: Princeton University Press, 1976); G. H. Snyder and P. Diesing, eds., *Conflict Among Nations: Bargaining, Decision Making, and System Structure in International Crisis* (Princeton, N.J.: Princeton University Press, 1977); Y. Y. I. Vertzberger, *The World in Their Minds: Information Processing, Cognition, and Perception in Foreign Policy Decision-Making* (Stanford, Calif.: Stanford University Press, 1990); P. Tetlock, "Cognitive Style and Political Ideology," *Journal of Personality and Social Psychology* 45 (1983): 118–25; R. K. White, *Nobody Wanted War* (New York: Doubleday, 1976).

BIBLIOGRAPHY

Aho, J. A. (1994). *This Thing of Darkness: A Sociology of the Enemy.* Seattle: University of Washington Press.

Albonetti, M. E., & Farabollini, F. (1993). "Restraint Increases Both Aggression and Defence in Female Rats Tested Against Same-sex Unfamiliar Conspecifics." *Aggressive Behavior* 19, pp. 369–376.

Alder, J. (1996). "Just Following Orders" [review of D. Goldhagen, *Hitler's Willing Executioners: Ordinary Germans and the Holocaust*] *Newsweek,* April, p. 74.

Alford, C. F. (1990). "The Organization of Evil." *Political Psychology* 11, no. 1, pp. 5–27.

———. (1990). "Response to Dallmayr." *Political Psychology* 11, pp. 37–38.

———. (1997). "The Political Psychology of Evil." *Political Psychology* 18, no. 1, pp. 1–17.

Allman, W. F. (1994). *The Stone Age Present: How Evolution Has Shaped Modern Life—From Sex, Violence, and Language to Emotions, Morals, and Communities.* New York: Simon & Schuster.

Allport, G. (1954). *The Nature of Prejudice.* Cambridge, Mass.: Addison-Wesley.

Anderson, E. (1994). "The Code of the Streets." *Atlantic Monthly* May, pp. 81–92.

Andreopoulos, G.J. (Ed.) (1994). *Genocide: Conceptual and Historical Dimensions.* Philadelphia: University of Pennsylvania Press.

Archer, J. (1991). "The Influence of Testosterone on Human Aggression." *British Journal of Psychology* 82, pp. 1–28.

Ardrey, R. (1966). *The Territorial Imperative: A Personal Inquiry into the Animal Origins of Property and Nations.* New York: Atheneum.

Arendt, H. (1963). *Eichmann in Jerusalem: A Report on the Banality of Evil.* New York: Viking Press.

———. (1973). *The Origins of Totalitarianism.* New York: Harcourt Brace Jovanovich.

Aronson, E. (1988). *The Social Animal.* 5th ed. New York: W. H. Freeman.

Asch, S. (1952). *Social Psychology.* Englewood Cliffs, N.J.: Prentice-Hall.

Baker, S. L., & Kirsch, I. (1991). "Cognitive Mediators of Pain Perception and Tolerance." *Journal of Personality and Social Psychology* 61, pp. 504–10.

Bandura, A. (1973). *Aggression: A Social Learning Analysis.* Englewood Cliffs, N.J.: Prentice-Hall.

————. (1985). *The Social Foundations of Thought and Action: A Social Cognitive Theory.* New York: Paramus Prentice-Hall.

————. (1991). "Social Cognitive Theory of Moral Thought and Action." In W. M. Kurtines & J. L. Gewirtz (Eds.), *Handbook of Moral Behavior and Development* (pp. 71–129). Hillsdale, N.J.: Erlbaum Associates.

Bandura, A., Underwood, B., & Fromson, M. E. (1975). "Disinhibition of Aggression Through Diffusion of Responsibility and Dehumanization of Victims." *Journal of Research in Personality* 9, pp. 253–269.

Barash, D. P. (1994). *Beloved Enemies: Our Need for Opponents.* Amherst, N.Y.: Prometheus Books.

Barefoot, J. C., Dodge, K. A., Peterson, B. L., Dahlstrom, W. G., & Williams, R. B. (1989). "The Cook-Medley Hostility Scale: Item Content and Ability to Predict Survival." *Psychosomatic Medicine* 51, pp. 46–57.

Bargh, J. A., Chaiken, S., Raymond, P., & Hymes, C. (1996). "The Automatic Evaluation Effect: Unconditional Automatic Attitude Activation with a Pronunciation Task." *Journal of Experimental Social Psychology* 32, no. 1, pp. 104–28.

Barkow, J. H., Cosmides, L., & Tooby, J. (1992). *The Adapted Mind: Evolutionary Psychology and the Generation of Culture.* Oxford: Oxford University Press.

Barnett, M. A., Quackenbush, S. W., & Sinisi, C. S. (1995). "The Role of Critical Experiences in Moral Development: Implications for Justice and Care Orientations." *Basic and Applied Social Psychology* 17, pp. 137–52.

Baron, R. A., & Richardson, D. R. (1994). *Human Aggression.* 2nd ed. New York: Plenum Press.

Baron-Cohen, S. (1997). *Mindblindness: An Essay on Autism and Theory of Mind.* Cambridge, Mass.: MIT Press.

Bartov, O. (1996). "Ordinary Monsters" [review of D. Goldhagen, *Hitler's Willing Executioners: Ordinary Germans and the Holocaust*]. *New Republic,* April, pp. 32–38.

Bateson, P. (Ed.). (1991). *The Development and Integration of Behaviour: Essays in Honour of Robert Hinde.* Cambridge: Cambridge University Press.

Baucom, D., & Epstein, N. (1990). *Cognitive-Behavioral Marital Therapy.* Brunner/Mazel Cognitive Therapy Series. New York: Brunner/Mazel.

Baumeister, R. (1997). *Evil: Inside Human Cruelty and Violence.* New York: W. H. Freeman.

Baumeister, R. F., Stillwell, A. M., & Heatherton, T. K. (1994). "Guilt: An Interpersonal Approach." *Psychological Bulletin* 115, pp. 243–67.

Bavelas, J., Black, A., Chovil, N., Lemery, C., & Mullet, J. (1988). "Form and Function in Motor Mimicry: Topographic Evidence That the Primary Function Is Communication." *Human Communication Research* 14, pp. 275–99.

Beck, A. T. (1963). "Thinking and Depression: Idiosyncratic Content and Cognitive Distortions." *Archives of General Psychiatry* 9, pp. 324–33.

————. (1988). *Cognitive Therapy of Depression: A Personal Reflection.* Aberdeen, Scotland: Scottish Cultural Press.

————. (1988). *Love Is Never Enough.* New York: HarperCollins.

Beck, A. T., & Emery, G., with Greenberg, R. L. (1985). *Anxiety Disorders and Phobias: A Cognitive Perspective.* New York: Basic Books.

Beck, A. T., Freeman, A., & Associates. (1990). *Cognitive Therapy of Personality Disorders.* New York: Guilford.

Beck, A. T., Wright, F. W., Newman, C. F., & Liese, B. (1993). *Cognitive Therapy of Substance Abuse.* New York: Guilford.

Beck, R., & Fernandez, E. (1998). "Cognitive-Behavioral Therapy in the Treatment of Anger: A Meta-analysis." *Cognitive Therapy and Research* 22, no. 1, pp. 63–74.

Berger, P., & Luckmann, T. (1966). *The Social Construction of Reality: A Treatise in the Sociology of Knowledge.* New York: Doubleday.

Berke, J. H. (1986). *The Tyranny of Malice: Exploring the Dark Side of Character and Culture.* New York: Summit Books.

Berkowitz, L. (1989). "Frustration-Aggression Hypothesis: Examination and Reformulation." *Psychological Bulletin* 106, pp. 59–73.

————. (1994). *Aggression: Its Causes, Consequences, and Control.* McGraw-Hill Series in Social Psychology. New York: McGraw-Hill.

Birtchnell, J. (1993). *How Humans Relate: A New Interpersonal Theory.* Foreword by Russell Gardner Jr. Westport, Conn.: Praeger.

Bjorkqvist, K., Nygren, T., Bjorklund, A., & Bjorkqvist, S. (1994). "Testosterone Intake and Aggressiveness: Real Effect or Anticipation?" *Aggressive Behavior* 20, pp. 17–26.

Blackburn, R. (1989). "Psychopathology and Personality Disorder in Relation to Violence." In K. Howells and C. R. Hollin (Eds.), *Clinical Approaches to Violence* (pp. 187–205). New York: Wiley.

Blainey, G. (1973). *The Causes of War.* New York: Free Press.

Brock, T. C., & Buss, A. H. (1964). "Effects of Justification for Aggression and Communication with the Victim of Postaggression Dissonance." *Journal of Abnormal and Social Psychology* 68, no. 4, pp. 403–12.

Bronfenbrenner, U. (1961). "The Mirror Image in Soviet-American Relations." *Journal of Social Sciences* 17, pp. 45–56.

Brothers, L. (1997). *Friday's Footprint: How Society Shapes the Human Mind.* Oxford: Oxford University Press.

Browning, C. (1992). *The Path to Genocide.* Cambridge: Cambridge University Press.

Brunner, C. (1992). *The Tyranny of Hate: The Roots of Antisemitism: A Translation into English of Memsheleth Sadon.* Lewiston, N.Y.: Edwin Mellen Press.

Brustein, W. (1996). *The Logic of Evil: The Social Origins of the Nazi Party, 1925–1933.* New Haven, Conn.: Yale University Press.

Buchanan, G. M., & Seligman, M. E. P. (Eds.). (1995). *Explanatory Style.* Hillsdale, N.J.: Erlbaum Associates.

Burt, M. R. (1980). "Cultural Myths and Supports for Rape." *Journal of Personality and Social Psychology* 38, pp. 217–30.

Bush, J. (1995). "Teaching Self-risk Management to Violent Offenders." In J. McGuire et al. (Eds.), *What Works: Reducing Reoffending: Guidelines from Research and Practice* (pp. 139–54). Wiley Series in Offender Rehabilitation. Chichester, Eng.: Wiley.

Butterfield, F. (1998). "Southern Curse: Why America's Murder Rate Is So High." *New York Times,* July 26, 1998, pp. D1, D16.

Cahn, D. D. (Ed.). (1994). *Conflict in Personal Relationships.* Hillsdale, N.J.: Erlbaum Associates.

Calley, W. L., as told to J. Sack. (1970). *Lieutenant Calley: His Own Story.* New York: Viking Press.

Campbell, D. T. (1975) "On the Conflict Between Biological and Social Evolution." *American Psychologist* 30, pp. 1103–26.

Cannon, W. B. (1963). *Wisdom of the Body.* New York: Norton.

Caprara, G. V., Barbaranelli, C., Pastorelli, C., & Perugini, M. (1994). "Individual Differences in the Study of Human Aggression." *Aggressive Behavior* 20, pp. 291–303.

Caprara, G. V., Renzi, R., Alcini, R., D'Imperio, G., & Travaglia, G. (1983). "Instigation to Aggress and Escalation of Aggression Examined from a Personological Perspective: Role of Irritability and of Emotional Susceptibility." *Aggressive Behavior* 9, pp. 345–53.

Caprara, G. V., Renzi, P., D'Augello, D., D'Imperio, G., Rielli, G., & Travaglia, G. (1986). "Interpolating Physical Exercise Between Instigation to Aggress and Aggression: The Role of Irritability and Emotional Susceptibility." *Aggressive Behavior* 12, pp. 83–91.

Carr, W. (1979). *Hitler: A Study in Personality and Politics.* New York: St. Martin's Press.

Cassidy, K. W., Chu, J. Y., & Dahlsgaard, K. K. (1997). "Preschoolers' Ability to Adopt Justice and Care Orientations to Moral Dilemmas." *Early Education and Development* 8, pp. 419–34.

Chalk, F., & Jonassohn, K. (1990). *The History and Sociology of Genocide: Analyses and Case Studies.* New Haven, Conn.: Yale University Press.

Charny, I. W. (1984). *Toward the Understanding and Prevention of Genocide.* Boulder, Colo.: Westview Press.

Chen, M., & Bargh, J. A. (1997). "Nonconscious Behavioral Confirmation Processes: The Self-fulfilling Consequences of Automatic Stereotype Activation." *Journal of Experimental Social Psychology* 33, no. 5, pp. 541–60.

Chirot, D. (1994). *Modern Tyrants: The Power and Prevalence of Evil in Our Age.* New York: Free Press.

Chorover, S. L. (1979). *From Genesis to Genocide: The Meaning of Human Nature and the Power of Behavior Control.* Cambridge, Mass.: MIT Press.

Choucri, N., & North, R. C. (1987). "Roots of War: The Master Variables." In R. Vayrynen (Ed.), *The Quest for Peace* (pp. 204–16). London: Sage.

Christie, R., & Geis, F. L. (1970). *Studies in Machiavellianism.* New York: Academic Press.

Cigar, N. (1995). *Genocide in Bosnia: The Policy of "Ethnic Cleansing."* College Station: Texas A&M University Press.

Clark, D. A., & Beck, A. T., with Alford, B. (1999). *The Scientific Foundations of Cognitive Theory of Depression.* New York: Wiley.

Cleckley, H. (1950). *The Mask of Sanity: An Attempt to Clarify Some Issues About the So-called Psychopathic Personality.* St. Louis: Mosby.

Cohen, D., & Nisbett, R. E. (1994). "Self-protection and the Culture of Honor: Explaining Southern Violence." *Personality and Social Psychology Bulletin* 20, pp. 551–67.

Cohen, R. (1998). "Yes, Blood Stains in the Balkans; No, It's Not Just Fate." *New York Times,* October 4, p. D1.

Cohn, N. (1975). *Europe's Inner Demons: An Enquiry Inspired by the Great Witch-Hunt.* New York: Basic Books.

———. (1996). *Warrant for Genocide: The Myth of the Jewish World-Conspiracy and the Protocols of the Elders of Zion.* London: Serif.

Connolly, C. (1982). *The Unquiet Grave.* New York: Persea Books.

Corrado, R. R. (1981). "A Critique of the Mental Disorder Perspective of Political Terrorism." *International Journal of Law and Psychiatry* 4, pp. 1–17.

Craig, G. (1989). "Making Way for Hitler" [review of D. C. Watt, *How War Came: The Immediate Origins of the Second World War, 1939–1939*]. *New York Review of Books,* October, pp. 11–12.

Crawford, C. J. (1993). "Basque Attitude Towards Political Violence: Its Correlation with Personality Variables, Other Sociopolitical Attitudes, and Political Party Vote." *Aggressive Behavior* 19, pp. 325–46.

Crenshaw, M. (1986). "The Psychology of Political Terrorism." In M. G. Hermann (Ed.), *Political Psychology* (pp. 317-342). San Francisco: Jossey-Bass.

Crick, N. R., & Dodge, K. A. (1994). "A Review and Reformulation of Social Information-Processing Mechanisms in Children's Social Adjustment." *Psychological Bulletin* 115, pp. 74–101.

Cronin, H. (1991). *The Ant and the Peacock: Altruism and Sexual Selection from Darwin to Today.* New York: Cambridge University Press.

Crooker, J., Luhtanen, R., Elaine, B., et al. (1994). "Collective Self-esteem and Psychological Well-being Among White, Black, and Asian College Students." *Personality and Social Psychology Bulletin* 20, pp. 503–13.

Dahlen, E. R., & Deffenbacher, J. L. (in press). "A Partial Component Analysis of Beck's Cognitive Therapy for the Treatment of General Anger." *Cognitive Therapy and Reasearch.*

Dallmayr, F. (1990). "Political Evil: A Response to Alford." *Political Psychology* 11, pp. 29–35.

Daly, M., & Smuts, B. (1992). "Male Aggression Against Women: An Evolutionary Perspective." *Human Nature* 3, no. 1, pp. 1–44.

Daly, M., & Wilson, M. (1988). *Homicide.* New York: Reed Elsevier.

———. (1993). "Spousal Homicide Risk and Estrangement." *Violence and Victims* 8, pp. 3–16.

Darley, J. M., & Shultz, T. R. (1990). "Moral Judgments: Their Content and Acquisition." *Annual Review of Psychology* 41, pp. 525–56.

Darwin, C. (1859). *The Origin of Species.* New York: Collier Press.

———. (1877). *The Descent of Man and Selection in Relation to Sex.* New York: New York University Press.

Davison, G. C., Williams, M. E., Nezami, E., Bice, T. L., & DeQuattro, V. L. (1991). "Relaxation, Reduction in Angry Articulated Thoughts, and Improvements in Borderline Essential Hypertension." *Journal of Behavioral Medicine* 14, pp. 453–69.

Dawkins, R. (1976). *The Selfish Gene.* Oxford: Oxford University Press.

Dedijer, V. (1966). *The Road to Sarajevo.* New York: Simon & Schuster.

Deffenbacher, J. L., Oetting, E. R., Huff, M. E., Cornell, G. R., & Dallager, C. J. (1996). "Evaluation of Two Cognitive-Behavioral Approaches to General Anger Reduction." *Cognitive Therapy and Research* 20, no. 6, pp. 551–73.

Degler, C. (1991). *In Search of Human Nature: The Decline and Revival of Darwinism in American Social Thought.* New York: Oxford University Press.

Devine, P. G. (1995). "Prejudice and Outgroup Perception." In A. Tesser (Ed.), *Advanced Social Psychology* (pp. 467–524). New York: McGraw-Hill, 1994.

Devine, P. G., Hamilton, D. L., & Ostrom, T. M. (1994). *Social Cognition: Impact on Social Psychology*. San Diego: Academic Press.

DeWaal, F. (1996). *Good Natured*. Cambridge, Mass.: Harvard University Press.

DeWaal, F. B. M., & Luttrell, L. M. (1988). "Mechanisms of Social Reciprocity in Three Primate Species: Symmetrical Mechanisms Characteristics or Cognition." *Ethological Sociobiology* 9, nos. 2–4, pp. 101–18.

Dobson, K. (1989). "A Meta-analysis of the Efficacy of Cognitive Therapy for Depression." *Journal of Consulting and Clinical Psychology* 57, no. 3, pp. 414–19.

Dodge, K. A. (1993). "Social Cognitive Mechanisms in the Development of Conduct Disorder and Depression." *Annual Review of Psychology* 44, pp. 559–84.

Doise, W. (1978). *Groups and Individuals: Explanations in Social Psychology*. Translated by D. Graham. Cambridge: Cambridge University Press.

Dreger, R. M., & Chandler, E. W. (1993). "Confirmation of the Construct Validity and Factor Structure of the Measure of Anthropocentrism." *Journal of Social Behavior and Personality* 8, pp. 189–202.

Du Preez, P. (1994). *Genocide: The Psychology of Mass Murder*. London and New York: Boyars/Bowerdean.

Eckhardt, C. I., Barbour, K. A., & Davison, G. C. (1998). "Articulated Thoughts of Maritally Violent and Nonviolent Men During Anger Arousal." *Journal of Consulting and Clinical Psychology* 66, no. 2, pp. 259–69.

Eckhardt, C. I., & Cohen, D. J. (1997). "Attention to Anger-Relevant and Irrelevant Stimuli Following Naturalistic Insult." *Personality and Individual Differences* 23, no. 4, pp. 619–29.

Eckhardt, C. I., & Dye, M. (in press). "The Cognitive Characteristics of Maritally Violent Men: Theory and Evidence." *Cognitive Therapy and Research*.

Ehrenreich, B. (1997). *Blood Rites: Origins and History of the Passions of War*. New York: Henry, Holt, and Co.

Eisenberg, N., & Miller, P. A. (1988). "The Relation of Empathy to Prosocial and Related Behaviors." *Psychological Bulletin* 103, pp. 324–44.

Eisenberg, N., & Mussen, S. (1989). *The Roots of Prosocial Behavior in Children*. New York: Cambridge University Press.

Eksteins, M. (1989). *Rites of Spring: The Great War and the Birth of the Modern Age*. Boston: Houghton Mifflin.

Elias, N. (1978). "On Transformations of Aggressiveness." *Theory and Society* 5, pp. 227–53.

Ellis, A. (1994). *Reason and Emotion in Psychotherapy*. New York: Carol Publishing. (Originally published in 1962)

———. (1985). *Anger: How to Live With and Without It*. New York: Carol Publishing.

Ellis, A., & Grieger, R. (1977). *Handbook of Rational-Emotive Therapy*. New York: Springer.

Enright, R. D. (1991). "The Moral Development of Forgiveness." In W. M. Kurtines & J. L. Gewirtz (Eds.), *Handbook of Moral Behavior and Development*, vol. 1, *Theory* (pp. 123–152). Hillsdale, N.J.: Erlbaum Associates.

Eron, L. D., Gentry, J. H., & Schlegel, P. (Eds.). (1994). *Reason to Hope: A Psychosocial Perspective on Violence and Youth*. Washington, D.C.: American Psychological Association.

Fabiano, E. A., & Ross, R. R. (1985). *Time to Think*. Ottawa: T3 Associate, Training & Consulting.

Fazio, R., Jackson, J., Dunton, B., & Williams, C. (1995). "Variability in Automatic Activation as an Unobtrusive Measure of Racial Attitudes: A Bona Fide Pipeline?" *Journal of Personality and Social Psychology* 69, no. 6, pp. 1013–27.

Fein, H. (1993). *Genocide: A Sociological Perspective.* London: Sage.

Ferguson, T. J., & Rule, B. G. (1983). "An Attributional Perspective on Anger and Aggression." In R. G. Geen & E. I. Donnerstein (Eds.), *Aggression: Theoretical and Empirical Reviews* (vol. 1, pp. 41–74). New York: Academic Press.

Feshbach, N. (1979). "Empathy Training: A Field Study in Affective Education." In S. Feshbach & A. Fraczek (Eds.), *Aggression and Behavior Change: Biological and Social Processes* (pp. 234–249). New York: Praeger.

Feshbach, N., Feshbach, S., Fauvre, M., & Ballard-Campbell, M. (1983). *Learning to Care: Classroom Activities for Social and Affective Development.* Glenview, Ill.: Scott, Foresman, 1983.

Feshbach, S. (1986). "Individual Aggression, National Attachment, and the Search for Peace." *Aggressive Behavior* 13, pp. 315–25.

Fest, J. C. (1973). *Hitler.* New York: Harcourt Brace Jovanovich.

Fischer, F. (1967). *Germany's Aims in the First World War.* London: Chatto & Windus.

Fisher, G. (1988). *Mindsets: The Role of Culture and Perception in International Relations.* Yarmouth, Mass.: Intercultural Press.

Fitzgibbons, R. P. (1986). "The Cognitive and Emotive Uses of Forgiveness in the Treatment of Anger." *Psychotherapy* 23, no. 4, pp. 629–33.

Fleming, G. (1994). *Hitler and the Final Solution.* Berkeley: University of California Press. (Originally published in 1984)

Ford, C. V. (1996). *Lies! Lies!! Lies!!!: The Psychology of Deceit.* Washington, D.C.: American Psychiatric Press.

Forgas, J. P. (1998). "On Being Happy and Mistaken: Mood Effects on the Fundamental Attribution Error." *Journal of Personal and Social Psychology* 75, pp. 318–31.

Forgas, J. P., & Fiedler, K. (1996). "Us and Them: Mood Effects on Intergroup Discrimination." *Journal of Personality and Social Psychology* 70, pp. 28–40.

Fox, R. (1992). "Prejudice and the Unfinished Mind: A New Look at an Old Feeling." *Psychological Inquiry* 3, pp. 137–52.

Freud, S. (1938). *The Basic Writings of Sigmund Freud,* trans. & ed. A. A. Brill. New York: Modern library.

Freud, S. (1994). "Mourning and Melancholia." In R. V. Frankiel (Ed.), *Essential Papers on Object Loss: Essential Papers in Psychoanalysis* (pp. 38–51). New York: New York University Press.

Friedlander, S. (1998). *Nazi Germany and the Jews: The Years of Persecution, 1933–1939.* Vol. 1. New York: HarperCollins.

Fromm, E. (1973). *The Anatomy of Human Destructiveness.* New York: Holt, Rinehart, & Winston.

Fursenko, A., & Naftali, T. (1997). *One Hell of a Gamble: The Secret History of the Cuban Missile Crisis.* New York: Norton.

Gable, M., Hollon, C., & Dangello, F. (1990). "Relating Locus of Control to Machiavellianism and Managerial Achievement." *Psychological Reports* 67, pp. 339–43.

Gabor, T. (1994). *Everybody Does It.* Toronto: University of Toronto Press.

Gagnon, V. P., Jr. (1994). "Serbia's Road to War." *Journal of Democracy* 5, pp. 117–31.

———. (1995). "Ethnic Nationalism and International Conflict: The Case of Serbia." *International Security* 19, pp. 130–66.

Gay, P. (1993). *The Cultivation of Hatred.* London: Fontanta.

Geen, R. G. (1990). *Human Aggression.* London: Open University Press.

George, J., & Wilcox, L. (1996). *American Extremists: Militias, Supremacists, Klansmen, Communists, and Others.* New York: Prometheus Books.

Gibson, J. W. (1994). *Warrior Dreams: Paramilitary Culture in Post-Vietnam America.* New York: Hill & Wang.

Gilbert, D. T., & Malone, P. S. (1995). "The Correspondence Bias." *Psychological Bulletin* 117, no. 1, pp. 21–38.

Gilligan, C. (1982). *In a Different Voice: Psychological Theory and Women's Development.* Cambridge, Mass.: Harvard University Press.

Girard, R. (1979). *Violence and the Sacred.* Baltimore: Johns Hopkins University Press.

Glass J. (1997). "Against the Indifference Hypothesis: The Holocaust and the Enthusiasts for Murder." *Political Psychology* 18, no. 1, p. 142.

Glenny, M. (1992). *The Fall of Yugoslavia: The Third Balkan War.* New York: Penguin.

Goldhagen, D. (1996). *Hitler's Willing Executioners: Ordinary Germans and the Holocaust.* New York: Alfred A. Knopf.

Goldstein, A. P. (1996). *The Psychology of Vandalism.* New York: Plenum Press.

Goldstein, A. P., & Keller, H. R. (1987). *Aggressive Behavior: Assessment and Intervention.* New York: Pergamon Press.

Goodall, J. (1992). *Through a Window: Thirty Years with the Chimpanzees of Gombe.* Boston: Houghton Mifflin.

Green, E. P., Glaser, J., & Rich, A. (1998). "From Lynching to Gay Bashing: The Elusive Connection Between Economic Conditions and Hate Crimes." *Journal of Personality and Social Psychology* 75, pp. 109–20.

Greenstein, F. I. (1969). *Personality and Politics: Problems of Evidence.* Chicago: Markham.

Grennstein, F. (1992). "Can Personality and Politics Be Studied Systematically?" *Political Psychology* 13, pp. 105–28.

Grossarth-Maticek, R., Eysenck, H. J., & Vetter, H. (1989). "The Causes and Cures of Prejudice: An Empirical Study of the Frustration-Aggression Hypothesis." *Personality and Individual Differences* 10, no. 5, pp. 547–58.

Grossman, D. (1995). *On Killing: The Psychological Cost of Learning to Kill in War and Society.* Boston: Little, Brown.

Guerra, N. G., Huesmann, L. R., & Zelli, A. (1993). "Attributions for Social Failure and Adolescent Aggresssion." *Aggressive Behavior* 19, pp. 421–34.

Haas, J. (Ed.). (1990). *The Anthropology of War.* Cambridge and New York: Cambridge University Press.

Hall, G. C. N. (1995). "Sexual Offender Recidivism Revisited: A Meta-analysis of Recent Treatment Studies." *Journal of Consulting and Clinical Psychology* 63, pp. 802–9.

———. (1995). "Conceptually Driven Treatments for Sexual Aggressors." *Professional Psychology: Research and Practice* 24, pp. 62–69.

Hamilton, K. (1997). "The Winners; Newsmakers of 1996: Hero of the Year." *Newsweek,* Winter 1997 special edition, p. 40.

Hamilton, W. D. (1964). "The Genetical Evolution of Social Behavior." *Journal of Theoretical Biology* 7, pp. 1–52.

Haney, C., Banks, C., & Zimbardo, P. (1973). "Interpersonal Dynamics in a Simulated Prison." *International Journal of Criminology and Penology* 1, no. 1, pp. 69–97.

Hare, R. D., McPherson, L. M., & Forth, A. E. (1988). "Male Psychopaths and Their Criminal Careers." *Journal of Consulting and Clinical Psychology* 56, pp. 710–14.

Haritos-Faturos, M. (1988). "The Official Torturer: A Learning Model for Obedience to the Authority of Violence." *Journal of Applied Social Psychology* 18, pp. 1107–20.

Harris, M. (1975). *Cows, Pigs, Wars, and Witches: The Riddles of Culture.* New York: Random House.

Hassard, J., Kibble, T., & Lewis, P. (Eds.). (1989). *Ways out of the Arms Race: From the Nuclear Threat to Mutual Security.* Teaneck, N.J.: World Scientific.

Hastorf, A., & Cantril, H. (1954). "They Saw a Game: A Case Study." *Journal of Abnormal and Social Psychology* 49, pp. 129–34.

Hatfield, E., Cacioppo, J. T., & Rapson, R. L. (1994). *Emotional Contagion.* New York: Cambridge University Press.

Heider, F. (1958). *The Psychology of Interpersonal Relations.* New York: Wiley.

Helm, C., & Morelli, M. (1979). "Stanley Milgram and the Obedience Experiment: Authority, Legitimacy, and Human Action." *Political Theory* 7, pp. 321–46.

Henning, K. R., & Frueh, B. C. (1996). "Cognitive-Behavioral Treatment of Incarcerated Offenders: An Evaluation of the Vermont Department of Corrections' Cognitive Self-change Program." *Criminal Justice Behavior* 23, no. 4, pp. 523–41.

Hersen, M., Ammerman, R. T., & Sisson, L. A. (Eds.). (1994). *Handbook of Aggressive and Destructive Behavior in Psychiatric Patients.* New York: Plenum Press.

Hilberg, R. (1992). *Perpetrators, Victims, and Bystanders: The Jewish Catastrophe, 1933–1945.* New York: Aaron Asher Books.

Hinde, R. A. (1979). *Towards Understanding Relationships.* London: Academic Press.

———. (1987). *Individuals, Relationships, and Culture.* Cambridge: Cambridge University Press.

———. (1992). *The Institution of War.* New York: St. Martin's Press.

Hinde, R. A., & Watson, H. E. (1995). *War, A Cruel Necessity?: The Bases of Institutionalized Violence.* New York: St. Martin's Press.

Hinsley, F. H. (1973). *Nationalism and the International System.* London: Hodder & Stoughton.

Hirshberg, M. S. (1993). "The Self-perpetuating National Self-image: Cognitive Biases in Perceptions of International Interventions." *Political Psychology* 14, pp. 77–98.

Hoffman, M. L. (1990). "Empathy and Justice Motivation." *Empathy and Emotion* 14, pp. 151–72.

Hofstadter, R. (1967). *The Paranoid Style in American Politics and Other Essays.* New York: Vintage Books.

Hollander, P. (1995). "Models and Mentors." *Modern Age* 37, pp. 50–62.

———. (1997). "Revisiting the Banality of Evil: Political Violence in Communist Systems." *Partisan Review* 64, no. 1, p. 56.

Holsti, O. R. (1972). *Crisis, Escalation, War.* Montreal: McGill-Queen's University Press.

Holtzworth-Munroe, A., & Hutchinson, G. (1993). "Attributing Negative Intent to Wife Behavior: The Attributions of Maritally Violent Versus Nonviolent Men." *Journal of Abnormal Psychology* 102, pp. 206–11.

Holtzworth-Munroe, A., Stuart, G. L., & Hutchinson, G. (1997). "Violent Versus Nonviolent Husbands: Differences in Attachment Patterns, Dependency, and Jealousy." *Journal of Family Psychology* 11, no. 3, pp. 314–31.

Horne, A. D. (1983). "U.S. Says Soviets Shot Down Airliner." *Washington Post,* September 2, p. A1.

Horney, K. (1950). *Neurosis and Human Growth: The Struggle Toward Self-realization.* New York: Norton.

Hovannisian, R. G. (Ed.). (1992). *The Armenian Genocide: History, Politics, Ethics.* New York: St. Martin's Press.

Huesmann, L. R. (1994). *Aggressive Behavior: Current Perspectives.* New York: Plenum Press.

Hunt, M. (1991). *The Compassionate Beast: The Scientific Inquiry into Human Altruism.* New York: Anchor Books/Doubleday.

Huxley, A. (1946). *Ends and Means.* London: Chatto & Windus.

Ickes, W. (Ed.). (1997). *Empathic Accuracy.* New York: Guilford.

Jacobson, E. (1968). *Progressive Relaxation: A Physiological and Clinical Investigation of Muscular States and Their Significance in Psychology and Medical Practice.* Chicago: University of Chicago Press. (Originally published in 1938)

Jacobson, N., & Gottman, J. (1998). *When Men Batter Women: New Insights into Ending Abusive Relationships.* New York: Simon & Schuster.

Janis, I. L. (1982). *Victims of Groupthink: A Psychological Study of Foreign Policy Decisions and Fiascoes.* Boston: Houghton Mifflin.

Jervis, R. (1976) *Perception and Misperception in International Politics.* Princeton, N.J.: Princeton University Press.

Jukic, I. (1974). *The Fall of Yugoslavia.* New York: Harcourt Brace Jovanovich.

Kahn, L. S. (1980). "The Dynamics of Scapegoating: The Expulsion of Evil." *Psychotherapy: Theory, Research, and Practice* 17, pp. 79–84.

Kapferer, J. (1990). *Rumors: Uses, Interpretations, and Images.* New Brunswick, N.J.: Transaction Publishers.

Kaplowitz, N. (1990). "National Self-images, Perception of Enemies, and Conflict Strategies." *Political Psychology* 11, pp. 39–82.

Kassinove, H. (Ed.). (1995). *Anger Disorders: Definition, Diagnosis, and Treatment.* Washington, D.C.: Taylor & Francis.

Katon, W. (1987). "The Epidemiology of Depression in Medical Care." *International Journal of Psychiatry in Medicine* 17, pp. 93–112.

Katz, J. (1988). *Seductions of Crime: Moral and Sensual Attractions in Doing Evil.* New York: Basic Books.

Kawachi, I., Sparrow, D., Spiro, A., Vokonas, P., & Weiss, S. (1996). "A Prospective Study of Anger in Coronary Disease." *Circulation* 94, pp. 2090–95.

Keen, S. (1986). *Faces of the Enemy: Reflections of the Hostile Imagination.* San Francisco: Harper & Row.

Kelman, H. C. (1973). "Violence Without Moral Constraint: Reflections on the Dehumanization of Victims and Victimizers." *Journal of Social Issues* 29, pp. 25–61.

Kelman, H. C., & Hamilton, V. L. (1989). *Crimes of Obedience: Toward a Social Psychology of Authority and Responsibility.* New Haven, Conn.: Yale University Press.

Kershaw, I. (1987). *The "Hitler Myth": Image and Reality in the Third Reich.* Oxford: Clarendon Press.

Kidder, L. H., & Stewart, V. M. (1975). *The Psychology of Intergroup Relations: Conflict and Consciousness.* New York: McGraw-Hill.

Kinderman, P., & Bentall, R. "The Clinical Implications of a Psychological Model of Paranoia." In E. Sanavio, (Ed.) (1998) *Behaviour and Cognitive Therapy Today: Essays in Honour of Hans J. Eysenck.* Oxford: Elsevier Press.

Klama, J. (1988). *Aggression: Conflict in Animals and Humans Reconsidered.* Harlow, Eng.: Longman.

Knutson, J. N. (1973). *Handbook of Political Psychology.* San Francisco: Jossey-Bass.

Kochanska, G. (1993). "Toward a Synthesis of Parental Socialization and Child Temperament in Early Development of Conscience." *Child Development* 64, pp. 325–47.

———. (1997). "Multiple Pathways to Conscience for Children with Different Temperaments: From Toddlerhood to Age Five." *Developmental Psychology* 33, pp. 228–40.

Koestler, A. (1970). *The Ghost in the Machine.* London: Pan Books. (Originally published in 1967)

Kohlberg, L. (1984). *The Psychology of Moral Development: The Nature and Validity of Moral Stages.* San Francisco: Harper & Row.

Kohn, A. (1990). *The Brighter Side of Human Nature: Altruism and Empathy in Everyday Life.* New York: Basic Books.

Korn, J. H. (1997). *Illusions of Reality: A History of Deception in Social Psychology.* Albany: State University of New York Press.

Kosterman, R., & Feshbach, S. (1989). "Toward a Measure of Patriotic and Nationalistic Attitudes." *Political Psychology* 10, pp. 257–74.

Krebs, J. R., & Davies, N. B. (1993). *An Introduction to Behavioural Ecology.* 3rd ed. Oxford: Blackwell Scientific.

Kropotkin, P. (1995). *Evolution and Environment.* Montreal: Blackrose Books.

Kull, S. (1988). *Minds at War: Nuclear Reality and the Inner Conflict of Defense Policymakers.* New York: Basic Books.

Kull, S., Small, M., & Singer, J. D. (1982). *Resort to Arms: International and Civil Wars, 1816–1980.* Beverly Hill, Calif.: Sage Publications.

Kuper, L. (1977). *The Pity of It All: Polarization of Racial and Ethnic Relations.* Minneapolis: University of Minnesota Press.

———. (1985). *The Prevention of Genocide.* New Haven, Conn.: Yale University Press.

Kurtines, W., & Gerwitz, J. (Eds.). (1991). *Handbook of Moral Behavior and Development.* Vols. 1, 2, and 3. Hillsdale, N.J.: Erlbaum Associates.

LaCapra, D. (1994). *Representing the Holocaust: History, Theory, Trauma.* Ithaca, N.Y.: Cornell University Press,

Laqueur, W. (1987). *The Age of Terrorism.* Boston: Little, Brown.

Lau, R. R., & Sears, D. O. (Eds.). (1986). *Political Cognition: The Nineteenth Annual Carnegie Symposium on Cognition.* Hillsdale, N.J.: Elrbaum Associates.

Lazarus, R. S. (1991). *Emotion and Adaptation.* New York: Oxford University Press.

Leary, M. R. (1983). *Understanding Social Anxiety: Social, Personality, and Clinical Perspectives.* Beverly Hills, Calif.: Sage Publications.

Le Bon, G. (1952). *The Crowd: A Study of the Popular Mind.* London: E. Benn.

Lebow, R. N. (1981). *Between Peace and War: The Nature of International Crisis.* Baltimore: Johns Hopkins University Press.

———. (1987). "Is Crisis Management Always Possible?" *Political Science Quarterly* 102, pp. 181–92.

———. (1987). *Nuclear Crisis Management.* Ithaca, N.Y.: Cornell University Press.

Lederer, G. (1982). "Trends in Authoritarianism: A Study of Adolescents in West Germany and the United States Since 1945." *Journal of Cross-cultural Psychology* 13, pp. 299–314.

Lerner, M. (1980). *The Belief in a Just World: A Fundamental Delusion.* New York: Plenum Press.

Lerner, R. M. (1992). *Final Solutions: Biology, Prejudice, and Genocide.* University Park: Pennsylvania State University Press.

Levenson, M. R. (1992). "Rethinking Psychopathy." *Theory and Psychopathy* 2, pp. 51–71.

Lévi-Strauss, C. (1963). "Do Dual Organizations Exist?" In *Structural Anthropology.* New York: Basic Books.

Levy, J. S. (1983). "Misperceptions and the Causes of War: Theoretical Linkages and Analytical Problems." *World Politics* 35, pp. 75–99.

———. (1987). "Declining Power and the Preventive Motivation for War." *World Politics* 40, pp. 82–107.

———. (1989). "The Causes of War: A Review of Theories and Evidence." In P. E. Tetlock, J. L. Husbands, R. Jervis, P. C. Stern, & C. Tilly (Eds.), *Behavior, Society, and Nuclear War* (vol. 1, pp. 209–333). New York: Oxford University Press.

Lewis, M., & Saarni, C. (1993). *Lying and Deception in Everyday Life.* New York: Guilford.

Lifton, R. J. (1986). *The Nazi Doctors: Medical Killing and the Psychology of Genocide.* New York: Basic Books.

Lifton, R. J., & Markusen, E. (1990). *The Genocidal Mentality: Nazi Holocaust and Nuclear Threat.* New York: Basic Books.

Lippmann, W. (1922). *Public Opinion.* New York: Harcourt, Brace, & Co.

Lore, R. K., & Schultz, L. A. (1993). "Control of Human Aggression: A Comparative Perspective." *American Psychologist* 48, pp. 16–25.

Lorenz, K. (1966). *On Aggression.* New York: Routledge.

Lykken, D. T. (1995). *The Antisocial Personality.* Hillsdale, N.J.: Erlbaum Associates.

———. (1996). "Psychopathy, Sociopathy, and Crime." *Society* 34, no. 1, pp. 29–38.

Malamuth, N. M., & Brown, L. M. (1994). "Sexually Aggressive Men's Perceptions of Women's Communications: Testing Three Explanations." *Journal of Personality and Social Psychology* 67, no. 4, pp. 699–712.

Mann, C. R. (1982). *When Women Kill.* Albany: State University of New York Press.

Marazziti, D., Rotondo, A., Presta, S., Pancioli-Guadagnucci, M. L., Palego, L., & Conti, L. (1993). "Role of Serotonin in Human Aggressive Behaviour." *Aggressive Behavior* 19, pp. 347–53.

Martin, L. L., & Tesser, A. (Eds.). (1992). *The Construction of Social Judgments.* Hillsdale, N.J.: Erlbaum Associates.

Masters, R. D., & McGuire, M. T. (Eds.). (1994). *The Neurotransmitter Revolution: Serotonin, Social Behavior, and the Law.* Carbondale, Ill.: Southern Illinois University Press.

Maxwell, M. (1990). *Morality Among Nations: An Evolutionary View.* Albany: State University of New York Press.

————. (Ed.). (1991). *The Sociobiological Imagination.* Albany: State University of New York Press.

Maybury-Lewis, D., & Alamagor, U. (Eds.). (1989). *The Attraction of Opposites: Thought and Society in the Dualistic Mode.* Ann Arbor: University of Michigan Press.

Mayer, M. S. (1955). *They Thought They Were Free: The Germans, 1933–1945.* Chicago: University of Chicago Press.

Mazian, F. (1990). *Why Genocide?: The Armenian and Jewish Experiences in Perspective.* Ames: Iowa State University Press.

McCauley, C. (1989). "The Nature of Social-Influence in Groupthink: Compliance and Internalization." *Journal of Personality and Social Psychology* 57, no. 2, pp. 250–60.

McGuire, M. T. (1991). "Moralistic Aggression and the Sense of Justice." *American Behavioral Scientist* 34, pp. 371–85.

————. (1993). *Human Nature and the New Europe.* Boulder, Colo.: Westview Press.

Meichenbaum, D. H. (1975). *Stress Inoculation Training.* New York: Pergamon Press.

Miedzian, M. (1992). *Boys Will Be Boys: Breaking the Link Between Masculinity and Violence.* London: Virago.

Mikulincer, M. (1994). *Human Learned Helplessness: A Coping Perspective.* New York: Plenum Press.

Miller, A.G. (Ed.). (1982). *In the Eye of the Beholder: Contemporary Issues in Stereotyping.* New York: Praeger.

————. (1986). *The Obedience Experiments: A Case Study of Controversy in Social Science.* New York: Praeger.

Miller, P. A., & Eisenberg, N. (1988). "The Relation of Empathy to Aggressive Behaviour and Externalising/Antisocial Behaviour." *Psychological Bulletin* 103, pp. 324–44.

Miller, T. C. (1993). "The Duality of Human Nature." *Politics and the Life Sciences* 13, pp. 221–41.

Monteith, M. J. (1993). "Self-regulation of Prejudiced Responses: Implications for Progress in Prejudice-Reduction Efforts." *Journal of Personality and Social Psychology* 65, pp. 469–85.

Morris, D. (1967). *The Naked Ape: A Zoologist's Study of the Human Animal.* New York: McGraw-Hill.

Moscovici, S. (1993). *The Invention of Society: Psychological Explanations for Social Phenomena,* trans. W. D. Halls. Cambridge: Polity Press.

Nagayama Hall, G. (Ed.). (1993). *Sexual Aggression: Issues in Etiology, Assessment, and Treatment.* Washington, D.C.: Taylor & Francis.

Nathan, O., & Norden, H. (Eds.). (1968). *Einstein on Peace.* New York: Schocken Books. (Originally published in 1960)

Newman, J. P., Schmitt, W. A, & Voss, W. D. (1997). "The Impact of Motivationally Neutral Cues on Psychopathic Individuals: Assessing the Generality of the Response Modulation Hypothesis." *Journal of Abnormal Psychology* 106, no. 4, pp. 563–75.

Nicolai, G. F. (1919). *The Biology of War.* New York: Century Co.

Nisbett, R., & Cohen, D. (1996). *Culture of Honor: The Psychology of Violence in the South.* Boulder, Colo.: Westview Press.

Noddings, N. (1984). *Caring: A Feminine Approach to Ethics and Moral Education.* Berkeley: University of California Press.

Noller, P. (1984). *Nonverbal Communication and Marital Interaction.* New York: Pergamon Press.

Novaco, R. W. (1975). *Anger Control: The Development and Evaluation of an Experimental Treatment.* Lexington, Mass.: D. C. Heath.

Ofshe, R., & Watters, E. (1996). *Making Monsters: False Memories, Psychotherapy, and Sexual Hysteria.* Berkeley: University of California Press.

O'Leary, K. D. (1996). "Physical Aggression in Intimate Relationships Can Be Treated Within a Marital Context Under Certain Circumstances." *Journal of Interpersonal Violence* 11, no. 3, pp. 450–52.

Oliner, S., & Oliner, P. (1988). *The Altruistic Personality: Rescuers of Jews in Nazi Europe.* New York: Free Press; London: Collier Macmillan.

Olweus, D. (1984). "Development of Stable Aggressive Reaction Patterns in Males." In R. J. Blanchard & D. C. Blanchard (Eds.), *Advances in the Study of Aggression* (vol. 1, pp. 103–37). New York: Academic Press.

Olweus, D., Mattsson, A., Schalling, D., & Low, H. (1988). "Circulating Testosterone Levels and Aggression in Adolescent Males: A Causal Analysis." *Psychosomatic Medicine* 50, pp. 261–72.

Orwell, G. (1949). *Nineteen Eighty-four.* New York: Harcourt, Brace & World.

Orwell, S., & Angus, I., eds. (1968). *The Collected Essays, Journalism, and Letters of George Orwell.* New York: Harcourt, Brace & World.

Ostow, M. (1996). *Myth and Madness: The Psychodynamics of Antisemitism.* New Brunswick, N.J.: Transaction Publishers.

Pagels, E. (1995). *The Origin of Satan.* New York: Random House.

Paret, P. (1985). *Clausewitz and the State: The Man, His Theories, and His Times.* Princeton, N.J.: Princeton University Press.

Paul, R. A. (1978). "Instinctive Aggression in Man: The Semai Case." *Journal of Psychological Anthropology* 1, pp. 65–79.

Pear, T. H. (1950). *Psychological Factors of Peace and War.* New York: Hutchinson.

Pearlstein, R. M. (1991). *The Mind of a Political Terrorist.* Wilmington, Del.: Scholarly Resources.

Pflanze, O. (1963). *Bismarck and the Development of Germany.* Vol. 1., 2nd ed. Princeton, N.J.: Princeton University Press.

Piaget, J., with the assistance of seven collaborators. (1960). *The Moral Judgment of the Child,* trans. M. Gabain. Glencoe, Ill.: Free Press. (Originally published in 1932)

Pincus, F. L., & Ehrlich, H. J. (Eds.). (1994). *Race and Ethnic Conflict: Contending Views on Prejudice, Discrimination, and Ethnoviolence.* Boulder, Colo.: Westview Press.

Pipes, D. (1996). *The Hidden Hand: Middle East Fears of Conspiracy.* New York: St. Martin's Press.

Plomin, R. (1994). *Genetics and Experience: The Interplay Between Nature and Nurture.* Thousand Oaks, Calif.: Sage Publications.

Plotkin, H. C. (1994). *Darwin Machines and the Nature of Knowledge.* Cambridge, Mass.: Harvard University Press.

Pluhar, E. B. (1995). *Beyond Prejudice: The Moral Significance of Human and Nonhuman Animals.* Durham, N.C.: Duke University Press.

Polaschek, D. L. L., Ward, T., & Hudson, S. M. (1997). "Rape and Rapists: Theory and Treatment." *Clinical Psychology Review* 17, no. 2, pp. 117–44.

Polk, K. (1994). *When Men Kill: Scenarios of Masculine Violence.* Cambridge: Cambridge University Press.

Pruitt, D. G., & Rubin, J. Z. (1986). *Social Conflict: Escalation, Stalemate, and Settlement.* New York: Random House.

Pryor, J. B., & Stoller, L. M. (1994). "Sexual Cognition Processes in Men High in Likelihood to Sexually Harass." *Personality and Social Psychology Bulletin* 20, pp. 163–69.

Ramirez, J. M., Hinde, R. A., & Groebel, J. (Eds.). (1987). *Essays on Violence.* Seville: Publicaciones de la Universidad de Sevilla.

Reich, R. (1997). *Locked in the Cabinet.* New York: Alfred A. Knopf.

Reich, W. (Ed.). (1990). *The Origins of Terrorism: Psychologies, Ideologies, Theologies, States of Mind.* Cambridge: Press Syndicate of the University of Cambridge.

Richardson, L. F. (1960). *Arms and Insecurity: A Mathematical Study of the Causes and Origins of War.* Pittsburgh: Boxwood Press.

———. (1960). *Statistics of Deadly Quarrels.* Pittsburgh: Boxwood Press.

Ridley, M. (1997). *The Origins of Virtue: Human Instincts and the Evolution of Cooperation.* New York: Viking.

Rieber, R. W. (1991). *The Psychology of War and Peace: The Image of the Enemy.* New York: Plenum Press.

Robins, R., & Post, J. (1997). *Political Paranoia: The Psychopolitics of Hatred.* New Haven, Conn.: Yale University Press.

Roes, F. L. (1995). "The Size of Societies, Stratification, and Belief in High Gods Supportive of Human Morality." *Politics and the Life Sciences* 14, pp. 73–77.

Roese, N. (1995). *What Might Have Been: The Social Psychology of Counterfactual Thinking.* Hillsdale, N.J.: Erlbaum Associates.

Rokeach, M. (1960). *The Open and Closed Mind: Investigations into the Nature of Belief Systems and Personality Systems.* New York: Basic Books.

Rosenbaum, R. (1998). *Explaining Hitler.* New York: Random House.

Rothbart, M., Evans, M., & Furlaro, S. (1979). "Recall for Confirming Events." *Journal of Experimental Psychology* 15, pp. 343–55.

Sanford, N., & Comstock, C. (1971). *Sanctions for Evil.* San Francisco: Jossey-Bass.

Sabini, J., & Silver, M. (1982). *The Moralities of Everyday Life.* New York: Oxford University Press.

Sagan, C., & Druyan, A. (1992). *Shadows of Forgotten Ancestors: A Search for Who We Are.* New York: Random House.

Santoni, R. E. (1991). "Nurturing the Institution of War: 'Just War' Theory's Justifications and Accommodations." In R. A. Hinde (Ed.), *The Institution of War* (pp. 99–120). London: Macmillan.

Sattler, D. N., & Kerr, N. L. (1991). "Might Versus Morality Explored: Motivational and Cognitive Bases for Social Motives." *Journal of Personality and Social Psychology* 60, pp. 756–65.

Saunders, N. (1991). "Children of Mars" [review of Haas, J., (Ed.), *The Anthropology of War*]. *New Scientist* 18, p. 51.

Scheff, T. J. (1994). *Bloody Revenge: Emotions, Nationalism, and War.* Boulder, Colo.: Westview Press.

Seligman, M. E., Abramson, L. Y., Semmel, A., & von Baeyer, C. (1979). "Depressive Attributional Style." *Journal of Abnormal Psychology* 88, no. 3, pp. 242–47.

Selman, R. (1980). *The Growth of Interpersonal Understanding: Developmental and Clinical Analyses.* New York: Academic Press.

Sempa, F. (1991). [Review of D. C. Watt, *How War Came: The Immediate Origins of the Second World War, 1938–1939*]. *Presidential Studies Quarterly* 21, pp. 621–23.

Sereny, G. (1995). *Albert Speer: His Battle with Truth.* New York: Alfred A. Knopf.

Serin, R., & Kuriychuk, M. (1994). "Social and Cognitive Processing Deficits in Violent Offenders: Implications for Treatment." *International Journal of Law and Psychiatry* 17, pp. 431–41.

Shapiro, D., Hui, K. K., Oakley, M. E., Pasic, J., & Jamner, L. D. (1997). "Reduction in Drug Requirements for Hypertension by Means of a Cognitive-Behavioral Intervention." *American Journal of Hypertension* 10, no. 1, pp. 9–17.

Shaw, R. P., & Wong, Y. (1989). *Genetic Seeds of Warfare: Evolution, Nationalism, and Patriotism.* Boston: Unwin Hyman.

Sherif, M., Harvey, O. J., White, B. J., Hood, W. R., & Sherif, C. W. (1988). *The Robbers Cave Experiment: Intergroup Conflict and Cooperation.* Middletown, Conn.: Wesleyan University Press.

Sigel, I. E., Stinson, E. T., & Kim, M. (1993). *Socialization of Cognition: The Distancing Model.* Hillsdale, N.J.: Erlbaum Associates.

Silber, L., & Little, A. (1996). *Yugoslavia: Death of a Nation.* New York: Penguin Books.

Silverstein, B. (1989). "Enemy Images: The Psychology of U.S. Attitudes and Cognitions Regarding the Soviet Union." *American Psychologist* 44, pp. 903–13.

Simner, M. L. (1971). "Newborn's Response to the Cry of Another Infant." *Developmental Psychology* 5, no. 1, pp. 136–50.

Simon, H. A. (1985). "Human Nature in Politics: The Dialogue of Psychology with Political Science." *American Political Science Review* 79, pp. 293–304.

Simon, R. I. (1995). *Bad Men Do What Good Men Dream: A Forensic Psychiatrist Illuminates the Darker Side of Human Behavior.* Washington, D.C.: American Psychiatric Press.

Simonton, D. K. (1990). "Personality and Politics." In L. A. Pervin (Ed.), *Handbook of Personality: Theory and Research* (pp. 670–92). New York: Guilford.

Singer, E., & Hudson, V. (1992). *Political Psychology and Foreign Policy.* Boulder, Colo.: Westview Press.

Singer, J. D. (1960). "International Conflict: Three Levels of Analysis." *World Politics* 12, no. 3, pp. 453–61.

———. (1961). "The Level-of-Analysis Problem in International Relations." In K. Knorr & S. Verba (Eds.), *The International System: Theoretical Essays* (pp. 77–92). Princeton, N.J.: Princeton University Press.

Singer, J. D., & Small, M. (1972). *The Wages of War 1816–1965: A Statistical Handbook.* New York: Wiley.

Small, M., & Singer, J. D. (1982). *Resort to Arms: International and Civil Wars, 1816–1980.* Beverly Hill, Calif.: Sage Publications.

Smith, A. (1976). *The Theory of Moral Sentiments.* Oxford: Clarendon Press. (Originally published in 1759)

Smith, D. N. (1998). "The Psychocultural Roots of Genocide: Legitimacy and Crisis in Rwanda." *American Psychologist* 53, no. 7, pp. 743–53.

Smith, M. B. (1968). "A Map for the Study of Personality and Politics." *Journal of Social Issues* 24, pp. 15–28.

Smoke, R. (1977). *War: Controlling Escalation.* Cambridge, Mass.: Harvard University Press.

Smuts, B. (1996). "Male Aggression Against Women: An Evolutionary Perspective." In D. M. Buss & N. M. Malamuth (Eds.), *Sex, Power, Conflict: Evolutionary and Feminist Perspectives* (pp. 231–68). New York: Oxford University Press.

Snyder, G. H., & Diesing, P. (Eds.). (1977). *Conflict Among Nations: Bargaining, Decision Making, and System Structure in International Crisis.* Princeton, N.J.: Princeton University Press.

Snyder, W. (1981). "On the Self-perpetuating Nature of Stereotypes." In D. Hamilton (Ed.), *Cognitive Processes in Stereotyping and Intergroup Behavior* (pp. 183–212). Hillsdale, N.J.: Erlbaum Associates.

Sober, E., & Wilson, D. S. (1998). *Unto Others: The Evolution and Psychology of Unselfish Behavior.* Cambridge, Mass.: Harvard University Press.

Sprinzak, E. (1990). "The Psychopolitical Formation of Extreme Left Terrorism in a Democracy: The Case of the Weathermen." In W. Reich (Ed.), *Origins of Terrorism: Psychologies, Ideologies, Theologies, States of Mind* (pp. 78–80). Cambridge: Press Syndicate of the University of Cambridge.

Staby, R. G., & Guerra, N. G. (1988). "Cognitive Mediators of Aggression in Adolescent Offenders." *Developmental Psychology* 24, no. 4, pp. 580–88.

Staub, E. (1979). *Positive Social Behavior and Morality.* Vols. 1–2. New York: Academic Press.

———. (1989). *The Roots of Evil: The Origins of Genocide and Other Group Violence.* New York: Cambridge University Press.

Stephan, W. (1977). "Stereotyping: The Role of Ingroup-Outgroup Differences in Causal Attribution for Behavior." *Journal of Social Psychology* 101, pp. 255–66.

Stern, F. (1987). *Dreams and Delusions.* New York: Alfred A. Knopf.

Stern, K. S. (1996). *A Force upon the Plain: The American Militia Movement and the Politics of Hate.* New York: Simon & Shuster.

Stern, J. P. (1975). *Hitler: The Führer and the People.* London: Fontana.

Stern, P. C. (1995). "Why Do People Sacrifice for Their Nations?" *Political Psychology* 16, no. 2, pp. 217–35.

———. (1996). "Nationalism as Reconstructed Altruism." *Political Psychology* 17, no. 3, pp. 569–72.

Stern, P. C., Axelrod, R., Jervis, R., & Radner, R. (Eds.). (1989). *Perspectives on Deterrence.* New York: Oxford University Press.

Stewart, J. R. (1980). "Collective Delusion: A Comparison of Believers and Skeptics." Paper presented at the Midwest Sociological Society meeting, Milwaukee (April).

Stoessinger, J. G. (1993). *Why Nations Go to War.* New York: St. Martin's Press.

Stone, M. H. (1993). "Antisocial Personality and Psychopathy." In Store, M. H., *Abnormalities of Personality: Within and Beyond the Realm of Treatment* (pp. 277–313). New York: Norton.

Stromberg, R. N. (1982). *Redemption by War: Intellectuals and 1914.* Lawrence: Regents Press of Kansas.

Strozier, C. (1994). *Apocalypse: On the Psychology of Fundamentalism in America.* Boston: Beacon Press.

Struch, N., & Schwartz, S. H. (1989). "Intergroup Aggression: Its Predictors and Distinctness from Ingroup Bias." *Journal of Personality and Social Psychology* 56, pp. 364–73.

Suadicani, P., Hein, H. O., & Gyntelberg, F. (1993). "Are Social Inequalities as Associated with the Risk of Ischaemic Heart Disease as a Result of Psychosocial Working Conditions?" *Atherosclersis* 101, pp. 165–75.

Sun, K. (1993). "Two Types of Prejudice and Their Causes." *American Psychologist* 1152, pp. 1152–53.

Sutherland, S. (1994). *Irrationality: Why We Don't Think Straight.* New Brunswick, N.J.: Rutgers University Press.

Sykes, S. (1991). "Sacrifice and the Ideology of War." In R. A. Hinde (Ed.), *The Institution of War* (pp. 87–98). London: Macmillan.

Tajfel, H. (1981). *Human Groups and Social Categories: Studies in Social Psychology.* Cambridge: Cambridge University Press.

Taylor, A. J. P. (1955). *Bismarck: The Man and the Statesman.* New York: Alfred A. Knopf.

Taylor, K. M., & Shepperd, J. A. (1996). "Probing Suspicion Among Deception Research." *American Psychologist* 51, no. 8, pp. 886–87.

Tejirian, E. (1990). *Sexuality and the Devil: Symbols of Love, Power, and Fear in Male Psychology.* New York: Routledge.

Tesser, A. (1994). *Advanced Social Psychology.* New York: McGraw-Hill.

Tetlock, P. (1983). "Cognitive Style and Political Ideology." *Journal of Personality and Social Psychology* 45, pp. 118–25.

Tetlock, P., Husbands, J., Jervis, R., Stern, P., & Tilly, C. (1991). *Behavior, Society, and Nuclear War.* Vol. 2. New York: Oxford University Press.

———. (Eds.). (1993). *Behavior, Society, and International Conflict.* Vol. 3. New York: Oxford University Press.

Thompson, J. (1989). "Perceptions of Threat." In J. Hassard, T. Kibble, & P. Lewis (Eds.), *Ways out of the Arms Race* (pp. 238–44). Singapore: World Scientific.

Todorov, T. (1996). *Facing the Extreme.* New York: Metropolitan Books.

Totten, S., Parsons, W. S., & Charney, I. W. (Eds.). (1995). *Genocide in the Twentieth Century: Critical Essays and Eyewitness Accounts.* New York: Garland Publishing.

Trivers, R. L. (1985). *Social Evolution.* Menlo Park, Calif.: Benjamin/Cummings.

Tuchman, B. (1962). *Guns of August.* New York: Macmillan.

Upham, C. W. (1959). *Salem Witchcraft,* Vol. 2. New York: Frederick Ungar.

Van der Dennen, J., & Falger, V. (1990). *Sociobiology and Conflict: Evolutionary Perspectives on Competition, Cooperation, Violence, and Warfare.* New York: Chapmann & Hall.

Van Goozen, S. H. M., Frijda, N. H., Kindt, M., & van de Poll, N. E. (1994). "Anger Proneness in Women: Development and Validation of the Anger Situation Questionnaire." *Aggressive Behavior* 20, pp. 79–100.

Van Praag, H. M., Plutchik, R., & Apter, A. (Eds.). (1990). *Violence and Suicidality.* New York: Brunner/Mazel.

Vayrynen, R. (1987). *The Quest for Peace: Transcending Collective Violence and War Among Societies, Cultures, and States.* Beverly Hills, Calif.: Sage Publications.

Vertzberger, Y. Y. I. (1990). *The World in Their Minds: Information Processing, Cognition, and Perception in Foreign Policy Decision-Making.* Stanford, Calif.: Stanford University Press.

Victor, J. (1993). *Satanic Panic: The Creation of a Contemporary Legend.* Chicago: Open Court.

Volavka, J. (1995). *Neurobiology of Violence.* Washington, D.C.: American Psychiatric Press.

Volk, R. J., Pace, T. M., & Parchman, M. L. (1993). "Screening for Depression in Primary Care Patients: Dimensionality of the Short Form of the Beck Depression Inventory." *Psychological Assessment* 5, pp. 173–81.

Volkan, V. D. (1988). *The Need to Have Enemies and Allies.* Northvale, N.J.: Jason Aronson.

Voss, J. F., & Dorsey, E. (1992). "Perception and International Relations: An Overview." In E. Singer & V. Hudson (Eds.), *Political Psychology and Foreign Policy* (pp. 3–30). Boulder, Colo.: Westview Press.

Wallimann, I., & Dobkowski, M. N. (Eds.). (1987). *Genocide and the Modern Age: Etiology and Case Studies of Mass Death.* New York: Greenwood Press.

Waltz, K. (1969). *Man, the State, and War: A Theoretical Analysis.* New York: Columbia University Press. (Originally published in 1959)

Walzer, M. (1992). *Just and Unjust Wars: A Moral Argument with Historical Illustrations.* New York: Basic Books.

Ward, T., Hudson, S. M., Johnston, L., & Marshall, W. L. (1997). "Cognitive Distortions in Sex Offenders: An Integrative Review." *Clinical Psychology Review* 17, no. 5, pp. 479–507.

Watt, D. C. (1989). *How War Came: The Immediate Origins of the Second World War.* New York: Pantheon Books.

Weiner, I. B., (Ed.) (1983) *Clinical Methods in Psychology.* New York: John Wiley & Sons.

Weiss, J. (1996). *Ideology of Death: Why the Holocaust Happened in Germany.* Chicago: I. R. Dee.

White, E. (1993). *Genes, Brains, and Politics: Self-selection and Social Life.* Westport, Conn.: Praeger Press.

White, R. K. (1976). *Nobody Wanted War.* New York: Doubleday.

———. (1984). *Fearful Warriors: A Psychological Profile of U.S.-Soviet Relations.* New York: Free Press.

———. (1990). "Why Aggressors Lose." *Political Psychology* 11, pp. 227–42.

Whiteman, M., Fanshel, D., & Grundy, J. (1987). "Cognitive-Behavioral Intervention Aimed at Anger of Parents at Risk of Child Abuse." *Social Work* 32, no. 6, pp. 469–74.

Wickless, C., & Kirsch, I. (1988). "Cognitive Correlates of Anger, Anxiety, and Sadness." *Cognitive Therapy and Research* 12, no. 4, pp. 367–77.

Williams, G. C. (1992). *Natural Selection: Domains, Levels, and Challenges.* Oxford: Oxford University Press.

Williams, R., & Williams, V. (1993). *Anger Kills.* New York: Times Books.

Wilson, E. O. (1975). *Sociobiology: The New Synthesis.* Cambridge, Mass.: Belknap Press of Harvard University Press.

———. (1998). *Consilience: The Unity of Knowledge.* New York: Alfred A. Knopf.

Wilson, G. (1988). "Navy Missile Downs Iranian Jetliner over Gulf; Iran Says 290 Are Dead." *Washington Post*, July 4, p. A1.

Wilson, M. I., & Daly, M. (1996). "Male Sexual Proprietariness and Violence Against Wives." *Current Directions in Psychological Science* 5, no. 1, pp. 2–7.

Winter, D. G. (1993). "Power, Affiliation, and War: Three Tests of a Motivational Model." *Journal of Personality and Social Psychology* 65, pp. 532–45.

Winter, J. (1991). "Imaginings of War: Some Cultural Supports of the Institution of War." In R. A. Hinde (Ed.), *The Institution of War* (pp. 155–77). London: Macmillan

Wolman, B. B. (1987). *The Sociopathic Personality*. New York: Brunner/Mazel.

Worchel, S., & Austin, W. G. (Eds.). (1986). *Psychology of Intergroup Relations*. Chicago: Nelson-Hall.

Wrangham, R., & Peterson, D. (1996). *Demonic Males: Apes and the Origins of Human Violence*. Boston: Houghton Mifflin.

Wright, R. (1994). *The Moral Animal: The New Science of Evolutionary Psychology*. New York: Pantheon Books.

———. (1995). "The Biology of Violence." *New Yorker*, March 13, pp. 68–77.

Yochelson, S., & Samenow, S. E. (1976). *The Criminal Personality: A Profile for Change*. New York: Jason Aronson.

Young, J. W. (1991). *Totalitarian Language: Orwell's Newspeak and Its Nazi and Communist Antecedents*. Charlottesville: University Press of Virginia.

Zahn-Waxler, C., Cummings, M., & Iannotti, R. (Eds.). (1986). *Altruism and Aggression: Biological and Social Origins*. Cambridge and New York: Cambridge University Press.

Zillman, D. (1988). "Cognition-Excitation Interdependencies in Aggressive Behavior." *Aggressive Behavior* 14, pp. 51–64.

Zillmer, E. A., Harrower, M., Ritzler, B. A., & Archer, R. P. (1996). "The Quest for the Nazi Personality" [review of J. W. Young, *Totalitarian Language: Orwell's Newspeak and Its Nazi and Communist Antecedents*]. *Psychological Record* 46, no. 2, pp. 399–402.

Zur, O. (1987). "The Psychohistory of Warfare: The Co-evolution of Culture, Psyche, and Enemy." *Journal of Peace Research* 24, pp. 125–34.

INDEX